Edited by
Beatrice Bressan

From Physics to Daily Life

Related Titles

Davies, H., Bressan, B. (eds.)

A History of International Research Networking
The People who Made it Happen

2010
Print ISBN: 978-3-527-32710-2, also available in digital formats

Razgaitis, R.

Valuation and Dealmaking of Technology-Based Intellectual Property

2 Edition
2009
Print ISBN: 978-0-470-19333-4, also available in digital formats

Touhill, C., Touhill, G.J., O'Riordan, T.

Commercialization of Innovative Technologies
Bringing Good Ideas into Practice

2008
Print ISBN: 978-0-470-23007-7, also available in digital formats

Speser, P.L.

The Art & Science of Technology Transfer

2006
Print ISBN: 978-0-471-70727-1, also available in digital formats

Ganguli, P., Prickril, B., Khanna, R. (eds.)

Technology Transfer in Biotechnology
A Global Perspective

2009
Print ISBN: 978-3-527-31645-8, also available in digital formats

Edited by Beatrice Bressan

From Physics to Daily Life

Applications in Informatics,
Energy, and Environment

WILEY Blackwell

Editor

Dr. Beatrice Bressan
700 Chemin du Pont du Loup
74560 Monnetier-Mornex
France

Cover

Center: Higgs boson sculpture. Artistic view of Higgs Boson event from ATLAS experiments data. Courtesy of CERN Author: Davide Angheleddu; www.davideangheleddu.net; davide.angheleddu@gmail.com; artzoneproject@gmail.com
The virtual sculpture is created as a single smooth surface obtained by a computer graphic process starting from the original tracks in the CERN 3D file. It represents the essential step for creating the real sculpture by 3d printing and lostwax casting process.

Top left: Artists impression of an antiproton annihilation in the ALPHA experiment: Courtesy of Chukman So, ALPHA Collaboration, 2011

Solar panels: © Eyematrix – fotolia.com

Cloud communication: © Julien Eichinger- fotolia.com

■ **Limit of Liability/Disclaimer of Warranty:** While the publisher and author have used their best efforts in preparing this book, they make no representations or warranties with respect to the accuracy or completeness of the contents of this book and specifically disclaim any implied warranties of merchantability or fitness for a particular purpose. No warranty can be created or extended by sales representatives or written sales materials. The Advice and strategies contained herein may not be suitable for your situation. You should consult with a professional where appropriate. Neither the publisher nor authors shall be liable for any loss of profit or any other commercial damages, including but not limited to special, incidental, consequential, or other damages.

Library of Congress Card No.: applied for

British Library Cataloguing-in-Publication Data
A catalogue record for this book is available from the British Library.

Bibliographic information published by the Deutsche Nationalbibliothek

The Deutsche Nationalbibliothek lists this publication in the Deutsche Nationalbibliografie; detailed bibliographic data are available on the Internet at <http://dnb.d-nb.de>.

© 2014 Wiley-VCH Verlag GmbH & Co. KGaA, Boschstr.12, 69469 Weinheim, Germany

Wiley-Blackwell is an imprint of John Wiley & Sons, formed by the merger of Wiley's global Scientific, Technical, and Medical business with Blackwell Publishing.

All rights reserved (including those of translation into other languages). No part of this book may be reproduced in any form – by photoprinting, microfilm, or any other means – nor transmitted or translated into a machine language without written permission from the publishers. Registered names, trademarks, etc. used in this book, even when not specifically marked as such, are not to be considered unprotected by law.

Print ISBN: 978-3-527-33286-1
ePDF ISBN: 978-3-527-68701-5
ePub ISBN: 978-3-527-68702-2
Mobi ISBN: 978-3-527-68704-6
oBook ISBN: 978-3-527-68703-9

Cover Design   Adam-Design, Weinheim, Germany
Typesetting   Thomson Digital, Noida, India
Printing and Binding   betz-druck GmbH, Darmstadt

Printed on acid-free paper.

Il piacere più nobile è la gioia di comprendere.
The noblest pleasure is the joy of understanding.
Leonardo da Vinci

Proof reading: Beatriz de Meirelles and Callum MacGregor

Contents

Contributors' CVs *XIII*
Foreword *XXI*
List of Acronyms *XXIII*
List of Units *XXXI*

1	Introduction *1*	
	Sergio Bertolucci	
	Part I Knowledge Management and Technology Transfer in an Organization *3*	
2	Knowledge Management: From Theory to Practice *5*	
	Beatrice Bressan and Daan Boom	
2.1	Knowledge-Based and Innovative Organization *5*	
2.2	The Theory of Knowledge *7*	
2.2.1	Tacit and Explicit Knowledge *8*	
2.2.2	The SECI Model and the Knowledge Creation Spiral *9*	
2.2.3	The Two Dimensions and the Two Spirals of Knowledge Creation *11*	
2.2.4	The Five Conditions and the Five Phases in Two Dimensions *13*	
2.3	The Core Processes of Managing Knowledge *18*	
2.3.1	Knowledge Outputs and Outcomes *19*	
2.4	The Knowledge Worker *31*	
2.4.1	The Individual Learning Process *33*	
2.4.2	Scientific, Technological and Social Processes *36*	
2.4.3	Concept Formation and the Hierarchical Levels of Conceptualization *37*	
2.5	The Knowledge Creation, Acquisition, and Transfer Model *39*	
2.6	Knowledge Management: A Case Study of CERN *41*	
2.6.1	The LHC Case Study Survey *47*	

Part II Examples of Knowledge and Technology Transfer 57

Section 1 Linking Information 59

3 WWW and More 61
Robert Cailliau
3.1 The First Page 62
3.2 Influences on the History of the Web 64
3.2.1 A Matter of Age 64
3.2.2 The Approach 64
3.3 CERN's Role 65
3.3.1 A Possible Definition 65
3.3.2 Making it Work 66
3.3.3 On Documents 66
3.3.4 The Director General 66
3.3.5 Al Gore, the LHC, and the Rest is History 67
3.4 What-if Musings 68
3.4.1 Money, Money, Money . . . 68
3.4.2 And if Not? 69
3.5 The Dark Sides of the Force 70
3.5.1 Techies 70
3.5.2 Global Heating 72
3.5.3 Sin by Omission 72
3.6 Good Stuff 73
3.6.1 Public Domain 73
3.6.2 The Conferences 74
3.6.3 The Consortium 75
3.7 On the Nature of Computing 76
3.7.1 Copy 76
3.7.2 See 77
3.7.3 Understand 77
3.7.4 Remember 77
3.7.5 Interact 78
3.7.6 Share 78
3.7.7 Think 78
3.8 Science 'Un-human' 79
3.9 Lessons to be Learned 80
3.10 Conclusions 80

4 Grid and Cloud 81
Bob Jones
4.1 Why a Grid? 82
4.2 A Production Infrastructure 85
4.3 Transferring Technology: Grids in Other Science Domains 86

4.4	How CERN Openlab has Contributed to the WLCG Grid	86
4.5	Four Basic Principles	87
4.6	Three-Year Phases	88
4.7	EGEE to EGI Transition	90
4.8	Lessons Learned and Anticipated Evolution	91
4.9	Transferring Technology: Grids in Business	92
4.10	Sharing Resources Through Grids	94
4.11	What are the Hurdles?	94
4.12	Philips Research: Scientific Simulation, Modelling and Data Mining Supports Healthcare	95
4.13	Finance: Stock Analysis Application	95
4.14	Multimedia: GridVideo	96
4.15	Imense: From Laboratory to Market	97
4.16	Total, UK	97
4.17	Seismic Imaging and Reservoir Simulation: CGG Veritas Reaping Benefits from the Grid	98
4.18	Societal Impact	99
5	**The 'Touch Screen' Revolution**	*103*
	Bent Stumpe	
5.1	The Birth of a Touch Screen	103
5.2	The Novelty for the Control Room of the CERN SPS Accelerator	106
5.3	A Touch Screen as Replacement for Mechanical Buttons	110
5.4	Attempts at Early Knowledge Transfer	111
5.5	Evolution Turned Into Revolution	113
5.6	Touch Screen and Human Behaviour	115
	Section 2 Developing Future	*117*
6	**Solar Thermal Electricity Plants**	*119*
	Cayetano Lopez	
6.1	The Four STE Technologies	120
6.2	Optical Issues in the STE Plant	124
6.2.1	Solar Concentrators	124
6.2.2	Selective and Anti-Reflective Coatings	124
6.2.3	Thermography	128
6.3	Thermodynamic Issues in the STE Plant	131
6.4	Issues in STE Plants Related to Heat Transfer	134
6.5	Thermal Storage of Energy	137
6.6	Fluid Mechanics	138

7 Computers and Aviation *141*
Antony Jameson

7.1 Computing in Structural and Aerodynamic Analysis *145*
7.2 Computer-Aided Design and Manufacturing *149*
7.3 Fly-By-Wire and Other On-Board Systems *151*
7.4 Airborne Software *154*
7.5 Ground-Based Computer Systems *155*
7.6 Conclusions *156*

8 Antimatter Pushing Boundaries *159*
Niels Madsen

8.1 Science and the Unknown *159*
8.2 Antimatter and CERN *162*
8.2.1 Antimatter at the LHC *164*
8.2.2 The CERN Antimatter Facility *164*
8.3 The Anti-World in Everyday Life *167*
8.4 Beyond the Present Day *169*

Section 3 Sustainability and Learning *171*

9 Towards a Globally Focussed Earth Simulation Centre *173*
Robert Bishop

9.1 A String of Disasters *174*
9.2 Now is the Time *176*
9.3 A Global Synthesis of Knowledge *176*
9.4 Modelling and Simulation as a Platform for Collaboration *177*
9.5 Advances in High-Performance Computing *178*
9.6 Creating Value from Massive Data Pools *179*
9.7 Interactive and Immersive 4D Visualizations *180*
9.8 Leveraging the Many Layers of Computing *182*
9.9 Getting a Complete Picture of the Whole Earth *183*
9.10 Influence of the Solar System *184*
9.11 Prediction and Uncertainty of Extreme Events *186*
9.12 Impact on Cities and Bioregions *189*
9.13 Towards Urban Resilience *190*
9.14 Modelling the Whole-Earth System: A Challenge Whose Time has Come! *191*

10 Radiation Detection in Environment and Classrooms *195*
Michael Campbell

10.1 The Origins of the Hybrid Pixel Detector *196*
10.2 Hybrid Pixel Detectors for High-Energy Physics *197*
10.3 Hybrid Pixel Detectors for Imaging: The Medipix Chips *199*
10.4 Applications *205*

10.4.1	Medical X-Ray Imaging	205
10.4.2	Biology	206
10.4.3	X-Ray Materials Analysis	207
10.4.4	Gas Detector Readout	208
10.4.5	Radiation Monitoring	209
10.4.6	Chemistry	210
10.4.7	Dosimetry in Space	210
10.4.8	Education	211
10.4.9	Art Meets Science	212
10.5	Back to High-Energy Physics	213
10.6	Collaboration, Organization and Serendipity	214

11 Theory for Development 215
Fernando Quevedo
11.1 The Importance of Theoretical Research Through History 216
11.2 Knowledge Management and Science for Peace 219

Part III Economic Aspects of Knowledge Management and Technology Transfer 227

12 Innovation and Big Data 229
Edwin Morley-Fletcher
12.1 The Wealth of Nations: Agriculture, the Division of Labour, or Profits? 230
12.2 Industrialization and/or Exploitation 231
12.3 Perfect Competition, the Disappearance of Profits, Economies of Scale 232
12.4 Creative Destruction 233
12.5 Risk and Uncertainty 235
12.6 Accumulation Without Innovation 236
12.7 The Real Engine of Economic Growth 237
12.8 Endogenous Technological Change 238
12.9 The Appropriate Set of Market and Non-Market Institutions 239
12.10 Limitless Knowledge 241
12.11 Post-Scarcity and Networks 242
12.12 Intellectual Property Rights 244
12.13 Governments' Support of Scientific Research 245
12.14 The Remaining Scarce Resource is Human Creativity 246
12.15 Different Organizational Modes for Overcoming Uncertainty 247
12.16 Information and Allocation Gains of Peer Production 248
12.17 An Ecosystem of Technologies Leading to the Singularity? 250
12.18 Big Data Analytics and Data-Intensive Healthcare 251

13	**Universities and Corporations: The Case of Switzerland**	*255*
	Spyros Arvanitis and Martin Woerter	
13.1	Background *255*	
13.2	KTT Activities in the Swiss Economy: The Main Facts from the *Firm's* Point of View *261*	
13.2.1	Forms and Partners of KTT Activities *262*	
13.2.2	Technological Fields of KTT-Active and R&D-Active Firms *267*	
13.2.3	Mediating Institutions and Motives for KTT Activities *268*	
13.2.4	Impact of KTT Activities as Assessed by the Firms *270*	
13.2.5	Obstacles to KTT Activities *272*	
13.3	KTT Activities in the Swiss Economy: The Main Facts from the *Science Institution* Point of View *276*	
13.3.1	Incidence and Forms of KTT Activities *276*	
13.3.2	Mediating Institutions and Obstacles of KTT Activities *276*	
13.4	Analytical Part: Exploration of KTT Activities in Switzerland *278*	
13.4.1	Drivers of KTT Activities from the Point of View of the *Firm* *279*	
13.4.1.1	Determinants of KTT Activities of Firms *280*	
13.4.1.2	Empirical Evidence *281*	
13.4.2	KTT Activities Determinants from the *University* Point of View *284*	
13.4.2.1	Determinants of the Propensity to KTT Activities of Universities *286*	
13.4.2.2	Empirical Evidence *287*	
13.4.3	Impact of KTT Activities on Innovation and Labour Productivity *288*	
13.4.3.1	Empirical Evidence *290*	
13.4.4	KTT Strategies Determinants and their Impact on Innovation Performance *291*	
13.4.4.1	Firm Characteristics, KTT Strategies, and Innovation Performance *291*	
13.4.5	Exploration and Exploitation *294*	
13.4.5.1	Empirical Evidence *297*	
13.4.6	Technological Proximity Between Firms and Universities and TT *299*	
13.4.6.1	Empirical Evidence *300*	
13.5	Conclusion *302*	
14	**Conclusion** *307*	
	Marilena Streit-Bianchi	

Author Index *311*

Index *317*

Contributors' CVs

Spyros Arvanitis is Senior Researcher at the KOF (Konjunkturforschungsstelle) Swiss Economic Institute at the ETH (Eidgenössische Technische Hochschule) Zurich, Switzerland. He is head of the Research Section Innovation Economics of the KOF. He holds a doctoral degree in Economics from the University of Zurich and a doctoral degree in Chemistry from the ETH Zurich. Spyros has served as expert in national and international advisory bodies on the evaluation of technology policy. His research interests cover several fields of industrial economics, with a strong emphasis on empirical studies. He has published extensively on the economics of innovation, technology diffusion, determinants of firm performance, economics of market dynamics as well as the international competitiveness of firms and industries.

E-mail: arvanitis@kof.ethz.ch

Sergio Bertolucci, Physics degree *cum laude* (University of Pisa, Italy), serves as Director for Research and Computing at CERN. He chaired the Large Hadron Collider (LHC) Committee and was a physics research committee member at DESY (Deutsches Elektronen-Synchrotron). He was vice-president and a member of the Board of the Italian INFN (Istituto Nucleare di Fisica Nucleare). A former Pisa scholar, Sergio has worked at DESY (Germany), Fermilab (US) and Frascati (Italy). He was a member of the group that founded the experiment CDF (Collider Detector at Fermilab). He has been involved in the CDF detector design, construction and operation, and has been technical coordinator of the team responsible for the design and construction of the KLOE (K LOng Experiment) detector at DAFNE (Double Annular Φ Factory for Nice Experiments) at the LNF (Laboratori Nazionali di Frascati). Sergio was appointed head of the LNF accelerator and the DAFNE project, becoming Director of the Laboratory in 2002. The 'top quark' discovery was amongst the obtained scientific results. He designed and built many innovative detectors that are still state-of-the-art today. He has co-authored over 370 scientific publications (global h-index 86). Sergio has been the Italian Delegate of the Programme Committee on the research infrastructure of the 6th European Union Framework Programme. His memberships include: ESA (European Space Agency), FAIR (Facility for

Antiproton and Ion Research), KEK (Kō Enerugī Kasokuki Kenkyū Kikō), RECFA (Restricted European Committee for Future Accelerators), J-PARC (Japan Proton Accelerator Research Complex), and SLAC (Stanford Linear Accelerator Centre).

E-mail: sergio.bertolucci@cern.ch

Robert Bishop spent 40 years in the technical, engineering and scientific computing business. He was responsible for building and operating the international aspects of Silicon Graphics Inc. (SGI), Apollo Computer Inc., and Digital Equipment Corporation. To accomplish this task, he lived with his family in five countries: US, Australia, Japan, Germany and Switzerland. He was SGI Chairman and CEO (1999–2005). Robert is a Fellow of the Australian Davos Connection and is an elected member of the Swiss Academy of Engineering Sciences, and serves on the advisory boards for NICTA (National Information and Communication Technology Australia), and the Human Brain Project of EPFL (École Polytechnique Fédérale, Lausanne, Switzerland). He served on the advisory boards of the Multimedia Super Corridor and UNITEN (Universiti Tenaga Nasional), Malaysia, and LONI (Laboratory Of Neuro Imaging) of UCLA (University of California, Los Angeles), US. Robert earned a B.S. (First Class Honours) in Mathematical Physics (University of Adelaide, Australia), an M.S. from the Courant Institute of Mathematical Sciences (New York University, US), and received his D.S. *Honoris Causa* (University of Queensland, Australia). In 2006, he was awarded the NASA (National Aeronautics and Space Administration) Distinguished Public Service Medal for his role in delivering simulation facilities that helped NASA's space shuttle fleet return-to-flight after the 2003 Columbia disaster. Robert is Founder and Chairman of BBWORLD Consulting Services Sàrl and Founder and President of the ICES (International Centre for Earth Simulation) Foundation, Switzerland.

E-mail: bbishop001@hotmail.com

Daan Boom graduated in Library and Information Science at The Hague University, The Netherlands. He has over 30 years' experience in advising and implementing knowledge management practices and IT solutions in private and public sectors. Daan is associated as Director of International Relations with CCLFI (Centre for Conscious Living Foundation Inc.), Manila, Philippines. He was Knowledge Management Programme Manager at ICIMOD (International Centre for Integrated Mountain Development), Nepal and Head of Knowledge Management at the Asian Development Bank (ADB) where he co-authored a paper on Knowledge Based Economies (2007) and ADB's experience implementing knowledge management (Knowledge Management for Development Journal, 2005). Daan was editor in charge of a special edition on Knowledge Management for Mountain Development (ICIMOD, Issue 58, 2011). He published an article on social media use to enforce democracy and transparency (Dutch Information Professional Journal, May, 2012). In Asia he organized knowledge management and communication workshops to strengthen local capacities of Non Governmental Organizations, and South-South knowledge exchange programs between universities and research institutes. Throughout his work, he has contributed to curriculum development on knowledge management for universities and has participated or chaired international networks.

E-mail: daan.boom@cclfi.org

Beatrice Bressan, science and technology writer and EUSJA (European Union of Science Journalists' Associations) member, is the Responsible for the Outreach activities of the CERN TOTEM (TOTal cross-section, Elastic scattering and diffraction dissociation Measurement at the Large Hadron Collider) experiment and expert consultant for associations and companies. After a degree in Mathematical Physics (La Sapienza University, Rome) and a Master in Science Communication (ISAS, International School for Advanced Studies, Trieste), Italy, she obtained a PhD, *magna cum laude*, and carried out a postdoctoral study in Knowledge Management and Technology Transfer for the Department of Physical Sciences of Helsinki University, Finland, within the CERN research programmes. She has worked several years as Technology Transfer Promotion Officer at CERN. Her managerial positions include: Head of Communications (SIB, Swiss Institute of Bioinformatics), Communications Manager (Physics Department, Geneva University), Communications Director (MaatG Cloud Computing Company), Switzerland, and Chief Business Development Officer (gnúbila Software Development Company), France. Beatrice has extensive experience in project management, business development, media publishing, public relations, mediation, and negotiation. She is author and editor in a wide range of subject areas; her publications include: *A History of International Research Networking* (Wiley, 2010), *Knowledge Management in an International Research Centre* (Lambert Academic Publishing, 2011).

E-mail: beatrice.bressan@cern.ch

Robert Cailliau graduated in Electrical and Mechanical Engineering (Ghent University, Belgium). In 1974 he started at CERN as Fellow in the PS (Proton Synchrotron) division. In 1987 he became group leader of Office Computing Systems. In 1989, both he and Tim Berners-Lee independently proposed a hypertext system for access to the documentation needed by CERN. He joined Berners-Lee, whose system became the 'World Wide Web' or 'WWW' (1990). As a key proponent of the early Web, he launched many related efforts, notably through the European Commission (1993: WISE – World-wide Information system for Support of R&D Efforts – Web for the least favoured regions, with the Fraunhofer Gesellschaft in Germany; 1994: Web for Schools) and helped with the transfer of WWW to the W3C (WWW Consortium). His work with the Legal Service led CERN to release its Web software into the public domain (1993). Robert organized the First International WWW Conference (1994, CERN), and was co-founder of the International WWW Conferences

Committee of which he was a president. From 2002 until his retirement from CERN in 2007 he devoted his time to public communication. He is a public speaker and delivers many keynotes at conferences. Robert was awarded a knighthood by King Albert II of Belgium in 2005.

E-mail: robert@cailliau.org

Michael Campbell obtained an Honours Degree and PhD at the Department of Electrical and Electronic Engineering, University of Strathclyde, Glasgow, Scotland. After working for a short period at the Battelle Institute in Carouge (Geneva, Switzerland), he joined CERN, also located near Geneva, in 1988. After an initial period of training in CMOS (Complementary Metal Oxide Semiconductor) circuit design, he started designing pixel detector electronics for High-Energy Physics (HEP) experiments. The first full hybrid pixel detector system was used at the WA97 (West Area 97) experiment in the mid-90s. Several generations of CMOS technology later Michael was a major contributor to the design of the silicon pixel detector of the ALICE (A Large Ion Collider Experiment) experiment and the pixel readout system of the RICH (Ring-imaging Cherenkov) detector of the LHCb (Large Hadron Collider beauty) experiment at the LHC (Large Hadron Collider). In parallel with his activity in HEP he founded the Medipix Collaborations which have used successive generations of CMOS technology to bring spectroscopic X-ray imaging closer to reality. Michael is the spokesman of the Medipix2 and Medipix3 Collaborations and has authored or co-authored over 100 scientific publications, mostly in the field of hybrid pixel detectors.

E-mail: michael.campbell@cern.ch

Rolf-Dieter Heuer is the Director General of CERN since January 2009. He graduated in Physics at the University of Stuttgart and obtained his PhD at the University of Heidelberg in Germany. He then worked as research scientist in the JADE (Japan, Deutschland, and England) Collaboration at the PETRA (Positron-Electron Tandem Ring Accelerator) in the DESY (Deutsche Elektronen-Synchrotron) Laboratory of Hamburg, Germany. Between 1984 and 1998, as a CERN staff member, Rolf-Dieter worked at OPAL (Omni-Purpose Apparatus for LEP – Large Electron Positron collider), where he was responsible for the coordination of design and construction of the tracking jet chamber. He was also the run coordinator during the start-up phase of LEP1 (1989–1992) and the OPAL spokesperson (1994–1998). Rolf-Dieter was appointed Professor at the University of Hamburg (1998) and Research Director at DESY (2004). He initiated the reconstruction and focus of German Particle Physics at the energy frontier, with particular emphasis on the Large Hadron Collider (LHC). Rolf-Dieter is a member of many scientific committees and advisory bodies.

E-mail: rolf.heuer@cern.ch

Antony Jameson is the Thomas V. Jones Professor of Engineering in the Aeronautics and Astronautics at Stanford University, US. He graduated with first class honours in Engineering and went on to obtain a PhD in Magnetohydrodynamics from Trinity Hall, Cambridge University, UK. Subsequently, he worked as an Economist for the Trades Union Congress, and then became Chief Mathematician at Hawker Siddeley Dynamics, UK. In 1966, he joined the Aerodynamics Section of Grumman Aerospace, in 1974, he moved to the Courant Institute of Mathematical Sciences at New York University, and in 1982, he joined Princeton University, US, where he was appointed James S. McDonnell Distinguished University Professor of Aerospace Engineering. Amongst his awards: Gold Medal (Royal Aeronautical Society), UK, NASA (National Aeronautics and Space Administration) Medal for Exceptional Scientific Achievement, Spirit of St. Louis Medal, Elmer A. Sperry Award, US; Fellow: Trinity Hall, Royal Society, Royal Aeronautical Society, Royal Academy of Engineering, UK, American Institute of Aeronautics and Astronautics; Honorary Professor North Western Polytechnic University, Xian, China, Doctor *Honoris Causa* Paris VI University, France and Uppsala University, Sweden, and National Academy of Engineering Foreign Associate, US. Antony has authored and co-authored over 400 scientific papers in a wide range of areas including computational fluid dynamics, aerodynamics and control theory. Antony is the principal developer of the 'flo' and 'syn' codes series, widely used in the aerospace industry.

E-mail: jameson@baboon.stanford.edu

Bob Jones is head of the CERN openlab project which facilitates collaboration between CERN and its industrial partners to study and develop data-intensive solutions for scientists working at the next-generation Large Hadron Collider (LHC). Following a B.Sc. Honours Degree in Computer Science from Staffordshire University, UK, he joined CERN in 1986 as a software developer with the IT department providing support for the LEP (Large Electron Positron collider) experiments. Bob completed his PhD thesis in Computer Science at Sunderland University, UK, while working at CERN. He held the position of the online software system leader for the ATLAS (A Toroidal LHC ApparatuS) experiment. His experience in the distributed computing arena includes mandates as the technical and project director of the European Commission co-financed EGEE (Enabling Grids for E-sciencE) project, which operated a production Grid facility for e-Science spanning 300 sites across 48 countries for more than 12 000 researchers. The work of EGEE was preceded as deputy project leader for the European Union DataGrid project. He is a member of the advisory board for several Grid related European and national and ESFRI (European Strategy Forum on Research Infrastructures) projects. Bob regularly acts as a reviewer for the IST (Information Societies Technology) programme and is a leader of the Helix Nebula, a public private partnership to explore the use of commercial Cloud services for science applications.

E-mail: robert.jones@cern.ch

Cayetano Lopez is Professor of Theoretical Physics at the Universidad Autónoma de Madrid, Spain. He has lead several research projects and published numerous papers on Elementary Particles Physics, including rigorous properties, phenomenology, unified theories and supersymmetry, on energy issues and on accelerator driven systems for nuclear waste treatment and energy production. He was the Rector of the Universidad Autónoma de Madrid, from 1985 to 1994. He was member of the Council of CERN from 1983 to 1995 and Vice-president of this organism from 1987 to 1990. At present, Cayetano is the Director General of the Research Centre in Energy, Environment and Technology, CIEMAT (Centro de Investigaciones Energéticas, Medioambientales y Tecnológicas), Madrid.

E-mail: cayetano.lopez@ciemat.es

Niels Madsen, Professor in Experimental Physics at Swansea University, UK, studied Physics at Aarhus University in Denmark from where he was also awarded a PhD He moved on to work on antihydrogen at CERN, securing a Danish Steno Fellowship before he finally took up a position at Swansea University in 2005. Niels is co-founder and group leader in the ALPHA (Antihydrogen Laser Physics Apparatus) Collaboration that was the first to trap antihydrogen and observe quantum transitions in it. He is now pursuing precision comparisons with hydrogen. Niels has been actively involved in antihydrogen research since 2001, playing a substantial role in the ATHENA (Antihydrogen Apparatus) team that first formed low energy antihydrogen in 2002. He plays a leading role in the ALPHA experiment, and led the effort to implement several key techniques leading to the first antihydrogen trapping. Niels has furthermore conceptualized, designed and built significant parts of the ALPHA apparatus. For this work he and his colleagues were awarded the 2011 James Dawson Award for Excellence in Plasma Physics Research. Niels has recently held a Royal Society Senior Leverhulme research fellowship and is a fellow of the Institute of Physics, UK.

E-mail: n.madsen@swansea.ac.uk

Edwin Morley-Fletcher is President of Lynkeus Consultancy and former Professor of Administration Science of the Faculty of Politics, University La Sapienza (Rome, Italy). Member of the Italian CNEL (Consiglio Nazionale dell'Economia e del Lavoro) and chairman of the CNEL Working Group on the Social Market from 1995 to 2000, Senior Fellow at the School of Public Policy (University of California, Los Angeles, US) in 1999, and Jean Monnet Fellow at the European University Institute (EUI) in 1989. A member of the Scientific Committee of the MedChild Foundation, Edwin has been advisor of the Gaslini Foundation for several years now, and has led the 'Healthcare Governance and Technology' working group of ASTRID (Analisi, Studi e Ricerche sulla Riforma delle Istituzioni Democratiche) in Italy. He is currently advisor of the Bambino Gesù Hospital in Rome in the area of Information and Communications Technology (ICT) for Health. He operated as Project Manager within Health-e-Child (2006–2010), the FP6 IP (6th Framework Programme Integrated Project) Sim-e-Child, and the FP7 STReP (Specific Targeted Research Project, 2010–2012). He is currently Project Manager of the FP7 IP MD-Paedigree (Model-Driven European Paediatric Digital Repository) and Co-ordinator of the FP7 STReP Cardioproof (Proof of Concept of Model-based Cardiovascular Prediction). Edwin is the author of over 100 publications of which the most recent one is: *Modelli sanitari a confronto: nuovi assetti istituzionali europei evidenziati dal dibattito sull'Obamacare*, in R. Nania (Ed.), Attuazione e sostenibilità del diritto alla salute, Chapter 5, Sapienza Università Editrice, 2013.

E-mail: emf@lynkeus.com

Marilena Streit-Bianchi obtained a degree in Biological Sciences at La Sapienza University (Rome, Italy). She has been working at CERN for 40 years, and has carried out research on biomedical applications and biological effects of high-energy radiations and on economic benefits from big-science collaboration with industry and in particular technological learning, innovation acquisition, market opening and knowledge transfer. Marilena is the co-author and author of more than 55 publications in international science journals, conferences and proceedings and has been the organizer of thematic conferences on fundamental science-industry relations, representing CERN at innovation day exhibitions. She has been the supervisor of many students and occupied managerial positions. She was the Deputy Group Leader of the Technology Transfer Group at CERN and Responsible for the Technology Promotion Section until 2007 and of Technology Promotion events in Latin-America carried out within the HELEN (High Energy Physics Latinamerican-European Network) project. Since 2014, Marilena is curator of exhibitions and educational activities for the arSciencia Association (Santiago de Compostela, Spain), a sociocultural entity dedicated to the world of Science and Art. As Honorary Staff Member, she is leading a CERN Oral History project.

E-mail: marilena.bianchi@bluewin.ch

Bent Stumpe, after several years in the Danish Air Force as Radio/Radar engineer and in the Danish television industry, came to CERN in 1961 working on flying-spot cathode-ray tube scanners for analysing bubble chamber films and cutting-edge techniques which later came into widespread use that is in the newspaper industry. Recruited into the design team for the control centre of the new 400 GeV CERN proton generator, Bent made important contributions such as the Capacitive Transparent Touch Screen, Micro Computer controlled stand alone Touch Terminals, graphical colour display drivers, pointing devices (Tracker Ball) and the Computer Controlled Knob. He was consultant to the World Health Organization (WHO) for which he developed an instrument for the early detection of Leprosy. After leaving CERN in 2003, he started the East West Trade Development company and, invited by the Danish Ministry of Research, became the CERN Danish ILO (Industrial Liaison Officer), creating the Danish DENCERN network. Bent received a Danish industrial prize for his technical developments at CERN and for his collaboration with industry.

E-mail: bent.stumpe@bluewin.ch

Fernando Quevedo is Director of the Abdus Salam ICTP (International Centre for Theoretical Physics) of UNESCO (United Nations Educational, Scientific and Cultural Organization) and Professor of Theoretical Physics at the University of Cambridge, UK. He is a theoretical particle physicist with wide-ranging research interests in string theory, phenomenology and cosmology. A Guatemalan citizen born in Costa Rica, was educated in Guatemala before obtaining a PhD from the University of Texas, US, under the supervision of Nobel Laureate Steven Weinberg. A string of research appointments followed at CERN, McGill University in Canada, the Institut de Physique at Neuchatel University in Switzerland and the Los Alamos National Laboratory in the US, as well as a brief term as Professor of Physics at the Mexican National Autonomous University. Fernando joined the Department of Applied Mathematics and Theoretical Physics at the University of Cambridge in 1998, where he is currently Professor of Theoretical Physics and Fellow of Gonville and Caius College. He has received several honours and awards, including honorary degrees from two Guatemalan universities, the 1998 ICTP prize, a Guggenheim Fellowship and the Royal Society Wolfson award. Fernando is also a fellow of TWAS (The World Academy of Sciences for the advancement of science in developing countries).

E-mail: fernando.quevedo@ictp.it

Martin Woerter, PhD in Economics and Social Sciences, joined the KOF (Konjunkturforschungsstelle) Swiss Economic Institute at the ETH (Eidgenössische Technische Hochschule) Zurich, Switzerland as senior researcher in 2002. He is experienced in applied research in the field of Economics, with a strong emphasis on empirical studies. Martin worked as a researcher at the Scientific Institute of Communication Services (WIK, Wissenschaftliches Institut für Kommunikationsdienste, Bad Honnef, Germany), where he carried out and lead projects in the field of Information and Communications Technologies (ICT). During his work as a research assistant in Economics at the University of Innsbruck, Martin was part of a research team at the Austrian Academy of Sciences focusing on policies for media and ICT. He also was carrying out parts of his PhD at SPRU (Science Policy Research Unit, now Science and Technology Policy Research, Brighton, UK). He has been a member of national and international expert groups evaluating national innovation promotion policies and policies related to ICT. His research mainly focuses on applied economics and industrial organization. Currently, he is conducting research in the field of economics of innovation, diffusion of new technologies, knowledge and technology transfer between universities and firms, innovation and competition. In Austria Martin was Lecturer at Innsbruck and Vienna Universities and now he teaches at the ETH.

E-mail: woerter@kof.ethz.ch

Foreword

As we advance further into the 21st century, we realise more and more that international scientific organisations such as CERN, the European Laboratory for Particle Physics, are concrete examples of science uniting people around common goals across national and cultural boundaries. Back in the 1940s and 50s, this intention to connect was prominent in the minds of those who came together to create at CERN an environment in which fundamental science has flourished. Their vision has resulted in 60 years of discoveries, 60 years of science in the service of peace.

CERN's commitment to streamlining the transfer of knowledge to different fields has strengthened during recent years. Forgetting for a moment our most cited example, the World Wide Web, and its role in globalization, there are many other success stories linked, for example, to medicine.

As is to be expected with any research centre, we are always content, but never satisfied. Therefore, in order to establish CERN as an important facilitator for medical physics in Europe, we have recently created a new Office for Medical Applications. This will provide an environment in which medical practitioners and members of the physics community can come together to pursue common goals.

Of course this is not something peculiar to CERN. Similar initiatives can be found at other research infrastructures, many of which have produced brilliant breakthrough innovations, demonstrating the subtle ways in which knowledge transforms from novel idea into novel application.

Knowledge travels mainly with people. Information dissemination, education and training are therefore key elements in connecting research to society. This constant osmosis between research facilities and society at large enables the translation of ideas into innovation.

This book provides a stimulating reflection on the growing role of knowledge as a strategic asset in responding to the epochal challenges in front of us. It exposes the crucial role that curiosity driven research plays in the long-term sustainability of our society. And it intrinsically expresses a message of hope: a common thread in the motivations that moved the contributing authors to pursue their ideas.

By improving our quality of life, they have at the same time contributed to the realization of a more equitable society. Might this not be the best vehicle to enable sustainable peace?

Geneva
July 2014

Rolf-Dieter Heuer

List of Acronyms

2D, 3D, 4D	Two-dimensional, Three-dimensional, Forth-dimensional
AA	Antiproton Accumulator
AAA	Authentication, Authorization, Accounting
ACAS X	Airborne Collision Avoidance System X
ACE	Antiproton Cell Experiment
AD	Antiproton Decelerator
ADB	Asian Development Bank
ADM	Air Data Monitor
AEGIS	Antihydrogen Experiment: Gravity, Interferometry, Spectroscopy
AIMS	African Institute of Mathematical Sciences
ALICE	A Large Ion Collider Experiment
ALPHA	Antihydrogen Laser Physics Apparatus
AOL	America Online
APO	Asian productivity Organization
APQC	American Productivity & Quality Center
APS	American Physical Society
APU	Auxiliary Power Unit
AR5	Assessment Report 5
ARC	Advanced Resource Connector
ART	Advanced Rotating Tomograph
ASACUSA	Atomic Spectroscopy And Collisions Using Slow Antiprotons
ASF	Apache Software Foundation
ASTRID	Analisi, Studi e Ricerche sulla Riforma delle Istituzioni Democratiche
AT&T	American Telephone & Telegraph Company
ATC	Air Traffic Control
ATHENA	Antihydrogen Apparatus
ATLAS	A Toroidal LHC ApparatuS
ATRAP	Antihydrogen Trap
AUEB	Athens University of Economics and Business
AUST	African University of Science and Technology
BC	Before Christ

BASE	Baryon Antibaryon Symmetry Experiment
BBC	British Broadcasting Corporation
BRAIN	Brain Research through Advancing Innovative Neurotechnologies
B-Spline	Basis Spline
C17	Cargo aircraft model 17
CAD	Computer-Aided Design
CADAM	Computer-Augmented Design And Manufacturing
CAM	Computer-Aided Manufacturing
CAMAC	Computer Automated Measurement And Control
Cardioproof	Proof of Concept of Model-based Cardiovascular Prediction
CATIA	Computer Aided Three-dimensional Interactive Application
CCLFI	Centre for Conscious Living Foundation Inc.
CC-BY	Creative Commons Attribution 3.0 License
CD	Compact Disc
CDC	Control Data Corporation
CDF	Collider Detector at Fermilab
CD-ROM	Compact Disc Read Only Memory
CEA	Commissariat à lÉnergie Atomique et aux Énergies Alternatives
CEO	Chief Executive Officer
CERN	Conseil Européen pour la Recherche Nucléaire
CEST	Centre dÉtudes de la Science et de la Technologie
CFD	Computational Fluid Dynamics
CFL3D	Computational Fluids Laboratory Three-Dimensional
CGG	Compagnie Générale de Géophysique
CHEP	Computing in High Energy Physics and Nuclear Physics
CIEMAT	Centro de Investigaciones Energéticas, Medioambientales y Tecnológicas
CII	Compagnie Internationale pour l'Informatique
CIS	CompuServe Information Service
CME	Coronal Mass Ejection
CMIP5	Coupled Model Inter-comparison Project 5
CMOS	Complementary Metal Oxide Semiconductor
CMS	Compact Muon Solenoid
CN	Computing and Networking
CNEL	Consiglio Nazionale dell'Economia e del Lavoro
CoI	Communities of Innovation
CoP	Communities of Practice
CPU	Central Processing Unit
CRC	Chemical Rubber Company
CSDS	Cargo Smoke Detector System
CT	Computed Tomography
CTA	Cherenkov Telescope Array
CTI	Commission for Technology and Innovation
CU	Cantonal University

DAFNE	Double Annular Φ Factory for Nice Experiments
DC	Douglas Commercial
DESY	Deutsches Elektronen-Synchrotron
DGC	Digicom Inc.
DLR	Deutsches Zentrum für Luft- und Raumfahrt e.V.
DNA	Deoxyribonucleic Acid
DoD	Department of Defense
DoE	Department of Energy
EAWAG	Eidgenössische Anstalt für Wasserversorgung, Abwasserreinigung und Gewässerschutz
EC	European Commission
ECM	Enterprise Content Management
ECMWF	European Centre for Medium-Range Weather Forecasts
ECP	Electronics and Computing for Physics
ECS	Environmental Control System
EDS	Electronic Data Systems
EEC	Electronic Engine Control
EFP	European Framework Programme
EGEE	Enabling Grids for E-sciencE
EGEODE	Expanding GEOsciences on DEmand
EGI	European Grid Infrastructure
EGI-DS	EGI Design Study
EGI InSPIRE	EGI Integrated Sustainable Pan-European Infrastructure for Researchers in Europe
ELENA	Extra Low-Energy Antiprotons
ELFINI	Finite Element Analysis System
EMBA	Executive Master of Business Administration
EMBL	European Molecular Biology Laboratory
EMI	European Middleware Initiative
EMPA	Eidgenössische Materialprüfungs- und Forschungsanstalt
EPFL	École Polytechnique Fédérale, Lausanne
EPIC	Electronic Preassembly Integration on CATIA
EPRI	Electric Power Research Institute
ERP	European Research Programme
ESA	European Space Agency
ESFRI	European Strategy Forum on Research Infrastructures
ESGF	Earth System Grid Federation
ESO	European Southern Observatory
ESRF	European Syncrotron Radiation Facility
ESTELA	European Solar Thermal Electricity Association
ETH	Eidgenössische Technische Hochschule
ETHD	ETH Domain
EU	European Union
EUDET	European Union project for Detector R&D towards the International Linear Collider

EUI	European University Institute
EUSJA	European Union of Science Journalists' Associations
FAA	Federal Aviation Administration
FAIR	Facility for Antiproton and Ion Research
FBW	Fly-By-Wire
Fermilab	Fermi National Accelerator Laboratory
FET	Future and Emerging Technologies
FLOPS	FLoating-point Operations Per Second
FOD	Foreign Object Debris
FP	Framework Programme
FUN3D	Fully Unstructured Navier-Stokes Three-Dimensional
FYRON	Former Yugoslav Republic Of Macedonia
GaAs	Gallium Arsenide
GBAR	Gravitational Behaviour of Antihydrogen at Rest
GDS	Global Distribution System
GEM	Gas Electron Multiplier
GIS	Geographic Information Systems
gLight	Lightweitht Middleware for Grid Computing
GPS	Global Positioning System
GPSSU	GPS Sensor Unit
GRC	Geoscience Research Centre
GSM	Global System for Mobile
HBP	Human Brain Project
HBP-PS	HBP Preparatory Study
HELEN	High Energy Physics Latinamerican-European Network
HEP	High-Energy Physics
HIDAC	High-Density Avalanche Chamber
HMI	Hahn Meitner Institut
HP	Hewlett-Packard
HPC	High Performance Computing
HTML	HyperText Markup Language
HTTP	HyperText Transfer Protocol
IAM	Integrated Assessment Model
IBM	International Business Machines
IBN	Internet Business Network
ICES	International Centre for Earth Simulation
ICFAI	Institute of Chartered Financial Analystsof India
ICI	Investment Company Institute
ICIMOD	International Centre for Integrated Mountain Development
ICT	Information and Communications Technology
ICTP	Abdus Salam International Centre for Theoretical Physics
ID	Identity Document
IEA	International Energy Agency
IEAP	Institute of Experimental and Applied Physics
IEEE	Institute of Electrical and Electronics Engineers

IGTF	International Grid Trust Federation
IHM	Institute for HyperMedia Systems
IICM	Institute for Information Processing and Computer-supported New Media
ILO	Industrial Liaison Officer
IMAX	Image Maximum
INFN	Istituto Nucleare di Fisica Nucleare
INRIA	Institut National de Recherche en Informatique et en Automatique
IO	Industrial Organization
IoT	Internet of Things
IP	Integrated Project
IPCC	Intergovernmental Panel on Climate Change
IPO	Initial Public Offering
ISAS	International School for Advanced Studies
ISS	International Space Station
IST	Information Society Technology
IT	Information Technology
ITU	International Telecommunication Union
IUPAC	International Union of Pure and Applied Chemistry
JADE	Japan, Deutschland, and England
JAMSTEC	Japan Agency for Marine-Earth Science and Technology
JET	Joint European Torus
J-PARC	Japan Proton Accelerator Research Complex
KEK	Kō Enerugī Kasokuki Kenkyū Kikō
KLOE	K LOng Experiment
KOF	Konjunkturforschungsstelle
KPMG	Klynveld, Peat, Marwick, Goerdeler
KPN	Koninklijke PTT Nederland
KTT	Knowledge and Technology Transfer
LEAR	Low-Energy Antiproton Ring
LEEM	Low-Energy Electron Microscopy
LEP	Large Electron Positron collider
LHC	Large Hadron Collider
LHCb	LHC beauty
Lidar	Light and radar
LIST	Laboratoire d'Integration des Systèmes et des Technologies
LKE	Lauritz Knudsen Electric
LMB	Line Mode Browser
LNF	Laboratori Nazionali di Frascati
LONI	Laboratory Of Neuro Imaging
LUCID	Langton Ultimate Cosmic ray Intensity Detector
MAKE	Most Admired Knowledge Enterprises
MD-Paedigree	Model-Driven European Paediatric Digital Repository
MIC	Many Integrated Core

MIT	Massachusetts Institute of Technology
MOOC	Massive Open Online Course
MoT	Management of Technology
MP3	Short for MPEG3, Moving Picture Experts Group Audio Layer 3
MRI	Magnetic Resonance Imaging
MSDOS	Microsoft Disk Operating System
NASA	National Aeronautics and Space Administration
NASTRAN	NASA STRucture ANalysis
NatCatSERVICE	Natural Catastrophes Service
NBER	National Bureau of Economic Research
NEC	Nippon Electric Company
NFP	Nationale Forschungsprogramm
NICTA	National ICT Australia
NII	National Information Infrastructure
NIKHEF	Nationaal Instituut voor Kernfysica en Hoge-Energiefysica
NIST	National Institute of Standards and Technology
NOAA	National Oceanic and Atmospheric Administration
NODAL	Network-Oriented Document Abstraction Language
NSF	National Science Foundation
NTNU	Norges Teknisk-Naturvitenskapelige Universitet
NURBS	Non-Uniform Rational B-Spline
OAK	Metropolitan Oakland International Airport
OECD	Organization for Economic Co-operation and Development
OLS	Ordinary Least Squares
OPAL	Omni-Purpose Apparatus for LEP
OS	Operating System
OSG	Open Science Grid
OVERFLOW	OVERset grid FLOW solver
PALS	Positron Annihilation Lifetime Spectroscopy
PC	Personal Computer
PCC	Photon Counting Chip
PDF	Portable Document Format
PEPT	Positron Emission Particle Tracking
PERL	Practical Extraction and Reporting Language
PET	Positron Emission Tomography
PETRA	Positron-Electron Tandem Ring Accelerator
PFC	Primary Flight Computer
PHOENICS	Parabolic Hyperbolic Or Elliptic Numerical Integrated Code Series
PHP	Hypertext Pre-processor, originally Personal Home Page
PI2S2	Progetto per l'Implementazione e lo Sviluppo di una e-Infrastruttura in Sicilia basata sulla Grid
PRISM	Protect Respond Inform Secure Monitor
PRT	Partial Ring Tomograph

Ps	Positronium
PS	Proton Synchrotron
PS10	Planta Solar 10
PSA	Plataforma Solar de Almería
PSI	Paul Scherrer Institute
PTT	Post, Telegraph, and Telephone
Qubit	Quantum bit
R&D	Research and Development
RAL	Rutherford Appleton Laboratory
RECFA	Restricted European Committee for Future Accelerators
RICH	Ring-imaging Cherenkov
RIM	Research In Motion
SABRE	Semi-Automated Business Research Environment
SATCOM	Satellite Communications
SCADA	Supervisory Controls and Data Acquisition
SECI	Socialization, Externalization, Combination and Internalization
SFO	San Francisco International Airport
SGI	Silicon Graphics Inc.
SI	International System of Units
SIB	Swiss Institute of Bioinformatics
SJO	Juan Santamaria International Airport
SKA	Square Kilometre Array
SLA	Service Level Agreement
SLAC	Stanford Linear Accelerator Centre
SLC	Social Learning Cycle
SME	Small and Medium Enterprise
SMS	Short Message Service
SNSF	Swiss National Science Foundation
SOAS	School of Oriental and African Studies
SolarPACES	Solar Power And Chemical Energy Systems
SPRU	Science Policy Research Unit, now Science and Technology Policy Research
SPS	Super Proton Synchrotron
STAR-CD	Simulation of Turbulent flow in Arbitrary Regions Computational Dynamics
STE	Solar Thermal Electricity
STReP	Specific Targeted Research Project
SUK	Schweizerischen Universitätskonferenz
SVG	Scalable Vector Graphics
TCAS	Traffic Alert and Collision Avoidance System
TED	Technology, Entertainment, Design
THISS	Technische Hochschulen und Innovation: Start-ups und Spin-offs

TNO	Nederlandse Organisatie voor Toegepast Natuurwetenschappelijk Onderzoek
TNT	Trinitrotoluene
TO	Transfer Office
ToT	Time over Threshold
TOTEM	TOTal cross-section, Elastic scattering and diffraction dissociation Measurement at the LHC
TWAS	The World Academy of Sciences
UAS	Universities of Applied Sciences
UAV	Unmanned Air Vehicles
UCLA	University of California, Los Angeles
UK	United Kingdom
UMTS	Universal Mobile Telecommunications System
UN	United Nations
UNESCO	United Nations Educational, Scientific and Cultural Organization
UNICORE	Uniform Interface to Computing Resources
UNITEN	Universiti Tenaga Nasional
UNIX	Originally UNICS, UNiplexed Information and Computing System
US	United States
USB	Universal Serial Bus
USI	Università della Svizzera Italiana
USM3D	Unstructured Mesh Three-Dimensional
UTC	Coordinated Universal Time
VG	Vortex Generator
VINE	Very Informal Newsletter
VO	Virtual Organisation
VOD	Video On-Demand
W3C	WWW Consortium
WA	West Area
WHO	World Health Organization
WIK	Wissenschaftliches Institut für Kommunikationsdienste
WinCC	Windows Control Centre
WISE	World-wide Information system for Support of R&D Efforts
WLCG	World-wide LHC Computing Grid project
WSL	Forschungsanstalt für Wald, Schnee und Landschaft
WTT	Wissen- und Technologietransfer
WWW	World Wide Web
Wysiwyg	What you see is what you get

List of Units

The term **pico** (symbol **p**) is a prefix in the International System of Units (SI) denoting a factor of 10^{-12}. The prefix is derived from the Italian *piccolo*, meaning *small*.

The term **nano** (symbol **n**) is a prefix in the SI denoting a factor of 10^{-9}. The prefix is derived from the Greek *nanos*, meaning *dwarf*.

The term **micro** (symbol **μ**) is a prefix in the SI denoting a factor of 10^{-6}. The prefix comes from the Greek *mikrós*, meaning *small*.

The term **milli** (symbol **m**) is a prefix in the SI denoting a factor of 10^{-3}. The prefix comes from the Latin *mille*, meaning *one thousand* (the plural is *milia*).

The term **kilo** (symbol **k**) is a prefix in the SI denoting a factor of 10^{3}. The prefix comes from the Greek *khilioi*, meaning *thousand*.

The term **Mega** (symbol **M**) is a prefix in the SI denoting a factor of 10^{6}. The prefix comes from the Greek *megas*, meaning *great*.

The term **Giga** (symbol **G**) is a prefix in the SI denoting a factor of 10^{9}. The prefix is derived from the Greek *gigas*, meaning *giant*.

The term **Tera** (symbol **T**) is a prefix in the SI denoting a factor of 10^{12}. The prefix comes derived from Greek *teras*, meaning *monster*.

The term **Peta** (symbol **P**) is a prefix in the SI denoting a factor of 10^{15}. The prefix is derived from the Greek *pente*, meaning *five*.

The term **Exa** (symbol **E**) is a prefix in the SI denoting a factor of 10^{18}. The prefix is derived from the Greek *hexa*, meaning *six*.

The **ampere** (symbol **A**) is the SI base unit of electric current. It is defined as the current which produces a specified force between two parallel wires which are 1 metre apart in a vacuum.

The **hertz** (symbol **Hz**) is the SI base unit of frequency. It is defined as the number of cycles per second of a periodic phenomenon.

The **kelvin** (symbol **K**) is the SI base unit of temperature. It is defined as $1/273.16^{th}$ of the thermodynamic temperature of the triple point of water.

The **kilogram** (symbol **kg**) is the SI base unit of mass. It is defined as the mass of an international prototype in the form of a platinum-iridium cylinder kept at Sèvres in France.

The **metre** (symbol **m**) is the SI base unit of length. It is defined as the distance light travels, in a vacuum, in $1/299792458^{th}$ of a second.

The **second** (symbol **s**) is the SI base unit of time. It is defined as the length of time taken for 9192631770 periods of vibration of the caesium-133 atom to occur.

The **unified atomic mass unit** (symbol **u**) is the SI base unit of mass on an atomic or molecular scale. It is approximately the mass of a nucleon (one of the particles in the atomic nucleus).

The **coulomb** (symbol **C**) is the SI derived unit of electric charge (symbol **Q** or **q**). It is defined as the charge transported by a constant current of one ampere in one second.

The **grey** (symbol **Gy**) the SI derived unit of absorbed ionizing radiation dose equivalent to absorption per unit mass of one joule per kilogram of irradiated material.

The **Joule** (symbol **J**) is the SI derived unit of energy. It measures the energy expended in applying a force of one newton through a distance of one metre.

The **newton** (symbol **N**) is the SI derived unit of force. It is equal to the amount of force needed to accelerate a one kilogram mass at a rate of one metre per second squared.

The **pascal** (symbol **Pa**) is the SI derived unit of pressure. It is a measure of force per unit area, defined as one newton per square metre.

The **sievert** (symbol **Sv**) is the SI derived unit of ionizing radiation dose to measure the: equivalent radiation dose (the radiation absorbed by a fixed mass of biological tissue for the different biological damage potential of different types of ionizing radiation); effective dose (the cancer risk to a whole organism due to ionizing radiation); and committed dose (the probabilistic health effect on an individual due to an intake of radioactive material).

The **Tesla** (symbol **T**) is the SI derived unit of magnetic field strength. It is equal to the magnitude of the magnetic field vector necessary to produce a force of one newton on a charge of one coulomb moving perpendicular to the direction of the magnetic field with a velocity of one metre per second.

The **volt** (symbol **V**) is the SI derived unit of electric potential. It is the difference of potential between two points of an electrical conductor when a current of 1 ampere between them dissipates a power of 1 watt.

The **watt** (symbol **W**) is the SI derived unit of power (one joule per second). It measures the rate of energy conversion or transfer.

The **electron volt** (symbol **eV**) is a unit of energy. It is equal to approximately 1.6×10^{-19} joules, the amount of energy gained (or lost) by the charge of a single electron moved across an electric potential difference of one volt.

The **bit** (symbol **b**) is a binary digit (taking a value of either 0 or 1) basic unit in computing. It is also the information capacity of one binary digit. Eight bits equal one **byte**.

The **pixel** (symbol **pel**) is a physical point in a dot matrix data structure image: the smallest controllable element represented on the screen in an all points addressable display device.

The **euro** (currency sign **€**) is the unit of currency used by the EU Institutions and is the official currency of the eurozone. Its name was officially adopted on 16 December 1995 and was introduced to world financial markets as an accounting currency on 1 January 1999.

The US **dollar** (currency sign **$**) is the unit of currency used by the US and its overseas territories. It is defined by the Coinage Act of 1792 to be between 371 and 416 grains (27.0 g) of silver (depending on purity). It is divided into 100 cents (200 half-cents prior to 1857).

1
Introduction

Sergio Bertolucci

This book, using the voices of the key players in several success stories, offers an important contribution to the debate on how (and how much) fundamental research connects to innovation and to societal issues at large.

In the following I will mainly refer to CERN, simply because I know it best, but most of the arguments would perfectly apply to any other research infrastructures, across many fields of science.

The unrestricted access to knowledge and expertise represents one of the strongest assets of organisations like CERN. By connecting the creativity of many different cultures and backgrounds, research laboratories, while pursuing their primary mission, also create one of the best environments in which scientific ideas can develop into new technologies.

Since its inception, CERN has been a centre of knowledge production and dissemination. Currently it has a staff of around 2400 people and hosts a community of more than 10 000 physicists and engineers, coming from 113 countries. In its sixty years of life it has established itself as the most global centre for High-Energy Physics and has set a reference in science organization, which has been replicated by several other research infrastructures.

CERN's origins can be traced back to the late 1940s, when a few visionary scientists in Europe and North America realized that a research laboratory was a powerful tool to re-unite Europe through science, while mitigating at the same time the brain drain to the United States of America, which had begun during World War II. In 1951 a provisional body was created, the 'Conseil européen pour la recherche nucléaire', in other words: CERN. Two years later, the Council decided to build a central laboratory near Geneva in Switzerland and on the 29th September 1954 the CERN convention was ratified.

Nowadays CERN researchers are mainly involved in the exploitation of its most complex scientific instrument: the world's most powerful system of accelerators, whose final component is the Large Hadron Collider (LHC). In the 27 kilometres of circumference of the LHC, counter-circulating beams of protons and/or heavy nuclei collide head-on in order to re-create the conditions that have existed in the earliest stages of the Universe. Gigantic detectors with hundreds of millions of sensors are surrounding the collision points and take

From Physics to Daily Life: Applications in Informatics, Energy, and Environment, First Edition.
Edited by Beatrice Bressan.
© 2014 Wiley-VCH Verlag GmbH & Co. KGaA. Published 2014 by Wiley-VCH Verlag GmbH & Co. KGaA.

electronic snapshots of the collisions at a rate of 40 millions a second, selecting and storing the information of the potentially interesting 'events' (about a thousand a second) for further analysis. The amount of raw data produced is massive, exceeding 30 petabytes (millions of gigabytes) every year.

This description, albeit sketchy, is nevertheless sufficient to expose the working mechanism of CERN: in order to pursue our mission (i.e. to understand the fundamental laws that govern matter), we need to develop and deploy cutting edge technologies (IT, precision mechanics, submicron electronics, sensors, control systems, materials, etc.) through a very well organized collaborative effort. Moreover, in order to keep it sustainable and to maximize our chances of paradigm-changing discoveries, we have to constantly train the next generations of physicist and engineers. The composition of these elements creates a very lively and coherent environment, based on competition and collaboration (we call it 'coopetition'), where charisma is by far more important than authority.

How does this community, which is clearly science driven, connect to innovation and society at large?

This relation has been examined by several scholarly studies, which have updated the by now insufficient linear model to more sophisticated interconnected representations.

I personally like to consider research infrastructures as gigantic 'knowledge libraries', where each and every actor, internal or external, can contribute a 'book' or use a 'book'. Many of these 'books' are in fact real people, with several of them going on permanent 'loan' to other 'libraries'.

If you allow unrestricted access to these libraries, both in the use and in the contributions, you create an agora where questions and answers meet in a dynamic way, and where no player has the rigid role of owning of all the answers or of all the questions.

Moreover, such an environment is very fertile in generating new and unpredictable developments, acting as a motor and amplifier of ideas.

The rules of this 'library' should be kept reasonably simple: key elements are Open Access, Open Source, Peer Review, Protection of the Intellectual Property, pretty much the same principles which are behind research infrastructures.

The model keeps the coexistence of competition and collaboration, which has proven fundamental in the field of research, but it can be much more effective in spreading Knowledge (and its use) across geographical and political boundaries.

In short you go from Open Science to Open Innovation.

In a world, which is facing increasingly urgent and global challenges, it might be the most powerful message that science can give for a sustainable future.

Part I
Knowledge Management and Technology Transfer in an Organization

Knowledge is the most important asset of humankind and society. Today, we live in a Knowledge Society that is governed by knowledge management and technology transfer demands. Those knowledge-management processes allowing more efficient knowledge transfer are highly valued by industry. Introducing knowledge management to science is a natural consequence, as large-scale international and multicultural scientific projects are often carried out in a collaborative global manner and involve many scientific disciplines.

Scientific organizations such as CERN, the European Laboratory for Particle Physics, have been centres of knowledge creation since their inception. The CERN laboratory employs over 2400 people, and each year welcomes around 10 000 visiting scientists from over 113 countries (half of the world's particle physicists) for their research. When leaving, some of these scientists transfer their acquired knowledge and experience to industry and, consequently, to society.

One of the most relevant forms of technology transfer to industry, institutions and society is this implicit transfer of the knowledge and know-how of people. Some industrial firms have also asked scientific research laboratories to host – at their own expense – engineers or applied physicists for training periods to work on various projects in order to benefit from more frequent and diversified exchanges.

This has been demonstrated to be a valuable mechanism for technological learning and innovation at both the individual and the organizational level. All people working at research centres have access to the rich programme of seminars and training courses, covering a wide range of state-of-the-art topics.

2
Knowledge Management: From Theory to Practice

Beatrice Bressan and Daan Boom

2.1
Knowledge-Based and Innovative Organization

Various observers describe today's global economy as one in transition to a Knowledge-Based Economy,[1] that is, an increasing interconnected globalized economy where knowledge assets such as know-how, expertise and capacity, innovation, knowledge networks, and creativity play a distinguished role towards sustainable growth. Although, in many countries and organizations around the world this works in different ways, research shows that countries and organizations capable of creating the enabling conditions for knowledge to be developed and shared are performing better, as well as showing higher social and economic progress. Research also shows that effective and sustainable organizations (both profit and nonprofit) tend to display a high degree of the following characteristics: efficiency; adaptability; flexibility; and financial prudence.[2] The first three characteristics are key dimensions of any knowledge-based and innovative organization.

- **Efficiency** allows an organization to implement and follow certain routines, whether production- or administration-based. Its business processes are well defined and documented, and the organization implements production, knowledge or administrative routines for delivering its core products and services at a defined high quality and price.
- **Adaptability** is the organizational capacity to learn, intervene and innovate in the business process and change its routines or delivery mechanism of its

---

1) Knowledge economy, knowledge society, knowledge management is a tautology, as societies have always been dependent on the interpretation and application of knowledge. Awareness of managing the pool of knowledge gained momentum during the early 1990s as distinct economic differentiator after the Organization for Economic Co-operation and Development published *The Knowledge-Based Economy*, 1996.
2) A. de Geus, *The Living Company: Habits for Survival in a Turbulent Business Environment*, Harvard Business School Press, 2002; P.E. Mott, *The Characteristics of Effective Organizations*, Harper & Row, 1972.

From Physics to Daily Life: Applications in Informatics, Energy, and Environment, First Edition.
Edited by Beatrice Bressan.
© 2014 Wiley-VCH Verlag GmbH & Co. KGaA. Published 2014 by Wiley-VCH Verlag GmbH & Co. KGaA.

products or services, and to define a new product or change the design and concept of the product. Those changes could be the result of client behaviour, substitution of materials, natural resources depletion, new market entry, and so on.[3] The so-called Toyota 'lean' principles can be mentioned as an example of efficiency and adaptability for their relentless attention for detail, their commitment to data-driven experimentation, and for the way in which they charge workers with on-going tasks of increasing efficiency and eliminating waste. These principles require from the staff, expertise and judgement that depend heavily on their 'knowledge' of the product and service. Adaptable organizations anticipate the changes in the external environment and often put scenarios in place to analyse potential changes of market conditions responding or adapting quickly to changes.

- **Flexibility** is an asset of organizations that have a culture and functions in place for adoption, sharing and testing of new ideas; including approaches to learning from failures.[4] Staff are continuously encouraged (or designated functions are put in place) to scan the external environment, to signal innovations, and to foster new thinking that may affect or change their business routines and their product–service portfolio rather than to reject them as disruptive.[5] While assets such as labour, capital, processes and technology continue to be important for any organization, the organizational capability to apply learning, to innovate, and to adapt and adapt are widely recognized as crucial pillars for sustainability and competitiveness as they are facing continuous changes in markets, demographics, technological developments and increased competition.[6]

The important role played by the stock and the application of knowledge for organizational growth is relatively well understood and has been addressed extensively by economic and social scientists.[7] The capacity and capability to absorb, create, recognize, share, and embody knowledge in new products, technologies and services is critical for any organization faced with shifting demands from markets, rapid product obsolescence, hyper-competition, and financial upheavals.[8] The active interest shown since the early 1990s by profit and non-profit organizations in putting in place mechanisms to manage knowledge, considering it as a critical activity, is largely a response to the challenges posed by an increasingly complex internal and external organizational environment.

---

3) M.E. Porter, *The Five Forces That Shape Strategy*, Harvard Business Review, 2008; C.K. Prahalad and G. Hamel, *The Core Competence of the Corporation*, Harvard Business Review, 1990.
4) J. Kim, President of the World Bank, *Speech on Science of Discovery*, World Bank, 2012.
5) *Once Dominant, Blackberry Seeks to Avoid Oblivion*, The New York Times, 12th August, 2013.
6) MAKE (Most Admired Knowledge Enterprises), Survey, 2011–2012, APQC (American Productivity & Quality Centre), Survey, 2011.
7) Marx, 1887; Shumpeter, 1942; Solow, 1956; Arrow, 1960; Kuznets, 1966; Polanyi, 1967; Machlup, 1980; Romer, 1986; Nonaka, 1991 and 1995; Stewart, 1997; Senge, 2000; Mokyr, 2002; Holsapple, 2003; Warsh, 2006; Bennet, 2010; World Bank, 2012.
8) I. Nonaka, *The Knowledge-Creating Company*, Harvard Business Review, 1991.

2.2
The Theory of Knowledge

Philosophers would claim that knowledge is nothing natural, while from a historian's perspective the accumulation of knowledge is the basis of the human evolution since its beginning thousands of years ago.[9] As early as the fifth century BC, the philosopher Socrates dealt already with the question of knowledge.[10] In Plato's *Meno*, Socrates says that once true beliefs are tied down, 'in the first place they become pieces of knowledge, and then they remain in place'. Aristotle, in the *Nicomachean Ethics*, describes the different virtues of thought (knowledge) in terms of: doxa, episteme, hypolepsei, nous, phronesis, sophia, and techne. Today, this would correspond to:

- Doxa-opinion
- Episteme, scientific knowledge, evidence-based knowledge
- Hypolepsei, conjectural knowledge
- Nous, intellectual reasoning
- Phronesis, practical and operational knowledge
- Sophia: good/bad practices
- Techne: technical knowledge, patents.

To which other forms of thought can be added such as:

- Religious knowledge
- Intuitive knowledge.

If you ask yourself the question 'What do I know?', you may come up with many things you know, but in reality you probably don't know what you know unless challenged in a discussion. In many cases you will not be able to articulate easily what you know, or to document your knowledge.[11]

The most common association we have with the term knowledge is that of scientific knowledge generated using scientific methods, tools and protocols. The protocols for scientific knowledge often include validation, testing, and extensive peer review processes before the results are disseminated. Another form of knowledge is operational and embedded knowledge generated as a result of organizational activities or feedback from consumers on products and services delivered. Messaging and other forms of personal related information shared through social media platforms are also used as knowledge in support of marketing, and social behavioural trend analyses. The recent disclosure of the US

---

9) C. McGinn, *Wie kommt der Geist in die Materie? Das rätsel des Bewusstseins*, Verlag C.H. Beck, 2001.
10) J. Mokyr, *The Gifts of Anthena: Historical Origins of the Knowledge Economy*, Princeton University Press, 2002.
11) C.G. Sieloff, *'If only HP knew what HP knows': the roots of knowledge management at Hewlett-Packard*, Journal of Knowledge Management, Volume 3, N. 1, p. 47, 1999.

Protect Respond Inform Secure Monitor (PRISM) programme learned that even private e-mail and SMS messages are analysed and used for intelligence purposes.[12]

However, the most important form of knowledge is the knowledge embedded in a person.[13] Comprehension, analytical skills, and the capacity to transfer and share knowledge are key competencies for the knowledge worker.

Managing knowledge is, therefore, a complex process that not only covers the human component, culture and capacities but also comprises other supporting processes, human resource policy, Information and Communications Technology (ICT) environment, communications, capacity to conduct research, culture for teamwork, and the capability to engage or partner with other organizations in the knowledge development cycle. With the rapid changes taking place in today's society, mastering quick access to knowledge, acquiring and fostering new knowledge and managing the resulting knowledge to respond vigorously to new challenges is considered a fundamental comparative advantage for nations and organizations.

2.2.1
Tacit and Explicit Knowledge

The main challenge for organizations is to protect and maximize the value derived from tacit knowledge held by employees, clients and stakeholders to the creation of new knowledge and the sharing of useful existing knowledge through the interaction between tacit and explicit knowledge. In general, there are three distinct forms of knowledge: *tacit*; *implicit*; and *explicit*. With the new technologies – especially decision support systems – the distinction between these three forms of knowledge becomes blurred, but for a good understanding of how knowledge can be organised the following clarification might be useful:

- **Implicit and tacit knowledge** are often used interchangeably in the literature, and many argue over the differences; even common dictionaries show a considerable overlap between the terminologies.[14] Some of the literature defines implicit knowledge as knowledge stored in the brain of which the bearer is not immediately aware. Although this information is not readily accessible, it can be 'restored', 'assembled' or 'associated' when triggered by

---

12) Contributors are not aware that their knowledge or observations are shared for specific objectives. The recent PRISM (Protect Respond Inform Secure Monitor) scandal is also an example of information to generate knowledge or insights.
13) In most cases, personal knowledge is used to advance society but in some cases knowledge is also used to gain advantage (business or military espionage, financial advantage through manipulation of information, or inner circle information shared to a few to gain a financial profit).
14) A. Bennet and D. Bennet, *Associative Patterning in the Unconsciousness Life of an Organization*, in J.P. Girard (Ed.), Building Organizational Memories: Will You Know What You Knew, ICI (Investment Company Institute) Global, 2009.

questioning, dialogue, or even during and after a good night-rest.[15] When implicit knowledge is discovered, the bearer may still not be able to share or transfer it easily to someone else.
- **Tacit knowledge** is also 'brain' or personal knowledge but is easier to 'restore or reproduce' by the holder. The bearer can write or share elements of knowledge, but the bearer will never be able to put all knowledge in writing. If the bearer shares knowledge, the receiver may have a better understanding of a subject, but will not in all cases be able to comprehend for effective action. Tacit knowledge can be broken down further into four categories: embodied; intuitive; affective; and spiritual. Tacit knowledge is person-bound and is often very specific to a particular work context. A person can reflect on tacit knowledge but may be unable to articulate or document the knowledge in full for sharing with other people. In this chapter, the term tacit includes within it also implicit knowledge.
- **Explicit knowledge** is knowledge documented or captured in print, video, databases or any other formats or standard commonly used in the printed or digital world that enables access, reproduction and transfer. In other words, explicit knowledge is expressed in words, numbers and images, and can be captured, stored, communicated, assembled or mixed and transferred. For that reason, explicit knowledge is easier to manage. During the early 1990s most knowledge management activities were aimed at improving the capturing, storing and searching of organizational information in repositories. This is one facet of what we now would label as 'information management'. The social interaction of knowledge came more into play in the mid-1990s.

The dynamic process of managing knowledge comes into play in the social interaction between tacit and explicit knowledge. This interaction – called 'knowledge conversion' – is a social process between individuals and is not confined within one individual. Through the social conversion process, tacit and explicit knowledge expand in terms of both quality and quantity, as outlined in the Socialization, Externalization, Combination and Internalization (SECI) model, introduced by Nonaka and Takeuchi.[16]

2.2.2
The SECI Model and the Knowledge Creation Spiral

Organizational knowledge is the knowledge shared by individuals, and is best highlighted by the four different modes of knowledge conversion according to the SECI model. Based on the SECI model, the conversion between tacit and

---

15) A. Bennet and D. Bennet, *Engaging tacit knowledge in support of organizational learning*, VINE (Very Informal Newsletter), The Journal of Information and Knowledge Management Systems, Volume 38, N. 1, 2008.
16) I. Nonaka and I.H. Takeuchi, *The Knowledge-Creating Company: How Japanese Companies Create the Dynamics of Innovation*, Oxford University Press, 1995.

explicit knowledge is carried out through four different modes, experienced by the individual, which are: 1. Socialization; 2. Externalization; 3. Combination; 4. Internalization:

1) **Socialization** is the process of sharing experiences, mental models and technical skills through observation and imitation, thereby creating tacit knowledge. The key element for acquiring tacit knowledge is experience. Without some shared experience, such as double loop learning (e.g. in brainstorming, informal meetings and discussions), motivated by the search for meaning, it is extremely difficult for one person to project her or himself into another individual thinking process.
2) **Externalization** is the process of articulating tacit knowledge into explicit concepts, and is typically seen in the process of concept creation triggered by dialogue or collective reflection. This mode holds the key to knowledge creation because it creates new, explicit concepts from tacit knowledge. It may however be difficult to find adequate verbal expression for a mental image through the use of analytical methods of deduction or induction. Metaphors or analogies, which are distinctive methods of perception, are often driving the externalization.[17] Using an attractive metaphor or analogies can be highly effective in fostering direct commitment to creative processes during the early stages of knowledge creation.
3) **Combination** is a process of systematising concepts into a knowledge system such as a repository or database environment. This mode of knowledge conversion involves combining different bodies of explicit knowledge. Individuals exchange and combine knowledge through such media as documents, meetings, telephone conversations or computerized communication networks. Reconfiguration of existing information through the sorting, adding, combining and categorising the explicit knowledge (as conducted in computer databases) can lead to new knowledge. Knowledge creation in formal education and training at school usually takes this form. The wide use of different modern technologies (Web, iPad, blogs, etc.) has resulted in many initiatives being focused within the combination mode.
4) **Internalization** is a process of embodying explicit knowledge into tacit knowledge. It is closely related to learning by doing. When experience through socialization, externalization and combination are internalized into individuals, tacit knowledge based in the form of shared mental models or technical know-how becomes valuable assets for knowledge. In order for organizational knowledge creation to take place, the tacit knowledge accumulated at the individual level needs to be socialized with other organization members. To facilitate the individual internalization, knowledge

---

17) Metaphor and analogy are often confused. The association of two things through metaphor is driven mostly by intuition and holistic imagery and does not aim to find differences between them. On the other hand, association through analogy is carried out through rational thought and focuses on structural or functional similarities between two things and, hence, their differences.

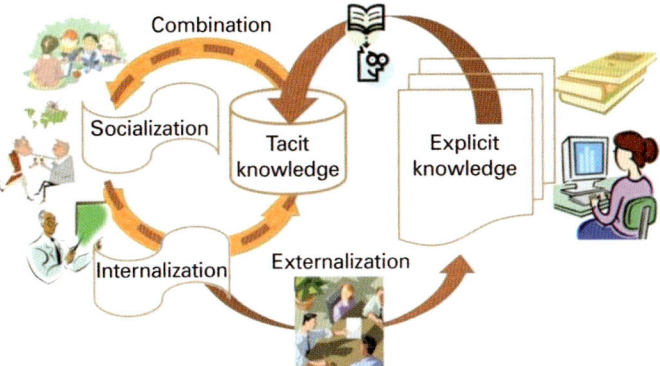

Figure 2.1 The spiral of knowledge.

shall be verbalized or drawn into documents or oral presentations, thus enriching individual tacit knowledge as well as contributing to the transfer of explicit knowledge to other people. The experience from the past may in addition become a tacit mental model.

Each mode creates, respectively: sympathized knowledge; conceptual knowledge; systemic knowledge; and operational knowledge. When most members of the organization share a mental model, the tacit knowledge becomes part of the organizational culture. Organizational knowledge creation is a continuous and dynamic interaction between tacit and explicit knowledge; innovation emerges thanks to this interaction. Each mode of knowledge conversion enriches a different part of the knowledge, and the interactions between them occur within the spiral of knowledge creation (see Figure 2.1).

This interaction is shaped by shifts between different modes of knowledge conversion, which are in turn induced by several triggers. First, the socialization mode usually starts with building fields of interactions that facilitate the sharing of people's experiences and mental models. Second, dialogue or collective reflection using metaphor and/or analogies to articulate hidden tacit knowledge triggers the externalization mode. Third, networking of the newly created knowledge and of knowledge existing in other sectors of the organization, triggers the combination mode, thereby crystallizing them into a new product, service or system. Finally, learning by doing is also an important trigger of the internalization process.

2.2.3
The Two Dimensions and the Two Spirals of Knowledge Creation

The theoretical framework of organizational knowledge creation contains two dimensions of knowledge formation: the epistemological and the ontological. The epistemological dimension is where conversion takes place between tacit

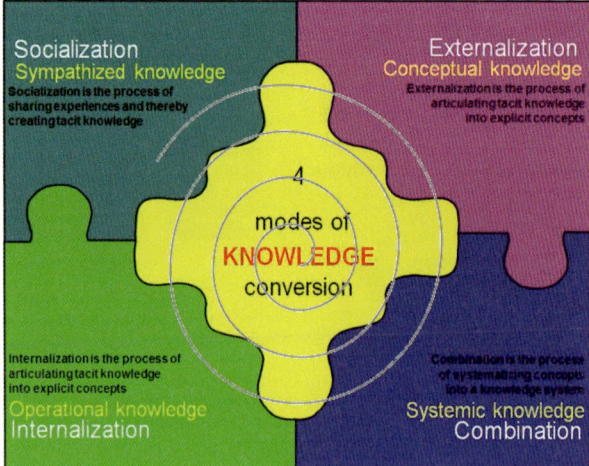

Figure 2.2 The knowledge creation spiral at the epistemological dimension.[18] (Reproduced with permission of Oxford University Press).

knowledge and explicit knowledge; the ontological is where knowledge created by individuals is transformed into knowledge at the group and organizational levels. In other words, the organization has to mobilize tacit knowledge created and accumulated at the individual level. The mobilized tacit knowledge is amplified through the four modes of knowledge conversion and crystallized at higher levels of the ontological dimension. Interaction between tacit knowledge and explicit knowledge at the epistemological dimension moves through all the levels of the ontological dimension, starting at the individual and by continuing through the organizational boundaries becomes ever larger in scale. The process leads to different cycles of knowledge creation fuelling innovation and is, in fact, its cornerstone.

The process is dynamic and produces two different types of knowledge spiral. The first spiral takes place at the epistemological dimension across the four modes of knowledge, as explained in Figure 2.2. Another spiral takes place at the ontological dimension, where knowledge developed at the individual level is transformed into knowledge at the group and organizational levels, as outlined in Figure 2.3.

Although each dimension produces a dynamic spiral, the truly dynamic nature of Nonaka's theory can be depicted as the interaction of these two knowledge spirals over time. In a three-dimensional chart (with time as the third dimension), the knowledge spiral at the epistemological dimension rises upwards, whereas the knowledge spiral at the ontological dimension moves from left to right and back to the left, in a cyclical motion.

---

18) I. Nonaka and I.H. Takeuchi, *The Knowledge-Creating Company: How Japanese Companies Create the Dynamics of Innovation*, Oxford University Press, 1995.

2.2 The Theory of Knowledge

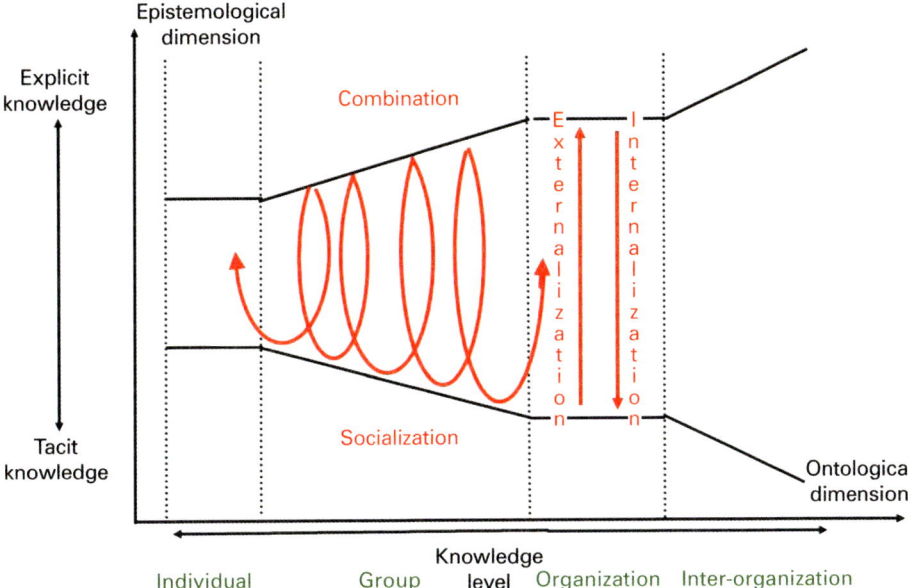

Figure 2.3 The knowledge creation spiral at the ontological dimension.[19] (Reproduced with permission of Oxford University Press).

2.2.4
The Five Conditions and the Five Phases in Two Dimensions

The four modes of knowledge conversion are not independent from each other, and interact continuously. At the epistemological dimension these interactions produce a five-condition knowledge spiral process, which enables the four modes to be transformed into this knowledge spiral. Neither of the organizational levels are independent of each other, but interact continuously when time is introduced. At the ontological dimension, these interactions produce another five-phase knowledge-creation spiral process, when knowledge developed at, for example, the project-team level, is transformed into knowledge at the divisional level and eventually at the inter-organizational level.

The five conditions of the knowledge spiral process at the epistemological dimension are Intention, Autonomy, Fluctuation and creative chaos, Redundancy, and Requisite variety.

- **Intention** is what drives and promotes knowledge in an organization. Efforts to achieve the intention take the form of a strategy to develop the organizational capability to acquire, create, accumulate and exploit knowledge.
- **Autonomy** increases the chance of unexpected opportunities arising, and the possibility that individuals will motivate themselves to create new

19) Ibidem.

knowledge. At the individual level, all members of an organization should be allowed to act autonomously as far as circumstances permit. Original ideas emanate from autonomous individuals, diffuse within a team, and then become organizational ideas.

- **Fluctuation and creative chaos** stimulate interaction between the organization and the external environment. Fluctuation is different from complete disorder, as it is characterized by an order without recursiveness, whose pattern is at first difficult to predict. When fluctuation is introduced into an organization, its members face a breakdown of routine, habits or cognitive frameworks and the opportunity to reconsider their fundamental thinking and perspective, and to hold dialogue by means of social interaction creating new concepts and knowledge. Some have called this phenomenon creating 'order out of noise' or 'order out of chaos'.[20] Chaos is generated naturally when the organization faces a real crisis. It can also be generated intentionally when the organization's leaders try to evoke a sense of crisis amongst organizational members by proposing challenging goals. This intentional creative chaos increases tension within the organization and focuses the attention of organizational members on defining the problem and resolving the crisis situation. This approach is in contrast to the paradigm of the information processing, in which a problem is simply given and the solution found through a process of combining relevant information based on a preset algorithm. Such a process ignores the importance of defining the problem. To attain definition, a problem must be constructed from the knowledge available at a certain point in time and context. The benefits of creative chaos can only be realized when organizational members are able to reflect upon their actions through After Action or Peer Review sessions. To

---

20) According to the principle of 'order out of noise', the self-organizing system can increase its ability to survive by purposefully introducing such noise into itself (H. von Foerster, *Principles of Self-Organization - in a Socio-Managerial Context*, in H. Ulrich and G.J.B. Probst (Eds), Self-Organization and Management of Social Systems, Springer-Verlag, 1984). Order in the natural world includes not only the static and crystallized order in which entropy is zero, but also the unstable order in which new structures are formed by the work of matter and energy. The latter is 'order out of chaos' according to the theory of dissipative structure (I. Prigogine and I. Stengers, *Order out of Chaos: Man's New Dialogue with Nature*, Bantam Books, 1984). In an evolutionary planning perspective, moreover, Jantsch argues: 'In contrast to widely held belief, planning in an evolutionary spirit therefore does not result in the reduction of uncertainty and complexity, but their increases. Uncertainty increases because the spectrum of options is deliberately widened; imagination comes into play' (E. Jantsch, *The Self-Organizing Universe: Scientific and Human Implications of the Emerging Paradigm of Evolution*, Pergamon Press, 1980). Researchers who have developed the chaos theory have found the creative nature of chaos (J. Gleik, *Chaos: Making a New Science*, Viking Penguin, 1987; M. Mitchell Waldrop, *Complexity: Life at the Edge of Order and Chaos*, Simon & Schuster, 1992). Nonaka also applied the chaos theory to management (I. Nonaka, *Creating Organizational Order Out of Chaos: Self-Renewal of Japanese Firms*, California Management Review, Volume 30, N. 3, p. 57, 1988; B.J. Zimmerman, *The Inherent Drive to Chaos*, in P. Lorange et al. (Eds), Implementing Strategic Processes: Change, Learning and Cooperation, Basil Blackwell, 1993).

make chaos truly creative, the knowledge-creating organization is required to institutionalize this reflection on action in order to induce and strengthen the subjective commitment of individuals.
- **Redundancy** means the existence of information that goes beyond the immediate operational requirements of organizational members. In order for organizational knowledge creation to take place, a concept created by an individual or group must be shared by the others, who may not need the concept immediately. Sharing redundant information promotes the sharing of tacit knowledge, through a sense what others are trying to articulate, and therefore also speeds up the knowledge creation process. Redundancy is especially important at the concept development stage, when it is critical to articulate images rooted into the tacit knowledge. At this stage, redundant information enables individuals to invade each other's functional boundaries, and offers advice or provides new information from different perspectives. Redundancy of information brings this learning by intrusion into each individual's sphere of perception. Even within a strictly hierarchical organization, redundant information helps to build unusual communication channels and to facilitate interchange between people's hierarchy and non-hierarchy. There are several ways to build redundancy into an organization. One way is to adopt the overlapping approach, in which different functional departments work together within a loosely defined organizational structure. Another way is to divide the product development team into competing groups, with each group developing a different approach to the same project; then, when the groups come together they argue over the advantages and disadvantages of their proposals. A third way is to implement a strategic rotation of personnel, especially when vastly different areas of technology or functions, such as R&D and marketing, are involved. Such rotation helps the members of the organization to understand their business from multiple and different perspectives, thereby making organizational knowledge more fluid and understandable.[21] It also enables the diversification of skills and information sources. The extra information held by individuals across different functions helps the organization to expand its knowledge creation capacity. Redundancy of information increases the amount of information to be processed. It can lead – at least in the short term – to information overload and an increase in the cost of knowledge creation. Therefore, a balance between the creation and processing of information is mandatory. One way to deal with the possible downside of redundancy is to make clear where information can be located and where knowledge is stored within the organization.
- **Requisite variety** is the complexity of the external environment which an organization's internal diversity must match in order to deal with challenges posed by its environment. Organizational members can cope with many contingencies if they possess requisite variety, which can be enhanced by combining information differently, flexibly and quickly, and providing equal

---

[21] This is a common practice at Royal Dutch Shell, Toyota, Wipro, and African Development Bank.

access to information throughout the organization. In order to maximize variety, everyone in the organization should be assured of direct and rapid access to the widest variety of needed information.

The five phases of the spiral process at the ontological level are: Sharing of tacit knowledge; Creating concepts; Justifying concepts; Building an archetype; and Cross-levelling of knowledge:

- **Sharing of tacit knowledge** is the starting point of organizational knowledge creation and corresponds to socialization, as the rich and untapped knowledge that resides in individuals must first be amplified within the organization. However, tacit knowledge cannot be communicated or passed to others easily, as it is acquired primarily through experience and is not easily expressible in words. The sharing of tacit knowledge amongst multiple individuals with different backgrounds, perspectives and motivations becomes critical for organizational knowledge creation.[22] Individuals' emotions, feelings and mental models have to be shared to build mutual trust. To make this sharing work, individuals from functionally separate organizational sectors need space to interact through dialogue and need to work together to achieve a common goal. Variety between the members, who experience a redundancy of information and share their interpretations of organizational intention, facilitates organizational knowledge creation. Management injects creative chaos by setting challenging goals and endowing team members with a high degree of autonomy. An autonomous team starts to set its own task boundaries and begins to interact with the external environment, accumulating both tacit and explicit knowledge. In the second phase, tacit knowledge shared through interactions is verbalized into words and phrases and finally converted and crystallized into explicit knowledge.
- **Creating concepts** is the phase in which the organization determines whether a new concept is truly worth pursuing. This process corresponds to externalization and is facilitated by the use of multiple reasoning methods such as deduction, induction and abduction, and by employing figurative language such as metaphors and analogies. Autonomy helps team members to diverge their thinking freely and then, when needed, to converge it. To create concepts, members have to fundamentally rethink their existing premises. Requisite variety helps in this regard by providing different angles or perspectives for a problem. Fluctuation and chaos, either from the outside or inside, also help members to change their way of thinking. Redundancy of information enables members to understand figurative language better and to crystallize a shared mental model. In the theory of organizational knowledge creation, knowledge is defined as justified true belief.

22) The rapid uptake of social media (Twitter, Facebook alike since 2007) changed the way in which organizations communicate internally and with their stakeholders. It is mostly the transfer of tacit knowledge, and it is maturing rapidly with new 'Big Data mining tools'.

- **Justifying concepts** is the phase in which the organization determines whether the new concept is truly worth pursuing. It is similar to a screening process, and corresponds to internalization. Individuals seem to be justifying or screening information, concepts or knowledge continuously and unconsciously throughout the entire organizational knowledge creation process. The organization, however, must conduct this justification more explicitly to ensure that the organizational intention is still intact and to ascertain whether the concepts being generated meet the needs of society at large. The most appropriate time for the organization to conduct this screening process is immediately after the concepts have been created. To avoid any misunderstanding of the organization's intention, redundancy of information helps to facilitate the justification process.
- **Building an archetype** occurs when, having been justified, the concepts are converted in the phase of what is tangible or concrete. It can take the form of a prototype in the case of hard product development, or an operating mechanism in the case of soft innovations, such as a novel managerial system or organizational structure. In either case, the archetype is built by combining newly created explicit knowledge with existing explicit knowledge. In building a prototype the explicit knowledge to be combined could take the form of technologies or components. Because justified concepts (which are explicit) are converted into archetypes (which are also explicit), this phase is akin to combination. This phase is complex, and therefore a dynamic cooperation of various departments within the organization is indispensable. Both, requisite variety and redundancy of information facilitate this phase. Organizational intention also serves as a useful tool to converge the various types of know-how and technologies that reside within the organization, as well as for promoting interpersonal and interdepartmental cooperation. On the other hand, autonomy and fluctuation are generally not especially relevant at this stage of the organizational knowledge-creation process. A knowledge-creating organization does not operate in a closed system but rather in an open system, in which knowledge creation is a never-ending process, upgrading itself continuously, and in which knowledge is constantly exchanged with the outside environment.
- **Cross-levelling of knowledge** is the phase which extends the created knowledge across all the organizational divisions (intra-organizationally by horizontal and vertical cross-fertilization) or outside (inter-organizationally through dynamic interaction). In order for this phase to function effectively, it is essential that each organizational division has the autonomy to take the knowledge developed elsewhere and apply it freely across different levels and boundaries. Internal fluctuation such as the frequent rotation of personnel, redundancy of information and requisite variety, all facilitate this knowledge transfer. The nature of the knowledge conversion behind the dynamic and interactive process occurring within the two knowledge creation spirals is the key to understanding knowledge formation.

The starting point in building a knowledge conversion is to recognize the need to transcend beyond dichotomies. For example, tacit knowledge and explicit knowledge, as well as an individual and an organization, are not opposing ends of two dichotomies but are mutually complementary entities, interacting dynamically and simultaneously with each other to create a synthesis, which is something new and different. The individual is the creator of knowledge while the organization is the amplifier of knowledge, and the context in which much of the conversion takes place is within the groups or team, which functions as the synthesizer of knowledge. The essence of knowledge creation is deeply rooted in building and managing synthesis through a knowledge conversion process. It is this interactive process that fuels innovation.[23]

2.3
The Core Processes of Managing Knowledge

Reviewing the practices of managing knowledge in profit and nonprofit organizations shows that each organization has set objectives, tasks and activities for its knowledge strategy, regardless of whether they label this as a knowledge strategy. Where a knowledge management strategy is applicable, it is observed that organizations struggle with the concept and face difficulties implementing the various moving parts of knowledge management and to make knowledge management or knowledge sharing the foundation of every day work. The reason for this implementation struggle might be grounded in the failure of organizations to make clear to staff what they want to achieve with knowledge management (business case); the organizational culture is not conducive and not attuned to knowledge sharing, and high-level governance and coordination oversight absent. Often quoted in literature of failed knowledge implementation projects are: (1) the organizational structure and accountabilities; (2) the inability to formulate a clear business case why managing knowledge is essential; (3) the lack of staff support; (4) the absence of a supportive technical environment; and (5) cultural barriers and active leadership.

The difficulty to make knowledge management mainstream in an organization is to formulate and define tangible and measurable outcomes in terms of gains, organizational improvements or exposure. This is another impediment for implementing knowledge-management processes. In order to make knowledge management more quantifiable, an analytical framework that spells out which knowledge activities can be defined on the one hand, and how these activities can be mapped to organizational outcomes in order to measure and quantify progress or achievements is adopted. The literature reviewed on this subject

---

23) I. Nonaka and I. H. Takeuchi, *The Knowledge-Creating Company: How Japanese Companies Create the Dynamics of Innovation*, Oxford University Press, 1995.

shows that the knowledge management cycle or value chain comprises four to eight knowledge-specific outputs to achieve up to eight objectives.

2.3.1
Knowledge Outputs and Outcomes

Davenport and Prusak distinguish three major knowledge management core processes: (i) knowledge generation; (ii) knowledge codification and coordination; and (iii) knowledge transfer.[24] The key characteristics of managing knowledge categorized by Probst received most recognition in the German language area, and comprise eight outputs: (i) identification of knowledge; (ii) acquisition of knowledge; (iii) development of knowledge; (iv) sharing; (v) utilization; (vi) retention; (vii) assessment; and (viii) setting knowledge and knowledge goals.[25] The SECI model of Nonaka and Takeuchi distinguishes four key processes. The outputs can be linked to specific business objectives or outcomes.

The outcome-based view is perhaps more relevant for organizations to look at as a guidance to avoid labelling specific outputs as part of the knowledge management strategy. Many organizations do not adopt formally a knowledge management strategy rather focus for instance on 'lean manufacturing or production principles', 'improving external communications', 'quality enhancement', 'improved business processes', 'organizational performance', or 'people strategy'. The Knowledgeable Ltd.[26] matrix, SECI, and Most Admired Knowledge Enterprises (MAKE) studies can be considered for that reason a succinct methodology as they describe outcomes (see Figure 2.4).

Knowledgeable Ltd defines five outcomes within which there are five maturity levels, with level 1 being the baseline and level 5 the mature or most ideal situation.[27] MAKE uses eight outcomes to assess the organizational activities to become a knowledge-based organization. The main advantage of the MAKE methodology is the availability of comparative data. To define more clearly what it takes to become knowledge-based, MAKE provides a comparative analysis of organizations based on the eight characteristics.[28] The comparison provides useful insights on what organizations actually do to become knowledge-based, regardless of whether they call it knowledge management. The eight MAKE characteristics comprise: Creating the culture for knowledge sharing; Institutionalizing knowledge management; Maximizing intellectual capital; Creating a

---

24) T.H. Davenport and L. Prusak, *Working Knowledge: How Organizations Manage What They Know*, Harvard Business School Press, 1998.
25) G. Probst *et al.*, *Wissen managen: wie Unternehmen ihre wertvollste Ressource optimal nutzen*, Gabler Verlag, 1998.
26) www.knowledgeableltd.com.
27) C. Collison and G. Parcell, *Learning to Fly: Practical Knowledge Management from Leading and Learning Organizations*, Capstone Publishing Limited, 2004.
28) MAKE, annual review on global, regional and national level of organizations, excels in achieving one or more of the eight characteristics: www.knowledgebusiness.com. The MAKE comparative analysis provides few details publicly as to what each characteristic of the core knowledge process really entails, and the interpretation that follows are the authors' reflections.

2 Knowledge Management: From Theory to Practice

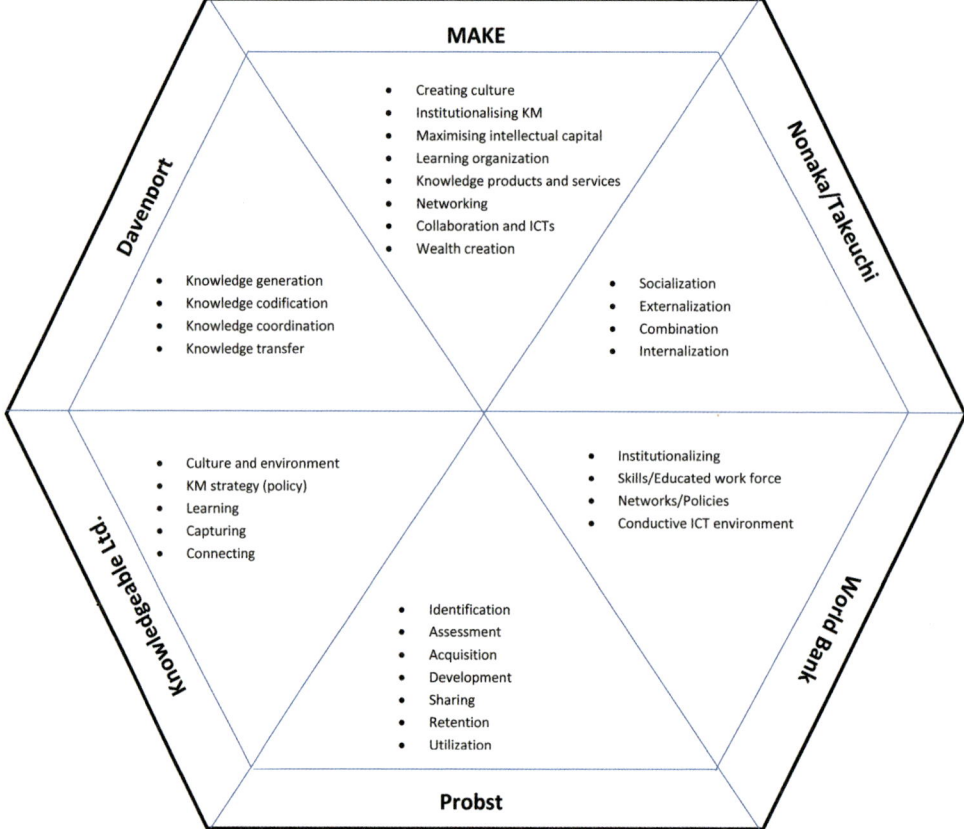

Figure 2.4 The Knowledgeable matrix.

learning organization; Creating knowledge; Networking; Creating an environment for collaborative knowledge sharing/ICTs; and Transforming knowledge into organizational wealth. The sum of all these contributes to making an organization knowledge-based. In reviewing the reports over the past years, the understanding of the eight characteristics can be summarized as follows:

- **Creating the culture for knowledge sharing:** Research identifies culture as one of the biggest impediments to knowledge sharing and transfer practices, citing the inability to change people's behaviour as the biggest hindrance to managing knowledge.[29] A 2011 evaluation of knowledge management from the World Bank shows the paradox between the core process of the bank (providing loans) and its knowledge activities. It was concluded that staff perception perceived knowledge sharing as secondary. Staff perception is

29) Ernst & Young, 1998; KPMG, 2001; APQC, MAKE, 2006 and 2007.

Figure 2.5 De Ocampo's knowledge-sharing image. (© Jovan de Ocampo, Ang Ilustrador ng Kabataan).

that completing a project is valued better and holds more potential for career advancement then devoting time and efforts to knowledge sharing.[30] Discussions on knowledge-sharing barriers focus primarily on organizational and national culture.[31] It is found that the barriers of knowledge sharing are a combination of soft and hard factors and are related to: the lack of interpersonal skills of the individuals; structures, processes and systems in the organization (e.g. deep-layered hierarchical structures); or inadequate tools and technologies (e.g. lack of software tools).[32] In 2012, Schein defines organizational culture as a set of implicit assumption held by members of a group that determines how the group behaves and responds to the environment.[33] The task of the management therefore is, amongst others, to advocate the importance of knowledge sharing and to create suitable conditions and incentives for staff participation (see Figure 2.5).

Creating a culture for sharing and collaboration is often a mix of various hard and soft investments. Some organizations apply soft activities such as creating spaces for knowledge sharing (in-house coffee shops, cafeteria, physical and brain fitness activities, changing the office landscape from closed to open door, etc.). In order to stimulate thinking or raising awareness on knowledge management and sharing issues, senior managers may leave their 'ivory towers' and join staff for lunch and coffee breaks, organize interactive breakfast sessions and promote speaker events. Some apply a blend of the foregoing with more formal activities such as organizing task teams or communities of practices to look into a specific organizational issue, random staff presentations on their work, stimulate corporate blogging, and so on. Culture changes in an organization will not occur overnight, as knowledge is a people-orientated process and forcing people by 'you must' will surely be counterproductive. Creating the conditions, minimizing

30) Knowledge work at the World Bank, 2011: www.worldbank.org.
31) McDermott and O'Dell, APQC, 2001 and 2013; Hofstede, 1993; Trompenaars, 1996 and 2004.
32) A. Riege, *Three-dozen knowledge-sharing barriers managers must consider*, Journal of Knowledge Management, Volume 9, Issue 3, p. 18, 2005.
33) E.H. Schein, *Organizational Culture and Leadership*, Jossey-Bass Publishers, 1992.

Table 2.1 The culture of managing knowledge.

Issue	Responses
Lack of trust	Build relationships and trust through face-to-face meetings
Different cultures, vocabularies, frames of reference	Create common ground through education, discussion, publications, teaming and job rotation
Lack of time and meeting places, narrow idea of productive work	Establish times and places for knowledge transfers for example fairs, talk rooms and reports
Status and rewards go to knowledge owners	Evaluate performance and provide incentives based on sharing
Lack of absorptive capacity in recipients	Educate employees for flexibility; provide time for learning, and hire for openness to ideas
Belief that knowledge is the prerogative of particular groups; 'not invented here' syndrome	Encourage non-hierarchical approach to knowledge; the quality of ideas is more important than the status of their source
Intolerance of mistakes or need for help	Accept and reward creative errors and collaboration; there is no loss of status from not knowing everything

barriers for sharing will help, especially when management shows engagement and gives the example.[34] The most identified impediments for sharing, and how to respond to them, are summarized in Table 2.1.[35]

- **Institutionalizing knowledge management:** This subject concerns the organizational capabilities to make knowledge management everybody's concern in the organization, and not just a senior management responsibility. Studies show that knowledge-based organizations often have an apostle at senior management level overseeing knowledge-orientated activities and coordination amongst various organizational departments or entities.[36] Knowledge management is a multidisciplinary and crosscutting activity in organizations, as it relates to departments responsible for human resource development, information-technology, marketing and communications, production, library, research and development, and even finance. For some departments the contribution to knowledge management may be marginal, but for others it may be significant. In any case, knowledge management requires coordination and a broad understanding of *why* it is needed, and *what* it will address or solve. For many organizations, the first step is to draft a plan, strategy or framework that clarifies what it is that needs to be

---

34) I. Nonaka, *Strategy as Distributed Phronesis: Knowledge Creation for the Common Good*, in S.D. Talisayon (Ed.), Proceedings of the International Productivity Conference, Knowledge Management: from Brain to Business, APO (Asian Productivity Organization), p. 9, 2007.
35) T.H. Davenport and L. Prusak, *Working Knowledge: How Organizations Manage What They Know*, Harvard Business School Press, 2000.
36) APQC, Knowledge Sharing Session and Survey, 2013.

resolved. The strategy should identify a clear business case, and how knowledge management can address this. MAKE assesses whether organizations have a clear plan, whether management is articulating the importance, whether there is a working group or a business owner to drive the change process, and whether staff know what their role is. Knowledge management can start as a top-down approach, or bottom-up or a mixed form through establishing a knowledge ecosystem in a single department before upscaling takes place. Senior management has an important role to play to introduce and promote knowledge management initiatives to the business owner, to the stakeholders, or to the governmental funding agencies.

- **Maximizing intellectual capital:** The intellectual capital school of knowledge management distinguishes three groups, for example the human capital, structural capital, and external capital.[37] These three forms of capital are collectively called 'intellectual capital' and are part of knowledge assets of any organization.[38] A seminal characterization of intellectual capital for organizations is what an organization's participants know that gives the organization a competitive advantage relative to others in its marketplace.[39] Every organization – including research organizations – possesses all three forms of intellectual capital. Human capital is the primary source of innovation and customization, and consists of the skills, competencies and abilities of individuals and groups. These range from specific technical skills to 'softer' skills, such as salesmanship or the ability to work effectively in a team. An individual's human capital cannot be owned by an organization; the term thus refers not only to individual talent but also to the collective skills and aptitudes of the workforce. One challenge faced by executives is how to manage the talent of truly outstanding members of their staff, or how to encourage star performers to share their skills with others. Skills that are irrelevant to an organization may be part of the individual's human capital but do not necessarily represent an added value for employers.

  Structural capital comprises knowledge assets that are developed by the organization such as: intellectual property rights in the form of patents, copyrights, and trademarks; business processes, methodologies, models; engineering; administrative systems, business software, and so on. Managing knowledge is converting human capital into structural capital, so it becomes accessible and shareable for use by others. This happens, for example, when a team documents a lesson or good practice from a project so that others can apply the acquired knowledge. Some structural capital can be a public good such as open-source software. Some research organizations, funded by public sources also make their findings publicly available. A typical example of such an organization is CERN.

---

37) Sullivan, 2000; Stewart, 1997; Edvinssons, 2003; Sveiby, 1997.
38) K.E. Sveiby, *The New Organizational Wealth: Managing and Measuring Knowledge-Based Assets*, Berrett-Koehler Publishers, 1997.
39) T.A. Stewart, *Intellectual Capital: the New Wealth of Organizations*; *The Wealth of Knowledge: Intellectual Capital and the Twenty-first Century Organization*, Doubleday, 1997 and 2003.

External capital is the value of relationships with financiers, suppliers, allies, and customers. An organization that takes a strategic approach to intellectual capital should examine its business or functional model and the economics of its industry to manage the combination of human, structural, and external capital in such a way as to create value that others in the marketplace cannot offer.

- **Creating a learning organization:** Managing knowledge and organizational learning are two sides of the same coin. Organizational learning is the sum of all learning processes within an organization, and that process is clearly articulated in the SECI model mentioned earlier. This includes not only individual learning (personal development) but also social learning from conversations, to team dialogues, to community meetings. Within the context of managing knowledge, learning is considered the creation or acquisition of the ability (potential and actual) for people to take effective action. The concept of the learning organization is based on Chris Argyris' and Donald Schön's work,[40] but in the early 1990s further research on the learning organization and its relationship to knowledge management appeared more prominently, focussing on the notion that a learning organization is a place where people are continually discovering how they create reality and how they can change it. Organizations, as such, should be considered as dynamic systems and in a state of continuous adaptation and improvement to meet external demands.[41] Further research on the evolution of organizations found that organizations that made learning part of their development usually 'live longer' because of their capacity to innovate and change. According to this analysis, long-living organizations have the following key characteristics:[42]
  - The ability to sense, learn, and adapt to changes in the environment in which they operate.
  - Cohesiveness, a strong sense of identity and a sense of belonging.
  - Avoidance of centralized control and a tolerance of eccentricities and experimentations.
  - Financially conservatism and a sound financial balance practice.

Learning organizations are thus organizations where people are stimulated to expand their capacity to create the results they truly desire, where new and expansive patterns of thinking are nurtured, where collective aspiration is set free, and where people are continually learning to see the whole together. In essence, organizational learning requires capacities to absorb and comprehend signals and turning these signals into action. The first type of learning is obtaining know-how or mastering a technique to solve specific

---

40) C. Argyris and D.A. Schön, *Organizational Learning: a Theory of Action Perspective*, Addison-Wesley Publishing, 1978.
41) P.M. Senge, *The Fifth Discipline: The Art & Practice of The Learning Organization*, Doubleday, 1990.
42) A. de Geus, *The Living Company: Habits for Survival in a Turbulent Business Environment*, Harvard Business School Press, 1997.

problems, based on existing premises. The second type of learning is establishing new premises (i.e. paradigms, schemata, mental models or perspectives) to override existing thinking. The creation of knowledge clearly involves the interaction between these two types of learning, and this interaction forms a type of dynamic spiral. Some of the specific ways that organizations learn are single-loop and double-loop learning, or mind-mapping exercises.[43] To sustain themselves, organizations need to have the capacity to transform and reinvent themselves by destroying existing knowledge systems and inventing new ways of thinking and doing. In the corporate world, the process of transforming is often expedited through forming alliances, joint ventures and mergers and acquisitions. The latter usually results in choosing the better of two information and knowledge bases.

> In Colombia, as in Korea, cultural norms tend to dictate that people avoid directly questioning authority, even if the authority is wrong. This culture was attributed to some fatal airline accidents, as described by M. Gladwell, in: Blink: The Power of Thinking Without Thinking, Little Brown and Company, 2005.

An organization cannot create knowledge on its own, and it needs its staff to take the initiative to generate ideas, new pathways of thinking. Organizations, as already pointed out, need to provide the framework and conditions in which individuals can excel and teams can interact within and beyond their organization.

Organizational learning is a dynamic process; it is a long journey that requires staff and teams to grow in awareness and self-responsibility continuously changing the nature of their organization, as exemplified by innovation, collaboration, and culture shifts.[44] Processes applied often entail single-loop learning (when mistakes in products or services are detected and corrected, perhaps an organizational manual or business process adjusted to reflect the new reality) or double-loop learning (when, in addition to the detection of mistakes and subsequent correction of errors, the organization is involved in the questioning and modification of existing norms, procedures, policies and objectives). Double-loop learning involves changing the organization's knowledge base competencies or routines. The organization of learning and the use of the Information space (I-space) framework created by Max Boisot has been analysed in ATLAS (A Toroidal LHC ApparatuS), the scientific collaboration of the Large Hadron Collider (LHC) at CERN.[45]

- **Creating knowledge:** The previous sections have explained the various interrelated dimensions that encompass managing knowledge in such a

---

[43] D. Bennet and A. Bennet, *Engaging tacit knowledge in support of organizational learning*, VINE, The Journal of Information and Knowledge Management Systems Volume 38, N. 1, 2008.
[44] Ibidem.
[45] M. Boisot et al. (Eds), *Collisions and Collaboration: The Organization of Learning in the ATLAS Experiment at the LHC*, Oxford University Press, 2011.

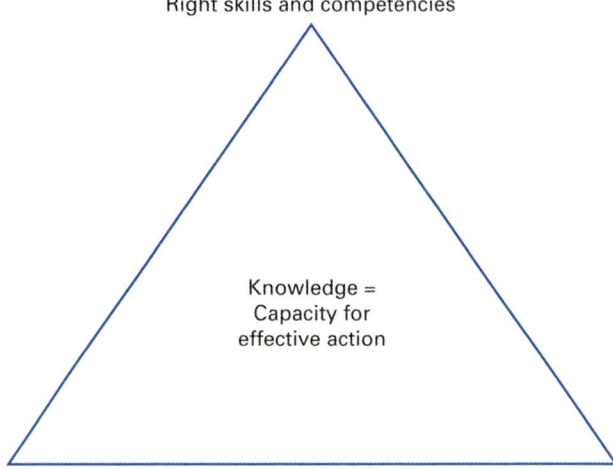

Figure 2.6 The managing knowledge mix of skills.

manner that organizations can create the capacities and conditions for staff to deal with the continuous flow of information and make decisions based on their understanding and the context of organizational objectives."[46],[47] While tacit and implicit knowledge may be useful in understanding the basic of knowledge, these terms here describe aspects of a fluctuating continuum (a range) rather than a strict classification system. In short, managing knowledge is a mix of skills, context relevant information, (decision) support systems and context, as outlined in Figure 2.6.[48]

Creating or developing new knowledge is a costly process. In the age of a superabundance of information made available on the Internet, scientific knowledge and practical useful knowledge (e.g. operational knowledge), that is or can be contextualized and applied relatively easily to solve a problem, is at a premium. This latter form of knowledge is often the result of executing

---

46) Justified belief that increases an entity's capacity for effective action (Nonaka, 1994); knowledge is a capacity to act (Sveiby, 1997); range of one's information or understanding (Webster New Collegiate Dictionary); Knowledge as a set or organizational statements of facts or ideas, presenting a reasoned judgement or an experimental result, which is transmitted to others through some communications medium in some systematic firm (this excludes news, entertainment, etc.). Knowledge consists of new judgements (research and scholarship) or new presentations of older judgments (Bell, 1974).

47) M. Polanyi, *The tacit dimension*, Anchor Books, 1967; E. Goldberg, *The Wisdom Paradox: How Your Mind Can Grow Stronger As Your Brain Grows Older*, Gotham Books, 2005.

48) The triangular representation (or semantic triangles related to metaphysic and triangular representations) appears in many fields including religions. A basic premise in Wilber integrative epistemology is the fundamental distinction between *I*, *We*, and *It*: personal individual knowledge acquired through experience, group/phenomenological knowledge, and empirical/objective knowledge, respectively, 1998.

ordinary business processes. Other forms of new knowledge are the results of R&D. The Frascati Manual distinguishes three forms of R&D in the main fields of science and technology: Basic research; Applied research; and Experimental development:

- **Basic research** is experimental or theoretical work undertaken primarily to acquire new knowledge of the underlying foundation of phenomena and observable facts, without any particular application or use in view.
- **Applied research** is also original investigation undertaken in order to acquire new knowledge. It is, however, directed primarily towards a specific practical aim or objective.
- **Experimental development** is systematic work, drawing on existing knowledge gained from research and/or practical experience, which is directed to producing new materials, products or devices, to installing new processes, systems and services, or to improving substantially those already produced or installed. Experimental development R&D comprises all creative work undertaken to increase the stock of knowledge of humankind and society, including culture.[49]

- **Networking:** Because of the costly nature of developing new knowledge or insights, organizations (and for that matter educational institutes as well) tend to focus on developing or maintaining specific core areas of knowledge and seek partnerships for the codevelopment of knowledge. In some cases, the organic appearance of research clusters in specific industrial areas that supports the flow of knowledge between educational institutes and organizations (Silicon Valley, Seattle, Boston 101, Stuttgart, Amsterdam) is observed, while in others there is an active intervention of government in creating knowledge clusters, for example Kuala Lumpur multi-corridor, Singapore, Dubai, London City (banking), and the Municipality of Amsterdam. Recently, organizations have found it advantageous to create particular entities or units where staff can generate, share and utilize knowledge, both internally and with external stakeholders. The advancement of information

---

Amsterdam's prestigious new science institute is to be developed by the Massachusetts Institute of Technology, Delft University and Wageningen University, the city announced on Friday. The project to set up the 'Institute for Advanced Metropolitan Solutions' was head and shoulders above the other short-listed entries, Robbert Dijkgraaf, chairman of the jury, said in a statement. 'Amsterdam will be getting something really new: a scientific institute of technology which will develop expertise in practical urban issues', Dijkgraaf said. The city council is pumping €50m into the new institute, which is expected to generate hundreds of jobs. Corporate sponsors such as Cisco, Shell, IBN, Accenture, TNO and Schiphol are also involved.

DutchNews.nl, Saturday 14 September 2013

---

49) Frascati Manual, *Proposed Standard Practice for Surveys of Research and Experimental Development*, Sixth Edition, OECD (Organization for Economic Co-operation and Development), 2002.

technology tools (organizational or public social media platforms) is propelling these forms of fluid networks. Reference is made to structures such as Communities of Practice (CoP), X-teams, and Communities of Innovation (CoI) that bypass the major building blocks in creating, transferring and applying organizational knowledge in order to create value. The organizations in which CoP and CoI are implemented can also be characterized as effective knowledge-based organizations. CoP and CoI are organizational forms able to anticipate, react, and lead change, complexity and uncertainty. In a networked organization, value is generated by nurturing informal and formal relations and encouraging a free, horizontal flow of knowledge across organizational boundaries, with the aim of opening new channels of communication and sustaining the propagation of new ideas. Examples of effective network organizations can be found in the airline and automotive industries, which maintain industrial co-supply chains as well as relationships with scientific institutes in the development and application of new materials. A networked organization promotes and maintains therefore native open innovation approach, explores, and exploits knowledge acquired from external sources (competitors, universities, partners), and retains the best talents.[50] Effective networking basic research organizations can also be found in High-Energy Physics (HEP), molecular biology, astronomy or space [in Europe, the European Organization for Nuclear Research (CERN), the European Molecular Biology Laboratory (EMBL), the European Southern Observatory (ESO), and the European Space Agency (ESA)] are typical examples. These types of networking have been promoted by the need to create infrastructures and tools that otherwise would be too costly for a single state, or by the necessity to join efforts and competences to contribute to substantial achievements of knowledge development. These comprise universities and institutions, often from all over the world, establishing synergies between individuals and between researchers and industries, fuelling R&D at the forefront of technology and innovation.

- **Creating an environment for collaborative knowledge sharing/ICTs:** Today, economic competitiveness depends on the ability to transform vast quantities of information into usable knowledge and to deploy it effectively across the organization and external stakeholders. To support knowledge management, organizations usually apply one or all of the following ICT support systems (see Figure 2.7):

  a) Enterprise Content Management (ECM). An ECM system enables the capture and storage of, and access to, organizational working, draft and official documents. An ECM supports collaborative work, access, editing and search. The ECM can be an accumulation of records (archived documents), working or active documents, notes, mails, photos and videos.

---

50) M. Grimaldi et al., *A methodology to assess value creation in communities of innovation*, Journal of Intellectual Capital, Volume 13, Issue 3, p. 305, 2012.

2.3 The Core Processes of Managing Knowledge

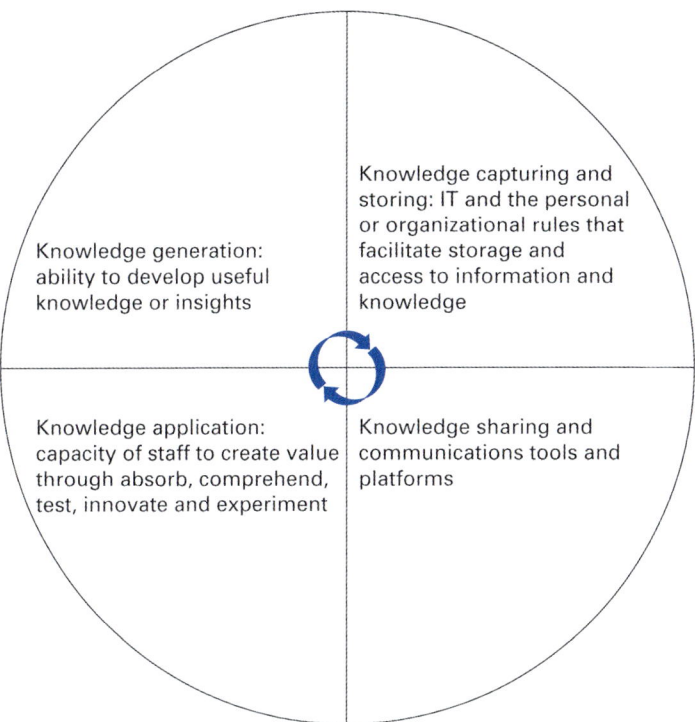

Figure 2.7 Information and Communications Technology (ICT) support systems for knowledge management.

    The CERN Invenio[51] free software system, with its navigable collection tree and powerful search engine, is an example of an ECM platform.

b) Enterprise portal. This a single entry point for staff or stakeholders to applications and information. The portal operates as a dashboard and syndicates information resources that meet a specific user profile. A portal empowers users also to publish information directly to the intranet or Internet, with little interference from a governance body that oversees quality of content.

c) Collaboration and social media platforms. These are dedicated platforms enabling groups and CoP to capture and share knowledge with other experts in the organization or beyond. Video-chat, presentation and e-learning applications are part of this platform.

d) Information robots applications. The amount of new information added daily to the existing stock of knowledge is impressive and it exponentially increases, with more and more people added to the World Wide Web from new markets. To select information that is useful, or which fits a

---

51) Invenio is a software suite that enables a personal digital library or document repository to be run on the Web.

certain information profile, information robots can play an important role in selecting what information will be displayed or disseminated to a user. Information robots have the capacity to roam specific information sources and define relevance based upon summary, abstract and references used. As early as the 1980s it was determined that a scientific user of information can probably select up to four of five journals to stay abreast of new scientific developments. What is missed will appear again through social and alumni networks, and conferences. Information robots are personal desktop tools that help a user to define journals, topics, and how the information will be delivered.

e) Wiki. Work-related wikis tend to focus on knowledge about an evolving project (strategic analysis, work progress, identification of unsolved issues, lessons learned, etc.) or the capture of information that might be typically be decentralized in an organization. Individuals of the organization, and also authorized outsiders, contribute their knowledge to the work-related wiki as they participate in the project or gain knowledge relevant to the wiki topic. Others in the organization viewing the work-related wiki can then reuse the accumulated knowledge to improve their own work experience. The business process is very similar to the face/face internalization and socialization of knowledge, and is especially useful in decentralized organizations or as team wiki to preserve the evolving team's knowledge of a task.[52]

- **Transforming knowledge into organizational wealth:** Amongst leading knowledge management academies and practitioners, knowledge is commonly understood as the capacity for effective action, which includes information useful for carrying out such actions. An action by an individual or an organization based on knowledge can be considered effective if the result is close to what is desired or intended. For the private sector, the desired result is often the creation of market value, exposure (free publicity) and sufficient profit, whereas for the public sector and civil society it is about the creation of social value and impact. For development organizations the desired result is its impact in alleviating poverty through scaling-up projects or through knowledge transfer of good practices, an achieved localization of knowledge, and capacity building efforts. For development organizations the desired result is its impact in alleviating poverty through scaling-up projects or through knowledge transfer of good practices, achieved localization of knowledge, and capacity building efforts. For research institutes, the desired results from the application of knowledge are new outstanding results, the validation and creation of a new scientific hypothesis, improved social exposure, valuable recognition and esteem also for members, the establishment of new synergies and larger networking, adequate policies, new and better deliverables, and patents registered and commercialized.

---

52) A. Majchrzak *et al.*, *The Impact of Shaping on Knowledge Reuse for Organizational Improvement with Wikis*, Management Information System Quarterly, Volume 37, Issue 2, p. 445, 2013.

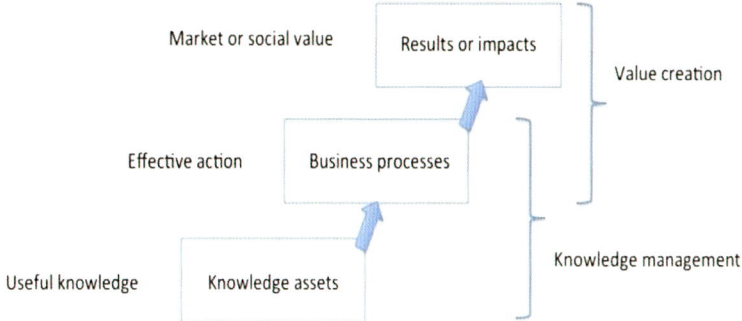

Figure 2.8 The ultimate aim of knowledge management value creation. (Courtesy of Serafin Talisayon, Centre for Conscious Living Foundation Inc., CCLFI).

Knowledge includes: embodied knowledge in people; embedded knowledge in technology, systems and processes; acculturated knowledge in work relationships, teams and networks; and actionable information and insights. The ultimate aim of knowledge management is to create value when used or applied for effective action as represented in the scheme in Figure 2.8.[53] Results, quality, the capacity to innovate, marketing, and revenues are measurable of knowledge management and deliverables of knowledge management practices.

Knowledge exists in the human brain in the form of stored or expressed neural patterns that may be activated and reflected upon through conscious thought. From this pattern, new patterns are created that may represent understanding, meaning and the capacity to influence actions. Through this process the mind is continuously growing, restructuring and creating an increased tacit knowledge base. Knowledge management is enabling and enhancing capabilities to perform the above processes, including the sourcing and deployment of the correct knowledge assets, in order to achieve the prefixed aims or better results.

2.4
The Knowledge Worker

Knowledge – much like living beings – is subject to 'selection' by humans or machines in the sense that an enormous amount is generated and needs to be processed, filtered, selected or rejected. In evolutionary epistemology, it is recognized that conscious identifiable actors carry out the selection of information,

---

[53] I. Nonaka, *Strategy as Distributed Phronesis: Knowledge Creation for the Common Good*, in S.D. Talisayon (Ed.), Proceedings of the International Productivity Conference, Knowledge Management: from Brain to Business, APO p. 9, 2007.

unlike in evolutionary biology where selection is a result of differential survival and reproduction, with no conscious selector operating it. The world of tacit and explicit knowledge is one in which the actors choose at the individual level. One of the agents is 'the knowledge worker', a term coined by Peter Drucker in 1959 to define a person who works primarily with information or who develops and uses knowledge in the workplace.[54] The term by itself is not very useful as most workers use information to perform their tasks, and it took Drucker almost 40 years to refine the six major elements that categorize (identify) the knowledge worker:[55]

1) Task management (better performance, better use of communication tools and networking capability to search for and gather strategic information).
2) The responsibility for personal productivity. Knowledge workers have to manage themselves, they have to have autonomy and mobility.
3) Continuing innovation has to be part of the work, the task and the responsibility of the knowledge workers.
4) Continuous learning is required on the part of the knowledge worker, but equally continuous teaching is required.
5) The productivity of the knowledge worker is not – at least not primarily – a matter of the quantity of output. Quality and efficiency is at least as important.
6) The knowledge worker is both seen and treated as an 'asset' rather than as a 'cost'. Furthermore they want to work for the organization in preference to all other opportunities.

The growing importance of knowledge workers and how to 'manage' them was further emphasized in the book *Thinking for a Living*.[56] Managing knowledge workers is a challenging task, but albeit is the factor driving the future of management because those who are managed as knowledge workers will be a substantially different group from the workers of the past.[57] Some companies have traditional workers coexisting with knowledge workers. It should be underlined that the eight specific changes described by Davenport (as reported below) resonate fairly well with the eight dimensions of the MAKE:

1) From overseeing work to doing it.
2) From organizing hierarchies to organizing communities.
3) From hiring and firing workers to recruiting them and retaining them.
4) From building manual skills to building knowledge skills.
5) From evaluating visible job performance to assessing invisible achievements.

54) P.F. Drucker, *Landmarks of Tomorrow*, Harper & Brothers, 1959.
55) P.F. Drucker, *Knowledge-Worker Productivity: The Biggest Challenge*, California Management Review, Volume 41, N. 2, p. 78, 1999.
56) T.H. Davenport, *Thinking for a Living: How to Get Better Performances And Results from Knowledge Workers*, Harvard Business School Press, 2005.
57) Ibidem.

6) From ignoring culture to building a knowledge friendly culture.
7) From supporting bureaucracy to fending it off.
8) From relying on interpersonal relations to considering a variety of sources.

More recent thinking from Davenport on the strategic approach to knowledge work stresses the importance of the approaches, benefits and shortcoming, for improving access to the information that lies at the core of knowledge work.[58]

As already mentioned, the primary qualification of knowledge workers is quality. If the level of quality output required is achieved, only then can the quantity of output (although difficult to assess in all its aspects) be measured and analysed.[59] Peer group teams and communities can be very effective tools in the management of knowledge workers. Making peer group or communities accountable for specific deliverables, and rewarding them as a team, will usually provide feedback if any one particular knowledge worker is inadequate. A team of knowledge workers will often carry the burden of an inadequate team member, but only for a short period of time. The changes in managing knowledge workers mean that individuals become slowly visible in major organizations, including the scientific ones. Companies which rely for their performance and innovation on knowledge workers will keep their doors open.

2.4.1
The Individual Learning Process

The understanding of how to manage knowledge starts from understanding the learning process at an individual level.[60] Individual learning is an expanding perception process that starts at birth. Perception refers originally to the creation of sensation from sensory excitation, which is the elementary process from which all learning starts.[61] The perceptional approach, which has been developed over the past 20 years by Kaarle Kurki-Suonio, is the foundation for a model and

---

58) T.H. Davenport, *Rethinking Knowledge Work: A Strategic Approach*, Mckinsey Quarterly, Volume 10, N. 7, p. 88, 2011.
59) B.D. Janz et al., *Knowledge worker team effectiveness: The role of autonomy, interdependence, team, development, and contextual support variables*, Personnel Psychology, Volume 50, Issue 4, p. 877, 1997.
60) B. Bressan, *A Study of the Research and Development Benefits to Society from an International Research Centre: CERN*, PhD thesis, Report Series in Physics HU-P-D112, University of Helsinki, Finland, 2004; *Knowledge Management in an International Research Centre*, Lambert Academic Publishing, 2011.
61) K. Kurki-Suonio and R. Kurki-Suonio, *The Concept of Force in the Perceptional Approach*, in H. Silfverberg and K. Seinelä (Eds), Ainedidaktiikan teorian ja käytännön kohtaaminen. Matematiikan ja luonnontieteiden opetuksen tutki- muspäivät 24–25.09.93, Reports from the Department of Teacher Education in Tampere, p. 321, University of Tampere, Finland, 1994.

provides insights for pedagogical applications.[62],[63] Concepts are processes – not products – and understanding arises from a perception of the concepts, and not from the concepts themselves. Perception is the primary creation of the concept's meaning; mental models or the individual understanding of nature call for subsequent verbal conceptualization. Conceptualization leads to terminology or language, which become the tools for further perception. In this way, concepts add new material to the perception process, leading to a hierarchy of concepts of increasing generality and abstraction. In particular, science is a highly structured perception process.[64],[65],[66],[67]

Perception, learning, studying and research are different stages of continuously ongoing process leading to the creation of knowledge, which is fundamentally everyone's personal process. Intuitive knowledge attained by perception is permanent. For Kurki-Suonio, both learning and research are conducted and controlled by intuitive sensitivity, and not by logical necessity.

Furthermore, according to the perceptional approach, understanding is based on the perception of empirical meaning, and a theory is the detailed quantitative representation of something that is understood. In this sense, observation and experiment form the basis of learning, where concepts are the representations of Nature as it is observed. When meaning ceases to be merely a formal representation, and is understood instead to be the essence of a concept, the nature of a concept belonging to scientific knowledge is seen under a new light. Concepts do not exist without their meanings, and the meanings cannot be separated from the process that creates them. Thus, concepts cannot be understood as the building blocks of a theoretical structure which maps on to the empirical structure of experimental results. The primary meanings of physical concepts stem from the basic perception of the relevant class of phenomena and results in a mental model of causal relationships of the perceived vision as a whole.

Concepts in sciences, such as physics or mathematics for example, are characterized by the quantification of observable properties. The quantification is the threshold process that transforms observable properties and their relationships

---

62) The perceptional approach constitutes the underlying framework of the physics teacher education programme at the Department of Physical Sciences of the University of Helsinki, Finland, and has been developed over 20 years. In Finland, this approach is the only substantial comprehensive, consistent, and well-known collection of ideas about physics education.
63) K. Kurki-Suonio and R. Kurki-Suonio, *Fysiikan Merkitykset ja Rakenteet* (The Meanings and Structures of Physics), Limes ry, 1994.
64) K. Kurki-Suonio and R. Kurki-Suonio, *Empiirinen Käsitteenmuodostus Fysiikassa*, in A. Airola et al. (Eds), Tieteen teitä, p. 51, University of Helsinki, Finland, 1992.
65) K. Kurki-Suonio and R. Kurki-Suonio, *Tutkimuksen ja Oppimisen Perusprosessit*, Dimensio 55, Volume 5, p. 18, 1991; Elämänkoulu, Volume 2, p. 4, 1993.
66) K. Kurki-Suonio and R. Kurki-Suonio, *Perceptional Approach in Physics Education*, in M. Ahtee et al. (Eds), Proceedings of the Finnish-Russian Symposium on Information Technology in Modern Physics Classroom, 1994.
67) K. Kurki-Suonio, *Principles Supporting the Perceptional Teaching of Physics: A 'Practical Teaching Philosophy'*, Science & Education, Volume 20, Issue 3–4, p. 211, 2011.

into quantities and laws. Above the levels of quantities and laws, there is the level of quantitative understanding, where the laws are combined into theories and models. Therefore, the knowledge and learning in such sciences includes, in addition to normal conceptual development, a quantification process. This threshold process transforms qualitative appreciation into quantitative formalization and builds a structure of quantitative concepts on the foundation of a qualitative system.

The primary evidence and the main motivation for the development of Kaarle Kurki-Suonio's study comes from the observation that a vast majority of students understand physical and mathematical sciences as an abstraction, without much relation to Nature, and that teaching reinforces this improper perception. In the analysis of the historical development of science, many examples demonstrate that perceptional approaches and experimentalism are necessary in teaching; it is not enough just to demonstrate and explore phenomena in the classroom. Kurki-Suonio's study focuses on the creation of knowledge through interaction between Nature and the human mind, starting from the hypothesis that this interaction has a fractal or self-similar repeating dynamic and cyclic structure. The basic features of his model link the elementary processes of perception of experimental and theoretical research to the science and technology processes. A Cartesian dualism prevails between empiry (an empirical approach) and theory.[68] Traditionally, separating dualism has prevailed in physics teaching and research. According to unifying dualism, the observation and the mind, or the experiment and theory, are inseparably coupled. There are neither purely experimental experiments nor purely theoretical theories. Separate experimental and theoretical processes do not drive the science, and perception is rather a one person, single expanding process.[69],[70] The basis for empirical conceptualization is the understanding of a concept through empirical perception, which includes the development process that creates the meaning of the concept. The meanings of concepts are created by a continuous process, which is fundamentally steered from perception to theory. Every scientific concept is a process in which experiment and theory are joined into one constantly developing concept formation. Therefore, conceptualization in science is neither purely a mental process nor a purely observational, and the interaction between theory and experiment creates continuous refinement. Theories constructed on the basis of current knowledge are tested by experiment and the outcomes used to refine or discard theories. Every concept, quantity, law or theory has its empirical basis, which forms the core of its meaning.

---

68) L. Palmer uses the term 'empiry' (i.e. observation and experimentation, as the counterpart of theory) in an electronic mail message to physl@atlantis.uwf.edu to embrace both classes of scientific activity, 1998.
69) K. Kurki-Suonio, *Ajatuksia Fysiikasta ja Todellisuudesta*, Dimensio 55, Volume 8, p. 38, 1991.
70) A. Hämäläinen, *An Open Microcomputer-based Laboratory System for Perceptional Experimentality*, PhD thesis, Report Series in Physics HU-P-D70, University of Helsinki, Finland, 1998.

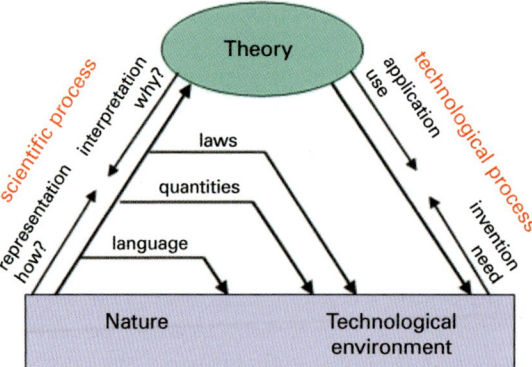

Figure 2.9 Basic processes of knowledge creation within the learning process.[71] (Courtesy of Kurki-Suonio University of Helsinki).

2.4.2
Scientific, Technological and Social Processes

According to Galileo Galilei, *how* is the only way to approach the question *why*. Understanding and use are the two basic motions of the knowledge creation process, and both have to answer the questions, why and how (Figure 2.9).

Two bidirectional branches – the *scientific process* and the *technological process* – are the two different types of interaction of Nature and the human mind or of empiry and theory. The primary subprocess of the scientific process, moving from *Nature* to *Theory*, is the *representation* that forms the whole perceived vision from the excitation of senses to the mental pictures generated to describe real entities and phenomena and their properties. The secondary subprocess of the scientific process, steered from *Theory* to *Nature*, is the *interpretation* controlled by hypotheses, predictions, verifications, and falsifications. The primary subprocess of the technological process is *application*, steered from *Theory* to the *Environment*, whereas the secondary subprocess, *invention*, is steered from *Environment* to *Theory*. The scientific and technological processes are mutually dependent and inseparably interconnected.

Nature and environment are also connected as the environment is the part of Nature that can be *technologically* manipulated, changed, and improved. Thus, together the scientific and the technological processes form a loop that is mediated by a third branch, the *social process*, which is the interaction between individuals and groups extending individual cognition into shared understanding. This loop primarily rotates clockwise, and all three processes are inseparably intertwined. Every concept contains a seed of application and has both a

71) Ibidem.

scientific and a technological meaning, agreed by the social process to create common concepts. This is not only a large-scale structure, but it is also a fractal feature of the knowledge creation process. The presence of these three processes and their interconnections can be identified at the early stages of knowledge acquisition. The entities, phenomena and properties perceived by a child as the start of the scientific process are often loaded with practical applications, which the child is constantly taking into account and making use of elsewhere. When a child – quite naturally and intuitively – adapts his or her behaviour, and when the child searches his or her possibilities and limits through trials and errors, that child is starting to develop the technological process. The significance of the social process, the child's interaction with his or her parents, is clearly a necessary condition for both the child and the parents. The scientific process builds conceptions of the world, aiming at an understanding of natural phenomena.

The technological process changes the structure of the world to satisfy *uses* and *needs*. The social process negotiates the meanings to find agreement about procedures and results within both the scientific and technological processes.

Technology products are the results of the technological process, whereas concepts and conceptual structures are the results of the scientific process. Inventions and unifying ideas, respectively, represent great achievements. For example, while the scientific process leads to the great unifying developments of physics and many other sciences, the technological process is responsible for changes in the environment, modifying living conditions and impacting on countries' technological capabilities. Ultimately, the scientific and technological processes which are the sources and driving forces of science and technology are interconnected.[72] Anyway, all these developments and changes have become possible only through agreement reached in the social process.

The role and the relationship of the three processes – as well as their action and interaction significance – are equally recognized in science as well as in learning. The scientific process dominates during the phase of planning and interpretation, while the technological process is required for the design and performance of the experiment in order to control Nature. The social process, activated by discussion, governs both the scientific and technological processes.

2.4.3
Concept Formation and the Hierarchical Levels of Conceptualization

In the learning process, understanding naturally comes first, while concept formation is cumulative and leads to a hierarchically layered and structured

---

[72] In the early development of the philosophy of technology as a discipline, Bunge considers 'technology' and 'applied science' as synonyms (M. Bunge, *Technology as Applied Science*, Technology and Culture, Volume 7, N. 3, p. 329, Society for the History of Technology, 1966).

knowledge. The concepts of different hierarchical levels are not end products but are active elements of the scientific process. The logic of conceptualization is a fractal, self-similar, bidirectional process which includes the three processes (scientific, technological and social), the inseparability of these three processes, the movements in two directions from Nature to Theory and from Theory to Nature, and the inseparability between these two movements. This means that all the basic features of the process structure are similar at any scale of the process and can be realized at all hierarchical levels.

There are three main approaches involved in the learning processes: the axiomatic-deductive approach; the empirical–inductive approach; and the perceptional approach. The first approach is a thinking method for learning based directly on well-understood structures of knowledge; it emphasizes the importance of deductive reasoning and begins with theory. This approach requires the ability for abstract reasoning and assumes that the empirical meanings of theoretical concepts are already well known. The second approach starts by questioning Nature to understand why. Its benefit is concreteness, in the sense that concepts have from the start an empirical meaning. Obviously, neither of these two approaches works alone. The third approach, the perceptional approach, emphasizes the development history of the concepts and the importance of intuition. This approach tries to present a true picture of the experimental basis of concept formation and bases the structure of physical knowledge on unifying theories, the great achievements of the scientific process.

All conceptualization starts from basic perception. The transition from qualitative to quantitative methods and concepts gives the perception process a new dimension. *Quantities*, *laws* and *theories* are the quantitative parallels of properties, phenomena and their resulting mental models. The primary direction of the conceptualization process is from concrete to abstract, from simple to structural, from observations through *language*, quantities and laws to theory. It puts the bidirectional dynamics of science into different directions: the *experimental approach* identified as the primary direction, and the *theoretical approach* as the opposite.

The hierarchical levels of conceptualization represent different abstraction levels. In studying each topic, the level of language is the most important hierarchical level, because upper levels can only be reached through it. This means that the main boundary is between the first level and the other three, because the first implicitly contains the other three levels. This level lays the foundation for the scientific language. The level of quantities is entered by *quantification*, based on experiments. The third connected level – laws – is an abstraction, a mathematical *representation* of the results of concrete experiments. The level of theories is the uppermost level. The theoretical meaning of a quantity comes about through *structurization*. This is a threshold process leading to the highest conceptual level of theories, the *interpretation*.

This hierarchical structure forms a spiral where a quantitative level of concepts is necessary to build up the basic perception for the next level of

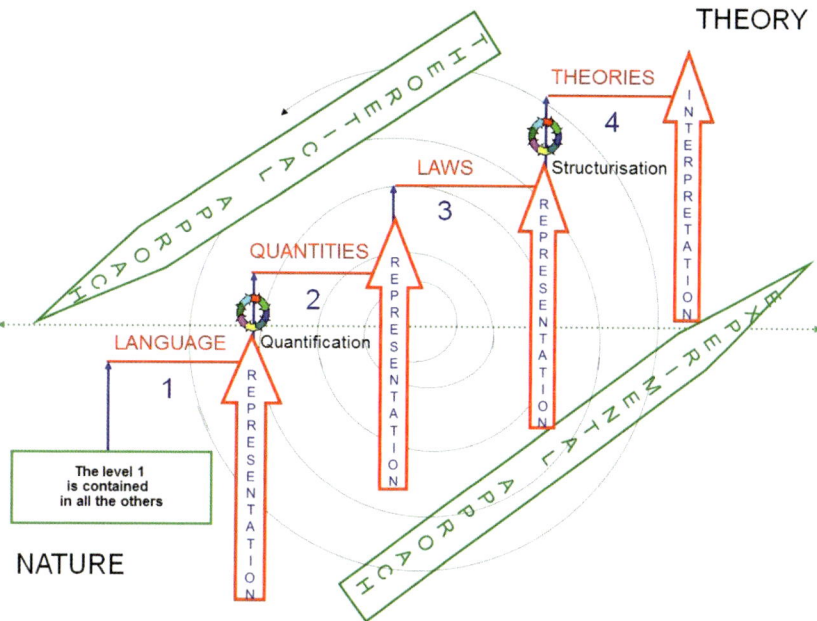

Figure 2.10 The spiral of the hierarchical levels of conceptualization.

qualitative appreciation of new phenomena, inventing new language and reinitializing the conceptualization process from first level (see Figure 2.10).

2.5
The Knowledge Creation, Acquisition, and Transfer Model

On the basis of the previous considerations, a model of creation, acquisition and transfer of knowledge (with particular regard to the scientific knowledge) is developed to describe the process path, from individuals to the organization, aiming at a better understanding and management of the acquired knowledge. This model absorbs the strengths of the two approaches described above: Kurki-Suonio's approach regarding individual knowledge perceptional creation; and Nonaka's approach focusing on organizational knowledge acquisition.

Considering that only individuals can create knowledge, and not an organization as such, it is very important for an organization to support and stimulate the knowledge of individuals and to aid knowledge acquisition by providing relationships and expanding opportunities for the development of new individual skills, absorbing and sharing them thereafter within the organization. This type of dynamic internal and external interaction and integration facilitates the transformation of individual knowledge into organizational knowledge, which can fuel innovation, leading to competitive advantage via enhanced technological distinctiveness, forefront developments and inventions.

The approach of Kurki-Suonio, which was developed originally for physics education purposes, based on scientific knowledge creation in individuals from universities and research organizations, places the main emphasis on the scientific process. This is considered more apt to generate innovative ideas while the roles of the technological process and the social process are discussed only in general terms. The approach of Nonaka is for industrial and commercial organizations, and is based on technological knowledge; it gives more weight to the details of the structure of the social process without paying too much attention to the scientific process.

In order to create a knowledge model applicable to international scientific nonprofit organizations or universities, where scientific knowledge and development is of paramount importance, both of these approaches are needed and the following associations have been identified to absorb them:

- The negotiation of meaning and the creation of scientific and technological knowledge in the social process of Kurki-Suonio have been matched with the creation of tacit knowledge, mental models and technical skills reported in Nonaka's description of the socialization mode.
- The transformation from empirical meaning to conceptualization has been correlated with the process of articulating tacit knowledge into explicit knowledge.
- The hierarchical development of knowledge, distinguishing the scientific process in the perceptional learning approach, became the process of systematizing concepts into a knowledge system typical of and specific to the organizational knowledge acquisition process.
- The empirical meaning, the starting point of knowledge in Kurki-Suonio's theory, and in Nonaka's theory the acquisition of tacit knowledge, is the experience.
- The relationship between Theory and Nature corresponds to the relationship between mental images and the creative process.

The spiral formed by the hierarchical structure of the conceptualization of science, composed of the four levels of language, quantities, laws and theories (see Figure 2.10), can be combined with the spiral that takes place at the epistemological dimension across the four modes of knowledge conversion, internalization, socialization, externalization, and combination (see Figure 2.2). The organizational knowledge acquisition as described by Nonaka can be interpreted as a description of the structure of the social process in Kurki-Suonio's individual learning process. Finally, the model describing the knowledge process path, from the individual learning process to the organizational knowledge acquisition, and transfer of the acquired knowledge (including the scientific one) to other organizations is represented in Figure 2.11, as a combination of the two different approaches.[73]

2.6
Knowledge Management: A Case Study of CERN[74]

By introducing knowledge management concepts (that have so far been limited to companies and information technology) to science organizations, the knowledge creation, acquisition and transfer model helps to reduce the gap between the scientific and technological worlds. The purpose of scientific research, in general, is to find the laws to describe Nature, and such purpose is pursued with the aim of understanding phenomena and explaining the world in which we live. To understand phenomena, a comprehension of the scientific process is important, but in doing that to satisfy our needs the technological process is essential. Increasingly, society asks scientists – and in particular physicists – to explain what advantages society has gained from their research, and which outcomes can be expected.

---

73) See the description of the original model combining Kurki Suonio's and Nonaka's approaches (see figure below) in B. Bressan, *Knowledge Management in an International Research Centre*, Lambert Academic Publishing, 2011.

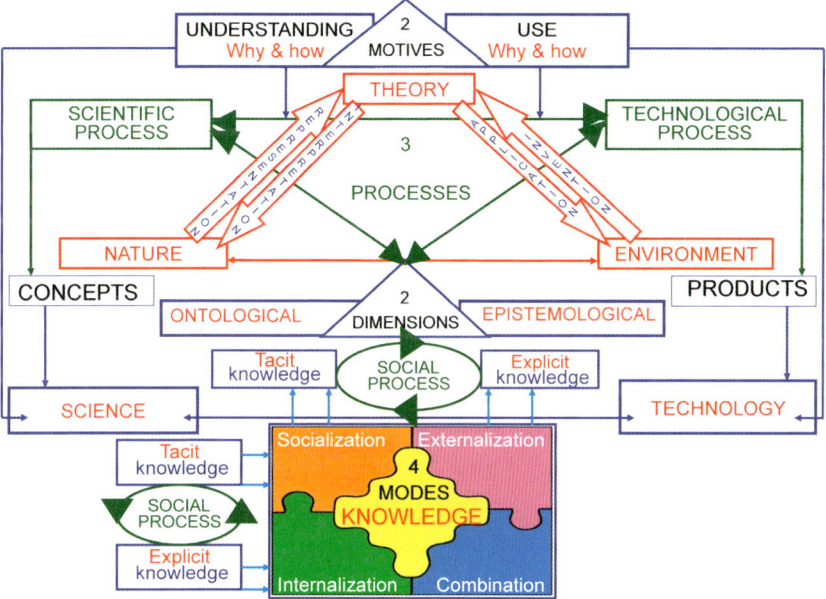

74) B. Bressan *et al.*, *Knowledge Creation and Management in the Five LHC Experiments at CERN: Implications for Technology Innovation and Transfer*, postdoctoral project, University of Helsinki, Finland, and CERN, 2008.

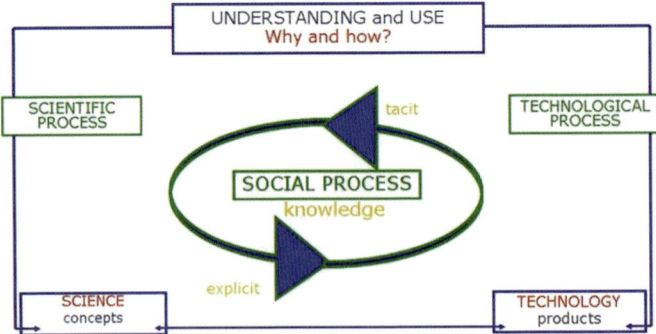

Figure 2.11 The knowledge creation, acquisition and transfer model. (Courtesy of CERN).

There are mainly three types of direct benefit to technology and society from HEP research: (i) entirely new fields of technology may be created; (ii) some of the pioneering technologies created may be used to create new products or services with an impact in other fields; and (iii) people trained as students or who have worked in HEP may incorporate industry or other institutions, thus transferring their acquired knowledge so as to produce innovation and contribute to the wealth of society. The lack or poorness of interaction between scientists and industry is a problem for the transfer of innovative technologies. It stems from a disparity of intents, the immediate applicability of technology, and the balance between investment and profit for developing a new application. It is important to remember that for many years (e.g. at CERN) a certain amount of knowledge and technology transfer that benefitted industry and society at large was achieved through procurement activities and interactions.[75],[76] Nowadays, the technology transfer structures made available by universities and scientific organizations are intended to facilitate, promote and publicize these interactions.[77] In addition, a widespread sharing of knowledge between scientists and industry exists in many countries and innovative R&D is increasingly globalized.

Knowledge creation has become an interesting and relevant research topic due to the impact it has on innovation and the creation of new ideas. Science and research organizations such as CERN are good places to conduct such investigations, as every year they train thousands of new research workers who are qualified and well trained both for academia and industry. The longest observation and broader comparative analysis on scientific epistemic cultures which consider the knowledge generated from scientific and technical practices and structures in molecular biology and HEP[78] found that different

75) H. Schmied, *A Study of Economic Utility Resulting from CERN Contracts*, CERN 75–05, 1975.
76) M. Bianchi-Streit et al., *Economic Utility Resulting from CERN contracts*, (Second study), CERN 84–14, 1984.
77) B. Bressan and M. Streit-Bianchi, *CERN technology transfer to industry and society*, CERN, 2005.
78) The ATLAS Collaboration from 1987 to 1996.

scientific fields exhibit different epistemic cultures, and outlined the communitarian culture and importance of distributed cognition in HEP.[79]

The knowledge spillovers from fundamental science are also a source of beneficial outcome for the economy, and for society. CERN represents a good place to investigate how knowledge is created, acquired, and transferred by:

- providing an introduction to other knowledge available within and outside the Organization;
- expanding opportunities for the sharing of knowledge and the development of new knowledge;
- improving personnel competence;
- integrating existing and new skills and competences into existing or new projects;
- facilitating interactions within the Organization leading to technological innovation.

Indeed, the fertile environment of CERN and its collaboration experiments, building and managing multiple processes, involves a dynamic, interactive and simultaneous exchange of knowledge both inside and outside their organization, including with the industrial world. The organization of learning and use of the I-space framework[80] describes how knowledge is created and shared within ATLAS.[81,82] The Social Learning Cycle (SLC) of this collaboration is described by six phases (Scanning, Codification, Abstraction, Diffusion, Absorption, and Impacting). The facilitators and impediments, due to institutional and culture differences within the project, affecting the SLC and their position within the I-space, have been identified. The interlaced knowledge created across boundaries is a very important factor, providing understanding and trust and facilitating the coordination of primary importance in such a large-scale project. Four cognitive moves (Codify, Abstract, Absorb and Impact) and two social moves (Scan and Diffuse) are used by the individual learner in the I-space;[83] these, when combined, generate complex dynamic processes of knowledge social learning. The physics collaborations have been able to offer the possibility to build up the cognitive, relational and structural dimension of the individual.

Furthermore, an analysis has been carried out to investigate the knowledge creation, acquisition and transfer in the five Large Hadron Collider (LHC)

---

79) K. Knorr Cetina, *Epistemic Cultures: How the Sciences Make Knowledge*, Harvard University Press, 1999.
80) M. Boisot, *Information space: a framework for learning in organizations, institutions and culture*, Routledge, 1995.
81) M. Boisot et al. (Eds), *Collisions and Collaboration: The Organization of Learning in the ATLAS Experiment at the LHC*, Oxford University Press, 2011.
82) M. Boisot and M. Nordberg, *A Conceptual Framework: The I-Space*, in M. Boisot et al. (Eds), Collisions and Collaborations: The Organization of Learning in the ATLAS Experiment at the LHC, p. 28, Oxford University Press, 2011.
83) M. Boisot and B. Bressan, *The Individual in the ATLAS Collaboration: A Learning Perspective*, in M. Boisot et al. (Eds), Collisions and Collaboration: The Organization of Learning in the ATLAS Experiment at the LHC, p. 201, Oxford University Press, 2011.

physics experiments (ALICE, ATLAS, CMS, LHCb, and TOTEM),[84] using the model illustrated in Figure 2.11, as the process involves scientific, social, and learning aspects.[85]

From individual perception, assessment, and analysis of context and tools in which the five LHC experiments evolved, it has been possible to track the various aspects of knowledge creation, acquisition, and transfer occurring within the scientific and technological processes.

Stronger social interactions provide scientists and engineers with an insight into the specialized systems and CERN structures, and result in specialized information, language, and know-how. By intensifying the frequency, breadth and depth of information exchange, social interaction increases relation-specific common knowledge, especially in large-scale scientific collaborations such as the CERN Large Electron Positron (LEP) collider or the CERN LHC experiments. Such diversity is necessary for new knowledge creation; it exposes the users to a greater range of knowledge acquisition opportunities and also enhances the users' ability to value such opportunities. Common knowledge is required for learning to occur between two exchange partners; however, some diversity of knowledge is necessary for the transfer of new knowledge to occur. Indeed, exposure to many different external contacts is essential for learning acquisition in a competitive environment, such as large physics experiments. The importance of social capital for technological distinctiveness and the input of the diversity and frequency of interactions in the innovation process have been assessed at CERN between scientists, engineers, technicians and industry.[86]

We now live in a Knowledge Society[87] and, in contrast to our former industrial society – where knowledge used to be just a resource – knowledge is today considered unanimously as the key factor driving innovation.

Founded in 1954 in order to make available to European physicists installations, the costs of which would be prohibitive for a single nation, CERN systems and equipment have from the very beginning employed cutting-edge technologies in many fields, from special materials to electronics, data acquisition and analysis. CERN has played a role in Europe over the past 60 years as a leading organization for the creation of knowledge and know-how, not only in the field of HEP but also in related technological fields, providing in particular a strong push towards information technology developments. Indeed, the importance of creating individual and organizational knowledge for the multicultural scientific and technological environment provides clear and factual evidence of CERN's input. Each year, many hundreds of young people join CERN as students, fellows, associates or staff members in their first employment, and become

---

84) ALICE (ALarge Ion Collider Experiment), ATLAS (A Toroidal LHC ApparatuS), CMS (Compact Muon Solenoid), LHCb (Large Hadron Collider beauty), and TOTEM (TOTal cross-section, Elastic scattering & diffraction dissociation Measurement at the LHC).
85) The results obtained are reported and discussed in Section 2.6.1.
86) E. Autio et al., Technology Transfer and Technological Learning through CERN's Procurement Activity, CERN-2003-005, 2003.
87) P.F. Drucker, Post-Capitalist Society, HarperCollins, 1993.

embedded in a scientific atmosphere that is characterized not only by scientific nonprofit and free sharing aims, but also a high technological distinctiveness. Subsequently, the newcomers are given the opportunity to confront and to interact with a vast array of technical and scientific specialists, thereby creating a strong common scientific identity. CERN as an organization has its own epistemology, with its own tacit and explicit knowledge and the creation of various entities (individuals, groups, and their organizations), where the mode of knowledge creation and innovation supersedes the national context.

Whilst the educational impact of this intergovernmental-funded scientific organization is evident, such impact was estimated only recently in a quantitative manner and analysed for the competitive knowledge and skills that people develop, and for the market value of these skills in relation to the CERN Member States' industries.[88],[89],[90]

Several quantitative and comparative assessments of the knowledge acquired by students, physicists and engineers who have worked at CERN for the four

[88] B. Bressan, *A Study of the Research and Development Benefits to Society from an International Research Centre: CERN*, PhD thesis, Report Series in Physics HU-P-D112, University of Helsinki, Finland, 2004.

[89] T. Camporesi, *Statistics and Follow-up of DELPHI Students Career*, CERN/PPE, 1996; *Physics acts as a career stepping stone*, CERN Courier, Volume 40, N. 8, p. 20, 2000.

[90] CERN is run by 21 European Member States, but many non-European countries are also involved. Member States have special duties and privileges, and make a contribution to the capital and operating costs of CERN's programmes. They are also represented in the council that is responsible for all important decisions regarding the organization and its activities. The CERN convention was signed in 1953 by the 12 founding states of Belgium, Denmark, France, the Federal Republic of Germany, Greece, Italy, The Netherlands, Norway, Sweden, Switzerland, the United Kingdom and Yugoslavia, and entered into force on 29th September 1954. The Organization was subsequently joined by Austria (1959), Spain (1961–1969, rejoined 1983), Portugal (1985), Finland (1991), Poland (1991), Czechoslovak Republic (1992), Hungary (1992), Bulgaria (1999), and Israel (2014). The Czech Republic and Slovak Republic rejoined CERN after their mutual independence in 1993. Yugoslavia left CERN in 1961. Today, CERN has 21 Member States, and Romania is a candidate to become a member state. Serbia is an associate member in the pre-stage to membership. Some states (or international organizations) for which membership is either not possible or not yet feasible are observers. Observer status allows Non-Member States to attend council meetings and to receive council documents, without taking part in the decision-making procedures of the organization. Over 600 institutes and universities around the world use CERN's facilities. Funding agencies from both Member and Non-Member States are responsible for the financing, construction and operation of the experiments on which they collaborate. Observer States and Organizations currently involved in CERN programmes include the European Commission, India, Japan, the Russian Federation, Turkey, United Nations Educational, Scientific and Cultural Organization (UNESCO), and the United States of America. Non-Member States with cooperation agreements with CERN include Algeria, Argentina, Armenia, Australia, Azerbaijan, Belarus, Bolivia, Brazil, Canada, Chile, China, Colombia, Croatia, Ecuador, Egypt, Estonia, Former Yugoslav Republic of Macedonia (FYROM), Georgia, Iceland, Iran, Jordan, Korea, Lithuania, Malta, Mexico, Montenegro, Morocco, New Zealand, Pakistan, Peru, Saudi Arabia, Slovenia, South Africa, Ukraine, United Arab Emirates, and Vietnam. CERN has scientific contacts with China (Taipei), Cuba, Ghana, Ireland, Latvia, Lebanon, Madagascar, Malaysia, Mozambique, Palestinian Authority, Philippines, Qatar, Rwanda, Singapore, Sri Lanka, Thailand, Tunisia, Uzbekistan, and Venezuela: http://home.web.cern.ch/about/member-states.

LEP and five LHC experiments, led to investigations as to how one's nationality – and therefore different academic curricula and cultural differences – can affect knowledge perception, learning, and acquisition.[91],[92]

Studies have also shown that young researchers participating in two of CERN's LEP experiments have ultimately started working in industry.[93],[94] There is no doubt that the continuous flow of workers who arrive at CERN, are trained by working with CERN's experts, and then return to their home countries, is an excellent example of technology and knowledge being transferred through people. Experience has shown that industry, universities and other private and public employers value these people very highly, and in particular the on-the-job training that they receive at CERN.

The interface between the industrial and public research domains is multifaceted, and different research institutions may possess a diversified range of distinctive potentials for knowledge creation with economic aims. CERN is a good example of knowledge being used to generate important economic outcomes to the benefit of society, even though such outcomes are within CERN's mandate. The results of previous research have shown that the potential of scientific centres may well be currently underutilized by industry. Specifically, paying more explicit attention to technological learning and knowledge management issues could enhance the spectrum of technological impact in industry. Economic returns have been monitored simply on the industrial return to member countries, and the total economic benefit resulting from technological learning in industry outweighs this aspect.[95],[96],[97],[98] It is to be noted that, in the LHC experiments, almost 50% of the participants are not from CERN Member States. In addition, the procurements from experiments are not entirely managed by CERN, or provided by European industry, which means that spill-overs of technological learning are generated worldwide.

---

91) H. Huuse and O.P. Nordahl, *Technology Transfer at CERN – a study on interorganizational knowledge transfer within multinational R&D collaborations*, Master's thesis, the Norwegian University of Science and Technology (NTNU, Norges Teknisk-Naturvitenskapelige Universitet), Norway, 2004.
92) E. Cruz et al., *From Purchasing Goods and Services to Proactive Technology Transfer and Partnerships*, Master's thesis, EMBA (Executive Master of Business Administration) in MoT (Management of Technology), the Swiss Federal Institute of Technology, Lausanne, EPFL (École Polytechnique Fédérale, Lausanne), Switzerland, 2004.
93) T. Camporesi, *Statistics and Follow-up of DELPHI Students Career*, CERN/PPE, 1996; *Physics acts as a career stepping stone*, CERN Courier, Volume 40, N. 8, p. 20, 2000.
94) OPAL (Omni-Purpose Apparatus at LEP) Management Collaboration, *private communication*, CERN, 2003.
95) H. Schmied, *A Study of Economic Utility Resulting from CERN Contracts*, CERN 75-05, 1975.
96) M. Bianchi-Streit et al., *Economic Utility Resulting from CERN contracts*, (Second study), CERN 84-14, 1984.
97) M. Nordberg, *Transaction Costs and Core Competence Implications of Buyer-Supplier Linkages: The Case of CERN*, PhD thesis, University of Tilburg, The Netherlands, 1997.
98) E. Autio et al., *Technology Transfer and Technological Learning through CERN's Procurement Activity*, CERN-2003-005, 2003.

2.6.1
The LHC Case Study Survey

The present survey assesses the dynamics of knowledge production and management within the LHC collaborations. It applies largely to the whole physics community, ranging from students to university professors and to CERN staff members involved in challenging, high-technology developments. The knowledge acquired represents the most important type of direct benefit to society from CERN, as it enables people to develop not only academic assets but also a market value for the Member States' industries. Communication and documentation barriers and enablers, and their impact in terms of efficiency and innovation, have been also assessed, as have the functioning of the project network and management and the importance of knowledge and industrial transfer. Knowledge acquisition plays a mediating role between the three social capital constructs (social interaction, relationship quality, and network ties) and the two competitive advantage outcomes (invention development and technological distinctiveness). Knowledge acquisition is measured by statements reflecting the scientific and technological knowledge that a user may acquire from CERN, while social interaction is measured by statements reflecting the extent to which the relationship between CERN users is characterized by personal and social ties.

The information presented below is derived from an analysis of 291 questionnaires, and is presented in the following figures either at the single collaboration level or with all experiments combined. The categorization of replies considers the aspects of knowledge acquisition, learning and management, as identified by the model in Figure 2.11.

The respondents represent a good cross-section of project managers, scientists, professors and students participating in the collaborations. Information is extracted from the value attributed by the respondents to each specific question. Information on the current position held by the respondents, as well as the position held while at CERN and information on scientific and technical functions and expertise – which were reasons for joining the collaboration – and technological developments performed has also been made available by the respondents.[99]

The results of the CERN case study have shown that social capital constructs in the multicultural environment of LHC experiments are associated with knowledge acquisition (Tables 2.2 and 2.3).

Figure 2.12 stresses the efficiency of communication amongst the members of the subprojects. Previous knowledge and culture sharing is also an important enabler within a shared framework where a community of practices is acting in facilitating knowledge transfer; participation in previous CERN collaborations amounted to 55%, and participation in physics collaborations outside CERN to 44%.

---

99) B. Bressan *et al.*, *Knowledge Creation and Management in the Five LHC Experiments at CERN: Implications for Technology Innovation and Transfer*, postdoctoral project, University of Helsinki, Finland, and CERN, 2008.

Table 2.2 Communication, new knowledge and skills in terms of importance.

Statement	Importance (Average value: 1 = low, 7 = high)
1. Telephone	6
2. Tele/video conference	5
3. E-mail	7
4. Newsgroup	4
5. Technical development meeting	5
6. Collaboration meetings	6
7. Taking part in the research activity: *planning activity*	6
8. Taking part in the research activity: *measuring*	6
9. Taking part in the research activity: *analysing data*	6
10. Taking part in the research activity: *writing a paper*	6
11. Taking part in the technology development: *planning activity*	6
12. Taking part in the technology development: *construction*	6
13. Taking part in the technology development: *evaluating*	6
14. Taking part in the technology development: *writing a paper*	5
15. Taking part in the problem-solving activity	6
16. Informal meeting space in cafeteria (lunch, coffee, etc.)	6
17. Informal meeting space in the evening (dinner, hostel)	5
18. Reading, writing and disseminating project documentation	6
19. Reading info from the general Web pages	6
20. Reading info from the Web pages (research article)	6
21. A course in particle physics (theoretical/experimental)	5
22. A course in engineering (mechanical, electrical)	4
23. A course in management of projects; innovation; finance	4

Table 2.3 Outcome of knowledge transfer, skills, network and industry: importance of the statement.

Statement	Importance (%)
1. Improved and widened social network	70
2. Increased multidisciplinary insight	71
3. Improved management skills	67
4. Enhanced scientific knowledge	83
5. Enhanced scientific skills: *planning skills*	69
6. Enhanced scientific skills: *measuring skills*	53
7. Enhanced scientific skills: *data-analysing skills*	62
8. Enhanced scientific skills: *paper-writing skills*	54
9. New technical skills: *planning skills*	55
10. New technical skills: *construction skills*	52
11. New technical skills: *evaluating skills*	50
12. New technical skills: *paper-writing skills*	40
13. Increased international network	79
14. Increased relation and knowledge of industry	50
15. New professional interests	51
16. Increased opportunity to find a job in industry	26

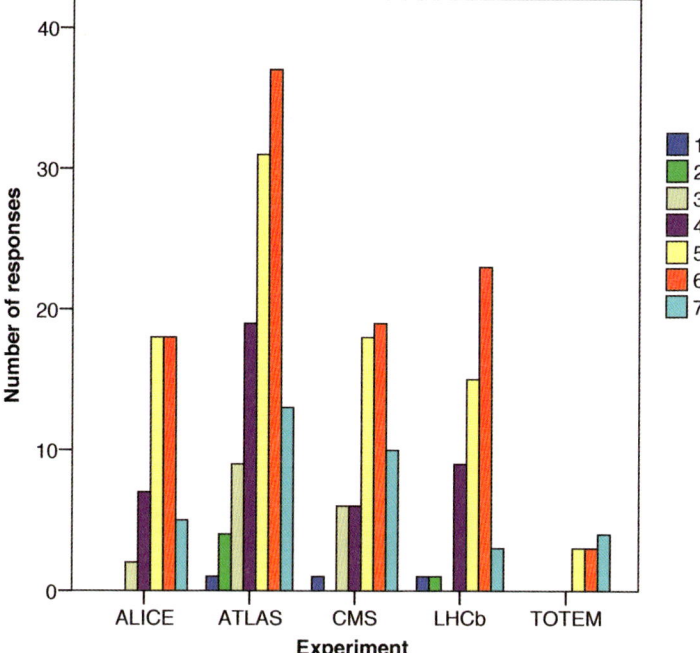

Figure 2.12 I have good communication with persons inside or outside the subproject of the projects (1 = disagree, 7 = agree). (Courtesy of CERN).

Communication, interactions, and shared context are important both for knowledge acquisition and transfer.[100] These aspects have, in general, been well managed by the LHC experiments, except for a minority of people that complained of not having been well enough informed (Figures 2.13 and 2.14).

Acquired knowledge appreciated by CERN LHC experimenters is a measure of the success of the social process in advancing the scientific and technological processes to create new knowledge and innovation (Figures 2.15 and 2.16).

The respondents underline that their most important experience at CERN was the opportunity to work in an international environment in a high-level research centre. Although they recognize that the multicultural and multifield interaction can be beneficial also outside HEP (Figure 2.17 and Table 2.4), only about 25% of the respondents could envisage leaving academic work, in spite of the greater attraction in terms of remuneration in a business environment.

The organizational learning – that is, the process whereby a group of people collectively enhance their capacities to produce the outcome – is a strong asset at CERN. This is due specifically to its longstanding multicultural, multifield environment characteristics.

100) E. Autio *et al.*, *Technology Transfer and Technological Learning through CERN's Procurement Activity*, CERN-2003-005, 2003.

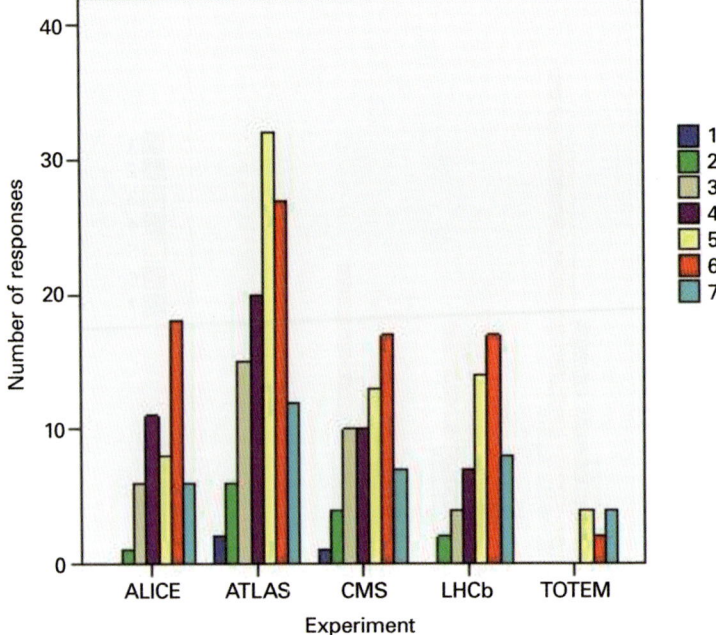

Figure 2.13 I am well enough informed about the development of the project (1 = disagree, 7 = agree). (Courtesy of CERN).

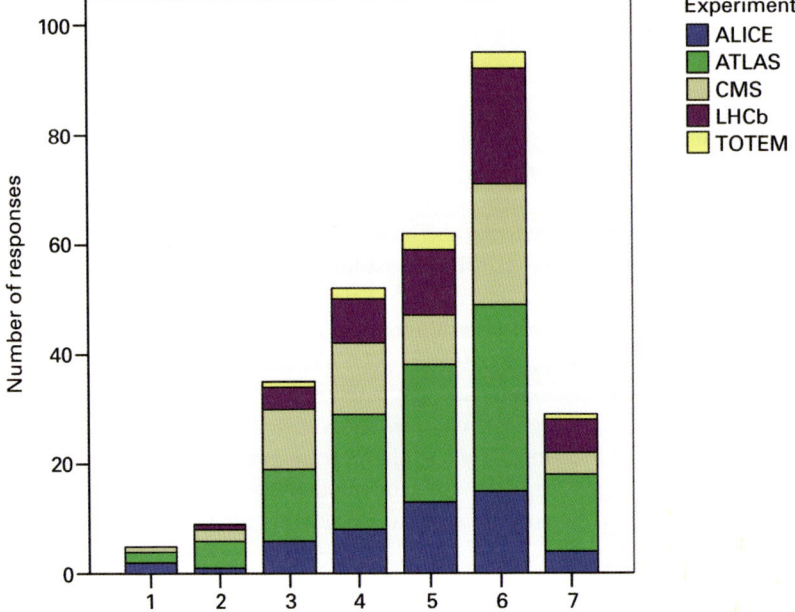

Figure 2.14 Project development flow of information in the team (1 = disagree, 7 = agree). (Courtesy of CERN).

2.6 Knowledge Management: A Case Study of CERN

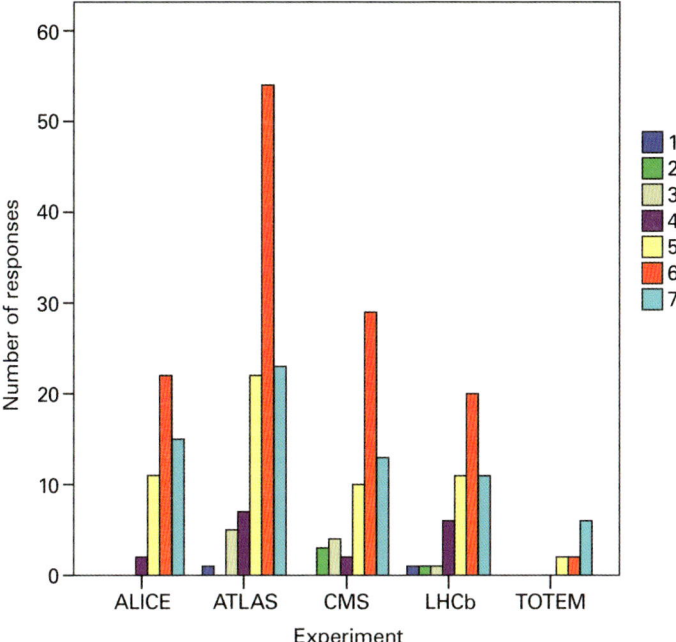

Figure 2.15 Knowledge transfer in the social process inside the project (1 = disagree, 7 = agree). (Courtesy of CERN).

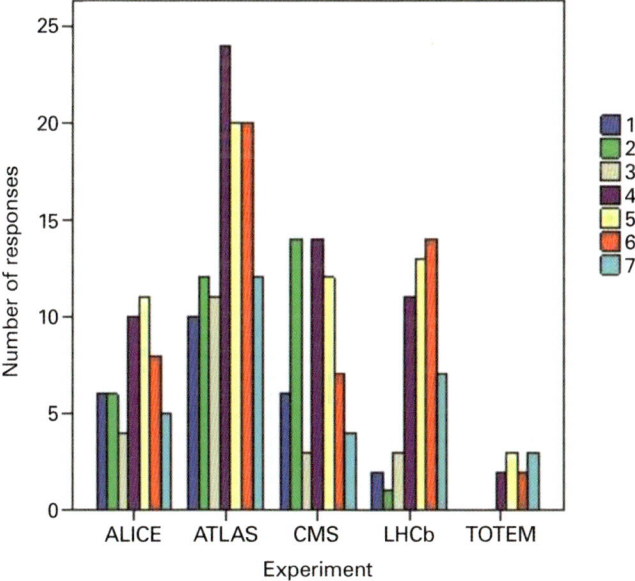

Figure 2.16 Knowledge transfer in the social process from other LHC experiments (1 = disagree, 7 = agree). (Courtesy of CERN).

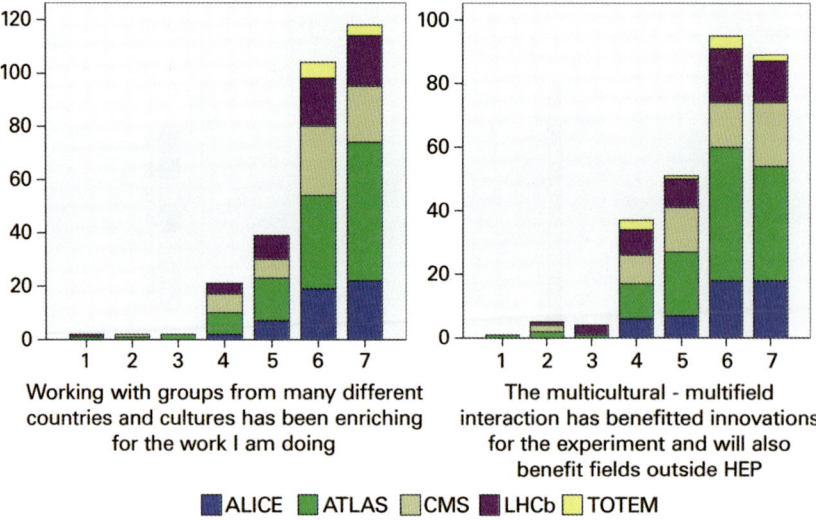

Figure 2.17 Importance of multicultural and multifield interaction (1 = disagree, 7 = agree). (Courtesy of CERN).

Table 2.4 Career development inside and outside HEP.

Career development	Percentage
Start a company	6
Go to work for a company	19
Continue (or start in academia)	72
No reply	3

The acquired skills of the participants of the LHC experiments enable them to develop market value for industry (Figure 2.18), as well as to motivate young researchers to work in industry or to start their own company (Tables 2.4 and 2.5).

The speed with which exchange takes place between the scientists themselves and between scientists and industrialists is a key factor for innovation. It must be remembered that the questionnaire refers to relations with the industrial world during the period when most of the relationships with industry for the LHC experiment construction were coming to an end (2006).

It is important to outline once more that, in HEP, the most efficient way of transferring knowledge is by transferring people, as demonstrated by previous studies carried out at CERN.[101),102),103),104)]

101) T. Camporesi, *Statistics and Follow-up of DELPHI Students Career*, CERN/PPE, 1996; *Physics acts as a career stepping stone*, CERN Courier, Volume 40, N. 8, p. 20, 2000.
102) OPAL Management Collaboration, *private communication*, CERN, 2003.
103) B. Bressan, *A Study of the Research and Development Benefits to Society from an International Research Centre: CERN*, PhD thesis, Report Series in Physics HU-P-D112, University of Helsinki, Finland, 2004.
104) E. Cruz et al., *From Purchasing Goods and Services to Proactive Technology Transfer and Partnerships*, Master's thesis, EMBA in MoT, EPFL, Switzerland, 2004.

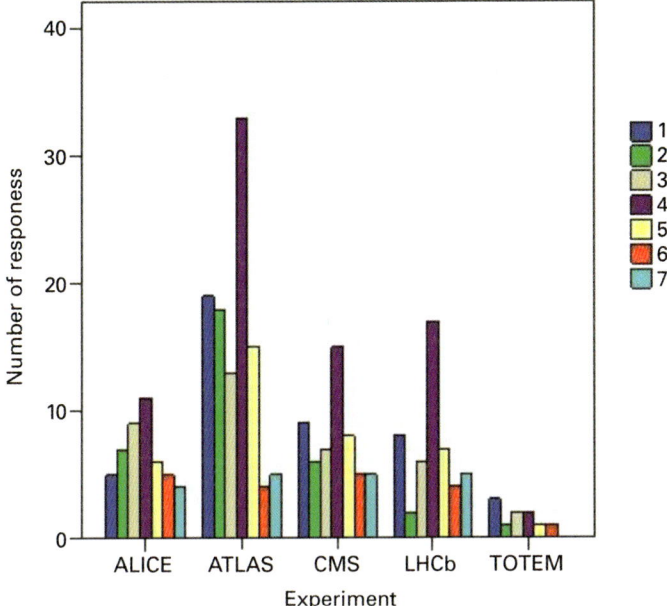

Figure 2.18 Increased opportunity to find a job in industry (1 = disagree, 7 = agree). (Courtesy of CERN).

Table 2.5 Innovative developments within the LHC experiments per domain.

Technological domain [a]	Frequency
Computing	64
Detector technology	79
Electronics	48
Mechanics	11
Other	8
Not known	6

a) Respondents = 194; missing replies = 97.

The scientific outcome has remained the main driving force in the project management as well as the motivation, as shown in Figures 2.19 and 2.20. The extent of the deepening and widening of scientific expertise has been assessed very positively by all five LHC experiments (Figure 2.20). This is not surprising considering the fact that HEP research is at the forefront of scientific knowledge.

In order to better understand the nature of knowledge learning and technological transfer, the data were analysed using specific tools (Exploratory Factor

54 | *2 Knowledge Management: From Theory to Practice*

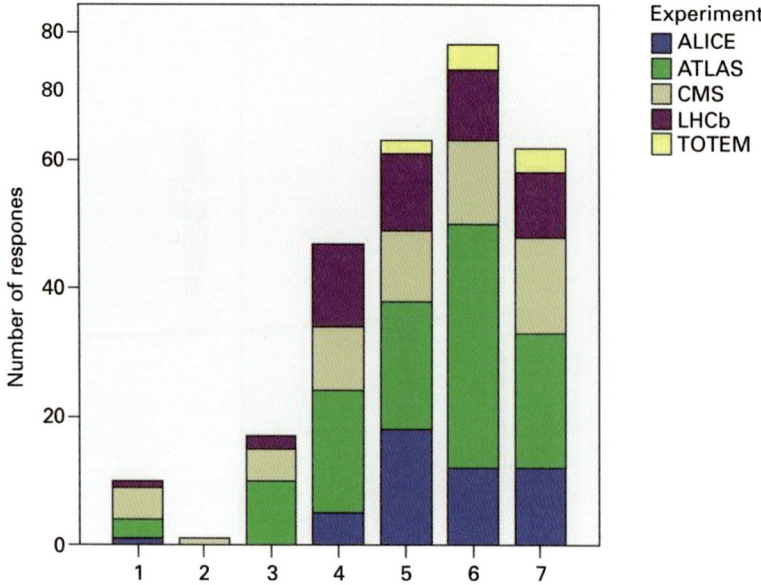

Figure 2.19 The scientific outcome determined in the project management (1 = disagree, 7 = agree). (Courtesy of CERN).

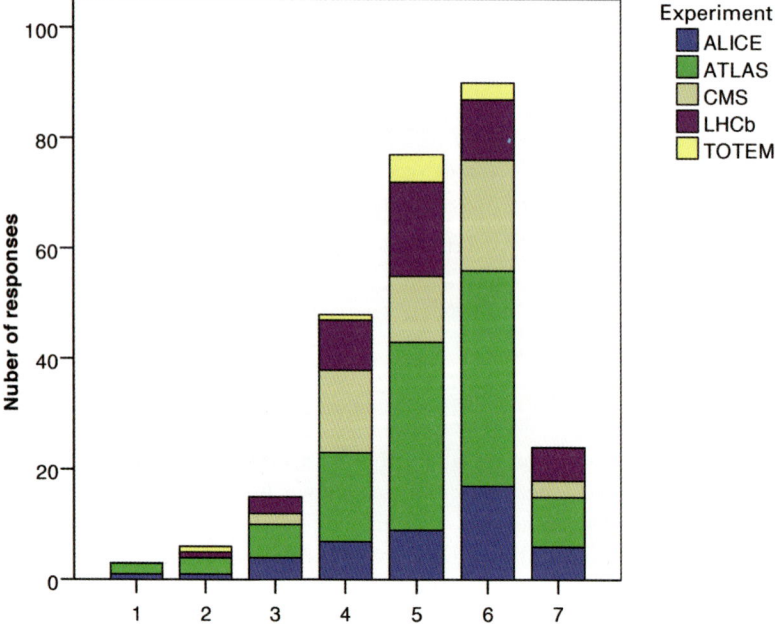

Figure 2.20 Motivation due to increase in expertise (1 = disagree, 7 = agree). (Courtesy of CERN).

Analysis), identifying four principal dimensions of personal outcome from the research activity in CERN:

1) The learning of new technical skills, such as construction skills, planning skills, evaluating skills, and skills needed in management.
2) The learning of new scientific knowledge and science-making skills, such as data analysis skills, paper-writing skills, and measuring skills.
3) The improvement and widening of social networks.
4) The improvement and widening of labour market competence.

Permanent staff and visitors evaluated their personal outcomes rather similarly. The most modern tools of communications were requested by those physicists who were more familiar with the latest developments in information technology. There were statistically significant differences in the value or importance of the communication type between young and old users in CERN. However, there was no statistically significant difference between physicists and engineers in the value or importance of the communication type.

The results of this case study provide evidence that the social process of participation in meetings, the acquisition of skills in different areas, and the development of interests by interaction with colleagues, represent key elements of the learning process. Science and technology outcomes are embedded and intertwined.

While the fundamental science mission of scientific centres should continue to dominate, greater attention should be paid to maximizing the technological impacts that scientific collaborations may potentially confer to industry and society. Scientific centres, as well as the contributing member countries, should encourage and prepare the terrain to make possible such types of collaboration with industrial companies and to make better-known the impact on society.

The present large physics collaborations have necessitated a change in approach, with a much greater importance being given to managerial aspects, and with about 30% of respondents having had a management and coordination role.

These positive outcomes of acquiring new technical skills and widening knowledge, observed in a population of 79% of users and 21% of staff members, was shown to span over a wide age range and thus benefitted both young and experienced physicists alike.

In accordance with the factor analysis presented in this study, these processes appear clearly as three different dimensions of the process, as illustrated by the model in Figure 2.11. An additional fourth dimension related to the acquisition of competencies for the labour market has been identified by the analysis.

In future, the development of personal skills, according to the four identified factors – that is, the learning of technical skills, the learning of science-making skills, improvements of social network, and increases in employment possibilities in the labour market – could be used to target individual development for improving opportunities for the labour market.

Scientific organizations are deemed to enable actions in order to achieve knowledge creation and diffusion at various levels (individual, group, and organization). Large-scale scientific projects – which often are trans-national – are, like any scientific research driven by the wish to understand why and how, governed by contextual situations.

The scale of working and interacting in the HEP scientific community has proved to be very communitarian and dynamic, and over the years the community has undergone many changes as a function of the projects increasing in both size and complexity. However, the project that will follow LHC (and its upgrades) will be carried out in a different contextual situation, with many structural and procedural changes having to be implemented in order to achieve the best scientific and technological outcomes.

The studies presented here may be of help, and hopefully will provide the tools to adapt and evolve the most appropriate methodologies to achieve the most efficient knowledge management. This, in turn, will help to transfer scientific and technical outcomes to society.

Part II
Examples of Knowledge and Technology Transfer

The contribution of Particle Physics to Informatics, Energy and Environment represents a longstanding adventure involving many domains of research and development. Initially pushed by the drive of enlightened pioneers, the transfer of forefront technologies to the benefit of society requires the correct context to foster inventiveness for the practical and cost-effective implementation of new receptiveness from research, medical, and industry practitioners.

Concrete examples exist of Knowledge and Technology Transfer achieved either through the dissemination and implementation policy established by scientific research laboratories, or through the conventional procurement mechanism. In particular, this has been occurring at CERN first through procurement mechanisms and, since 2000, with a properly established policy. High-technology suppliers now communicate results obtained through the conventional procurement mechanism, indicating product development benefits resulting from relationships with scientific research laboratories such as CERN.

In fact, CERN's technological requirements often exceed the available state-of-the-art technologies, thereby generating fruitful interactions, technological learning and innovation for industry. In turn, this impacts positively on society through new market products.

From Physics to Daily Life: Applications in Informatics, Energy, and Environment, First Edition.
Edited by Beatrice Bressan.
© 2014 Wiley-VCH Verlag GmbH & Co. KGaA. Published 2014 by Wiley-VCH Verlag GmbH & Co. KGaA.

Section 1
Linking Information

Information Technology (IT), which plays an essential role in scientific research achievements, has seen rapid development thanks to advances in Electronics and Network Technologies. Through the implementation of the World Wide Web (WWW) and of the Grid, IT has paved the way to the next generation of Computing. The WWW has become part of everyday modern communications, with tens of thousands of servers providing information to millions of users. It could be considered one of the most striking examples of Technology Transfer in the past three decades that has largely modified the functioning and behaviour both of modern society and of individuals.

The Internet, namely the technical infrastructure of the global network system, was born in 1969 with the first node at the University of California, Los Angeles, US. However, it would have been impossible to benefit from the advantages of the Internet without the research completed at CERN in the last decade of the 20thcentury. The WWW invented at CERN to enable the sharing of information between different computers was freely distributed to the Internet community and became a worldwide phenomenon. As the Web was CERN's response to a new wave of scientific collaboration at the end of the 1980s, the Grid is the answer to the need for data analysis to be performed by the world Particle Physics community.

With the Large Hadron Collider (LHC), the CERN experiments exploit petabytes of information. Many developments have been pursued both for the analysis and storage of the LHC data and for developing applications. The Grid is a very powerful tool linking Computing resources distributed around the world into one computing service for all requesting applications. Thanks to the Grid, a new way of interacting amongst scientists and different domains is possible with faster dissemination of data, better quality control, and more efficient use of information sources. These characteristics have allowed the rapid spread of the Grid in many different domains of application impacting in many ways people's daily life.

From Physics to Daily Life: Applications in Informatics, Energy, and Environment, First Edition.
Edited by Beatrice Bressan.
© 2014 Wiley-VCH Verlag GmbH & Co. KGaA. Published 2014 by Wiley-VCH Verlag GmbH & Co. KGaA.

3
WWW and More

Robert Cailliau

The World Wide Web (WWW), as it is known today, is almost commonplace. This incredible and powerful tool, the birth of which took place thanks to a number of players, is now taken for granted. This is one person's story of how the WWW came into being as the brainchild of Tim Berners-Lee, with the influence of many other people, at a very special place called CERN.

Public communication is the art of being accurate while leaving out all the essential details. As such it is, in the words of CERN theoretical physicist John Ellis: 'necessary but hopeless'. The tale is part historic, part sociological, and part private opinion. The author is alone to blame for any inaccuracies.

- **Caveat 1**: The Web is not based on any physical laws, its characteristics are not constrained by anything other than available computing power and network bandwidth. The Web is not based on any scientific discovery. Its development flowed from sociology, and therefore this chapter is much more about interacting people, and because the author was one of those actors it is quite personal.
- **Caveat 2**: The Web is on the Web, meaning that almost all documents of importance have been published there and do not exist in printed form. The two most important documents – the original proposal[1] and the public domain statement[2] – have never been published in the classical way; indeed, neither even carries an internal CERN reference. The proceedings of the Web conferences have been published mainly as HyperText Markup Language (HTML) files or on Compact Disc Read Only Memory (CD-ROM).[3]

---

1) T. Berners-Lee, *Information Management: A Proposal*, 1989, note: version of 1990: http://www.w3.org/History/1989/proposal.html.
2) http://cds.cern.ch/record/1164399.
3) The First International Conference on the World-Wide Web (WWW) was the first-ever conference about the WWW, and the first meeting of what became the International World Wide Web Conference. It was held at CERN on 25–27 May, 1994. The conference had 380 participants, accepted out of 600 applicants, and has been referred to as the 'Woodstock of the Web'.

From Physics to Daily Life: Applications in Informatics, Energy, and Environment, First Edition.
Edited by Beatrice Bressan.
© 2014 Wiley-VCH Verlag GmbH & Co. KGaA. Published 2014 by Wiley-VCH Verlag GmbH & Co. KGaA.

The Web Consortium publishes its documents on its site,[4] not in print, because the whole Web community obviously chose the Web as its medium.

3.1
The First Page

'What was the first Web page?' Journalists ask this question at almost every interview about the Web.

Unfortunately, there was no such thing. Web pages grow over time, then they keep changing, and finally they are in rare cases archived. In the case of the initial developments of the Web, archiving was simply not done. More importantly, the software itself was in constant flux: Tim Berners-Lee[5] kept adding and modifying to the browser/editor he developed for the NeXT[6] system (Figure 3.1). As is usual with software design, each time a page was produced, he found a feature that was either wrong or absent. None of these early pages was actually published officially (that came much later!), but they could be seen on the Internet. The pages and the software evolved almost in symbiosis.

As time passed, some more or less viable pages could be inspected by those who had a NeXT computer and had downloaded a copy of the browser.

Almost all of the early pages described what Tim Berners-Lee called 'the project', but as far as we know WWW never had official project status at CERN.

A few years later the number of pages on the world's first Web server had grown into a tangled amalgamate (a polite word for 'mess') of pages about CERN and pages about World Wide Web or WWW (and probably other things too!). This could of course not go on: the Web had become too visible to leave CERN's public image without any official supervision from the Organization. I decided to set up a separate server for publishing information about CERN and its many physics experiments, and carefully made it an endeavour supported by all departments.

The Web had come of age, but the earliest embryonic pages had long since been lost.

---

4) www.w3.org.
5) Tim Berners-Lee was awarded a knighthood by Queen Elizabeth II of the United Kingdom in 2004.
6) NeXT was a computer company formed by Steven Jobs, one of the founders of Apple Computer, and also the name of the advanced personal computer or workstation that the company developed and first offered in 1988. The NeXT computer was a black cube with a high-resolution display, a graphical user interface, and an operating system called 'NeXTStep'.

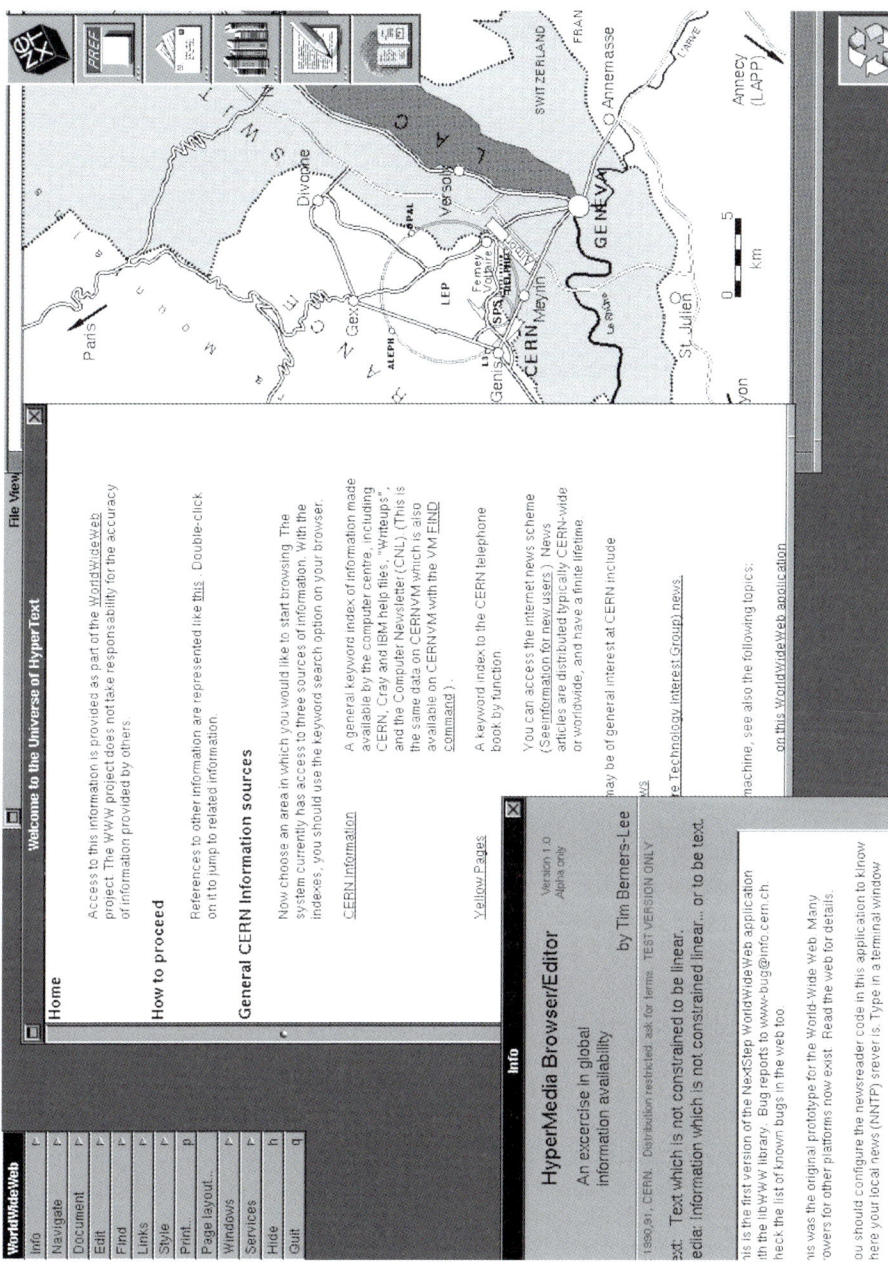

Figure 3.1 Appearance of the NeXT browser-editor: styled text and vector graphics. (Courtesy of CERN).

3.2
Influences on the History of the Web

In the sociological component of Web history, there are many influences that shaped the way the Web unfolded over time. Of these, two are of importance: age and, as a consequence, approach to life.

3.2.1
A Matter of Age

Looking back on what happened and why developments were what they were, the conclusion must be that most of it was perhaps a consequence of people's ages. There was a considerable factor of difference in life experience, attitudes and reflexes to circumstances. Within the group that factor was at least 2.5! No wonder that there was, in many respects, some stress built into the team.

3.2.2
The Approach

Another major factor that influenced developments was differences in approach to society. The perspectives of the group could not have differed more: on one side a post-war baby boomer (for today's young readers it may be necessary to identify the war in question as World War II), educated under the leaden cloud of the 'Cold War' in Europe. And on the other side were the very young programmers who considered that ancient history, if they even gave it a thought. The baby boomer was still concerned about loyalty to grand causes, to the organization he or she worked for, the paymaster, and the hierarchy. The youngsters, however, were highly independent spirits who unhesitatingly switched to the employer who provided them with the fastest Internet connection and the computer with the largest screen. Where the money to pay the pizza came from was of no interest to them, they had little or no loyalty to their place of work. But they did have a very strong social attitude: appreciation from their peers, and from their peers only, was the most important driving force in their lives.

For them the complex context of CERN was not easy to understand either: while our focus was on the nearly insoluble problem of convincing the management that working on the Web was an important activity (since of course CERN's primary occupation is particle physics), the newcomers were in disbelief that anyone could be so blind they could not see that the WWW was the most exciting happening of the century.

This almost surreal situation is the canvas on which the early Web's history was painted.

3.3
CERN's Role

To understand better why and how the Web came into being at CERN, one must understand the special type of environment generally present on its sites, as well as some of the specific historic happenings that coincided with the Web's inception.

3.3.1
A Possible Definition

It is important to try to define what is meant by the word 'CERN', because even when one has worked there for over 30 years and over that time has had four very different internal careers, it was not always clear what 'CERN' meant. Perhaps the best defining characteristic is that used by a visiting journalist. At one point, when drinking coffee in the cafeteria after having been on a detailed tour, he exclaimed: 'Amazing! Here are people who change their minds in the face of evidence!'

His name is not in the record, but his passionate exclamation has remained. That this attitude of the people who work at CERN struck him so much must be because, in the rest of society, 'evidence' is not something that influences people's behaviour much. No wonder humankind has planet-sized problems of all sorts.

CERN is indeed a most unique institute: it is one of the very few where people from all nations and backgrounds work on difficult, abstract problems with no apparent direct application to industry or society. They are trying to answer deep questions about the real world in which we live.

It is often said that research is exactly that: looking without really knowing where the answer lies. That requires an environment in which many people can explore different avenues, try out hypotheses and let the mind wander, so it can stumble over solutions.

This is not to say that at CERN people hang around staring at the ceiling. It is hard work, needing a lot of concentration and quiet. It also needs a complete absence of economic pressures: one should not have to worry about the minutiae of the immediate material future. That is why CERN has a housing service, on-site hotels, restaurants, post offices, banks and even a travel agency. These facilities are not provided with the intention of pampering the resident scientists; the environment at CERN is merely a necessary precondition for success in research (though it is of course not the only one).

There also needs to be an excellent communication infrastructure: blackboards, libraries, cafeterias, networks. And that infrastructure must be able to continue connecting the scientists when they have returned to their universities and laboratories.

3.3.2
Making it Work

A part of CERN's workforce (including the author) is present exclusively to support the scientists: these are the administrators and engineers keeping it all running. Engineers seek material problems and try to solve them, but they are not only concerned with making things happen – they also want to achieve some efficiency and quality.

No wonder then that it was worth observing the ways in which the physicists distributed documents. They spent a lot of attention on the content, of course, but much less on how the document was delivered to a colleague. During the early days they used to send printed versions through the paper mail, later switching to 'disquettes'.

3.3.3
On Documents

It was at the start of the 1980s that computer-based document editing systems sprang up, but these were incompatible with each other. This had already been signalled as a waste of effort, and a non-proprietary text editing system had been implemented and deployed.[7] When CERN decided to do something about document production on a laboratory-wide scale, via the Office Computing Systems, as the group leader I began to wonder about the relative virtues of centralized and distributed document systems, as well as the possible merits of computer-aided reading – that is, hypertexts. A future-proof format, independent of software vendors was also required. Having had the experience of making that prior system, we knew what was possible and what was difficult. However, the group was very busy with day-to-day support, and had little time for exploring new ideas; enthusiasm for investigating hypertexts and distributed document systems was at a low level.

3.3.4
The Director General

There came a change of CERN's Director General when Carlo Rubbia, Nobel laureate,[8] took the controls and immediately started restructuring CERN deeply, to prepare it for the future. Rubbia exploded the Data Handling Division, which had essentially become a sort of Information Technology (IT) department, and divided it into one department devoting its efforts to physicists, Electronics and

---

7) D. Bates et al., REPORT User's Guide for the Writing of Documents, CERN, PS/CCI Note, 1977.
8) In 1984, Carlo Rubbia was awarded the Nobel Prize in Physics jointly with Simon van der Meer '
    . . . for their decisive contributions to the large project, which led to the discovery of the field particles W and Z, communicators of weak interaction': http://www.nobelprize.org/nobel_prizes/physics/laureates/1984.

Computing for Physics (ECP), and another department to perform the more mundane infrastructure tasks, Computing and Networking (CN). Having 'emigrated' to ECP (at the suggestion of Mike Sendall) and not being well known at the time, I decided to seek a niche to devote at least some time to the study of distributed hypertext systems. Tim Berners-Lee was in CN. Although my half-page proposal to conduct such a study had been submitted to ECP, the reorganization of CERN meant that it did not float to the top of the list of priorities! However, one man had spotted it. Mike Sendall, who had also 'emigrated' to ECP, had been given a 14-page document titled *Information Management: A Proposal* by one Tim Berners-Lee.[9]

While CERN was fully engaged in the experiments at the Large Electron Positron (LEP) collider, it was also involved in considering the next major project, the Large Hadron Collider (LHC), a machine designed to compete with the US to obtain the Higgs boson. Mike Sendall understood very well the enormous computing power that the LHC would need, and the accompanying huge needs for networking and documentation.

Mike had qualified Tim's proposal by pencilling: 'Vague but exciting' on the front page, and passed it on to me for a second opinion. Since it was easily and immediately recognized as the potential solution, I went to see Tim Berners-Lee, whose ideas were subsequently adopted as a much better road to the information infrastructure needed for the LHC era.

3.3.5
Al Gore, the LHC, and the Rest is History

Later, when the Vice-President of the US, Al Gore, had pushed through his National Information Infrastructure (NII)[10] and the specification documents for it were distributed, we smiled and reflected: 'We have the solution to that, here and now!'

It may be said that the Web is one of the first spin-offs of the LHC. In that sense, the mere existence of the Web should silence all critics of the LHC's costs.

---

9) T. Berners-Lee, *Information Management: A Proposal*, 1989, version 1990: http://www.w3.org/History/1989/proposal.html.

10) The National Information Infrastructure (NII), as a telecommunications policy (a product of the *High Performance Computing Act* of 1991 popularized during the Clinton Administration under the leadership of Vice-President Al Gore), was a proposed, advanced, seamless Web of public and private communications networks, interactive services, interoperable computer hardware and software, computers, databases, and consumer electronics to place vast amounts of information at the users' fingertips. The NII was to have included more than just the physical facilities (cameras, scanners, keyboards, telephones, fax machines, computers, switches, compact disks, video/audio tape, cable, wire, satellites, optical fibre transmission lines, microwave nets, switches, televisions, monitors, and printers) used to transmit, store, process, and display voice, data, and images, and also to encompass a wide range of interactive functions, user-tailored services, and multimedia databases, interconnected in a technology-neutral manner that favored no one industry over any other.

The details of the history of the Web are provided in the book, *How the Web was Born*,[11] which has first-hand accounts of many of the people involved.

3.4
What-if Musings

Any momentous event leads people to examine the causes, and finding a cause prompts the question: 'What if something else had happened instead?' Trying to answer such questions is taboo to historians, but it is important in this context.

3.4.1
Money, Money, Money . . .

Perhaps the most frequent worry about the Web's history is: should CERN not have asked for money? A cent per page displayed? CERN would be immensely rich now, is the thought. This wishful thinking ignores the influence of price on sales. It assumes that the Web would have taken off no matter what, and that an opportunity was just missed. But that is not so.

First, putting a price on the service in whatever way would just have made the Web into yet another competitor in an already crowded information networks market. It would probably have had some share of that market, but the chances are that it would have been rather small.

Second, even if prices would have been low or zero, there was the fear of software intellectual property rights, patents and royalties. At the time those were not well-known terrain (are they even today?), so most developers and starting businesses were wary of anything that did not have a clear legal position.

The group perceived this too, especially after the University of Minnesota in the US had started charging for Gopher.[12] As a consequence, the use of Gopher instantly declined. To avoid making the same mistake, I worked with the CERN Legal Service, and after many months of reflection and argumentation, the Direction agreed to sign a document[13] whereby CERN relinquished all intellectual property and put the software into the public domain (see Section 3.6.1). That immediately made developers much more comfortable, and many useful applications sprang up.

The question, therefore, is not whether CERN would have become rich, but whether the Web would have taken off at all if there had been a cost associated.

---

11) J. Gillies and R. Cailliau, *How the Web was Born*, Oxford University Press, 2000.
12) Gopher is an Internet application protocol allowing file structures to be browsed remotely through the Internet. It was popular for several years, especially in universities. The Web with its hypertext links quickly transcended Gopher. It was developed by Mark McCahill's group, and then at the University of Minnesota, US, whose sports teams are called 'the Golden Gophers'.
13) http://cds.cern.ch/record/1164399.

Figure 3.2 A very rare photograph: from left to right: Tim Berners-Lee, Robert Cailliau, and Ted Nelson. (Courtesy of Håkon Wium Lie, Opera Software, 1997).

3.4.2
And if Not?

Another, equally important but much less asked question is: 'What would have happened if WWW had not been invented at all?'

It is highly likely that someone else would sooner or later have come up with a very similar idea. There were already other hypertext networked systems out there, like Hyper-G[14] from the University of Graz in Austria, and some visionaries such as Ted Nelson had already been thinking for a long time about networked hypertexts (see Figure 3.2).

The Web had even been foreshadowed as early as 1934 by Paul Otlet,[15] who had collected over 16 million references to documents spread over the planet's libraries. Otlet dreamed of also being able to read the documents themselves via TV links from the places where they were actually stored. His vision could be summed up as: 'Google and the Web, anno 1934'.

It might therefore have taken a few more years, though it is unclear how long. Working at the same organization as Tim Berners-Lee, I had been thinking of linking all CERNs documents over its internal networks. Supposing there had not been Tim's proposal, a much more restricted version could possibly have been implemented. It would have been highly likely that a colleague would soon after have pointed out that it would be nice to link across physics institutes.

---

14) Hyper-G is a publishing system with hypertext features more advanced than those available with the HyperText Transfer Protocol (HTTP) and today's Web browser. Hyper-G was inspired in part by the ideas of Ted Nelson (who coined the term 'hypertext'), and was developed in Graz, Austria, by a team of researchers at the Institute for Information Processing and Computer-supported New Media (IICM) and the Institute for HyperMedia (IHM) Systems of Joanneum Research.

15) P. Otlet, *Traité de Documentation: Le livre sur le livre*, Mundaneum, 1934.

Again, it would have taken more years – maybe many more – but we would have had something fairly similar to the Web.

The next question is what the evolution of other networked information systems would have been in the absence of WWW.

Indeed, one should not forget that there were commercial services such as Teletel (Minitel),[16] Ceefax,[17] and CompuServe,[18] which were widely available in their respective countries. These were powerful systems, arguably better thought out than the Web and, in the case of France's Minitel, were wildly popular, but they were run by the national telecom monopolies. And that was, at the end of the 1980s, precisely the problem: they could not operate outside their respective national boundaries. The question then is: 'What if the European Commission had been able to deregulate the telecoms a few years earlier, before the conception of the Web?' There is a fairly high probability that one would now be discussing 'Minitel version 5.3' or something similar, and that the whole world would be using it, in the same way that Global System for Mobile (GSM) phone networks spread.

But no one really has a crystal ball.

3.5
The Dark Sides of the Force

It is often asked: 'What might be done differently if we were to relive the WWW adventure again? Were there mistakes and omissions?' Certainly, and quite a few.

There are some technical mistakes, some managerial mistakes, and a few political mistakes, but perhaps the easiest to describe are the technical ones.

3.5.1
Techies

Tim Berners-Lee's Web was a computer application (Figure 3.3); to write it he used the most advanced software development system available (Interface Builder, by Jean-Marie Hullot of Paris, France) running on a NeXT computer.

---

16) Minitel was a Videotex (one of the earliest implementations of an end-user information system) online service accessible through telephone lines, and was considered one of the world's most successful pre-WWW online services. It was rolled out experimentally in 1978, and throughout France in 1982.

17) Ceefax (homonymous with 'See Facts') was the world's first teletext (a television information retrieval service created in the UK in the early 1970s) information service started by the BBC in 1974.

18) CompuServe (or CIS, CompuServe Information Service), the first major commercial online service in the US, dominated the field during the 1980s and through the mid-1990s. Since the purchase of CIS Division by AOL Inc. (previously known as America Online), it has operated as an online and an Internet service provider.

Figure 3.3 The first Web server. (Courtesy of CERN).

With the NeXT it was possible to edit and browse in one and the same seamless way, and Hullot's system allowed Berners-Lee to get the Web programs up and running in only a few months' time. However, by using Interface Builder it also meant that it was possible to share the software only with others who owned the same computers. A version was needed that anyone could install on any computing platform with little effort. The attempt to do that was called 'Line Mode Browser' (LMB). To distribute to virtually any system necessarily means adopting the largest common denominator, and that meant going back to bare basics: one could not assume much about the target machines. It was 1990, and computers ran Operating Systems (OS) such as Windows 3.0, Mac System-7 and IBM OS/2-1.3. Many people were still using dumb terminals! So, the LMB could not assume its user had cursor controls, let alone a mouse; type fonts, multiple windows, colours, raster graphics; but only a fixed 80-column-by-24-rows character grid, whereby the last line would be used as the command line. No active links (since there was no cursor anyway), and definitely no editing of pages.

The resulting software, written to the group's specifications by Nicola Pellow,[19] did work and could be termed as being useful, but was eclipsed by the NeXT software.

Any software developer who saw the LMB immediately thought: 'Brilliant idea, but I can do better!' And many went off and did exactly that: write what they thought was a better browser. But actually there was almost no contact from them and they spent their efforts on improved LMB versions. Although many browsers saw the light, none of them came close to the NeXT version, and none of them allowed editing. Many schemes were set up to overcome these

---

19) N. Pellow, *HyperText and HyperMedia*, Industrial Year Assignment II, BSc Mathematics III, 1991.

problems, all avoided the root of the evil: a 'wysiwyg' ('what you see is what you get') editor of HTML was a very difficult thing to implement (and still is).

3.5.2
Global Heating

There was a terribly turbulent activity around Web development, driven by a desire to be first or get credit. One participant at a meeting organized to calm the waters exclaimed: 'Manque total de sérénité!'[20]

The rational course of action would have been to pool efforts and make a portable version of what was on the NeXT, but one has to remember that this all happened before the Web and the Internet were widespread. Had more effort been put into coordinating the energies of all those young programmers, maybe it would have been possible to gain several years.

The LMB, for all the service it delivered and the good intentions of its developers, had another negative effect: it showed a single-window passive application. Later browsers followed that implicit model: the consumption of information by staring at a single window, with no possibilities to be creative. Converters quickly became the rage: write a page using a text processor (usually Microsoft Word) then push it through the converter to get an HTML page. This was not the way to go, but it was the way the world went.

Physics needs a lot of mathematics and graphics, yet here was a tool that could handle neither! One may now see the emergence of Scalable Vector Graphics (SVG),[21] at some 20 years after it was needed. Mathematics slowly gets into Web pages too, but even Wikipedia does not use a lot of it, as most of their equations are images.

3.5.3
Sin by Omission

Possibly the worst mistake of all was an omission. Web pages needed a programming language, but the team at CERN was adamant that everything should be declarative and no programming language should be incorporated, least of all an imperative one.

But Nature abhors vacuum: the hole left in the Web by the absence of a programming language was eventually filled in by arguably a not so user-friendly language: JavaScript.[22] A deluge of programming tools continue to fill the vacuum,

---

20) Total absence of serenity!
21) http://www.w3.org/Graphics/SVG.
22) JavaScript is a dynamic computer programming language most commonly used as part of Web browsers, whose implementations allow client-side scripts to interact with the user, control the browser, communicate asynchronously, and alter the document content that is displayed.

including Practical Extraction and Reporting Language (PERL)[23] and Hypertext Pre-processor (PHP).[24]

The pinnacle of absurdity is that multiple programming languages occur side by side in the same Web page, one being treated by the server before sending the page out, and the other being handled by the browser at the time the page is displayed to the reader. To this day, it is still difficult to set up a Web server unless one has a vast amount of detailed knowledge of computers, networks, programming languages and graphics.

3.6
Good Stuff

The Web is much more about people and society than it is about technology. Three important events are discussed here, all of which were crucial to the ultimate shape that the Web took: CERN donated it to the world, conferences brought the developers together, and the Consortium provided the common ground for industry.

3.6.1
Public Domain

Tim Berners-Lee had long been feeling that the Web should be made freely available to the world. At first, one idea had been to set up a company in neighbouring France at the industrial estate foreseen for such activities, located almost adjacent to CERN. WWW could be an interesting enterprise, and when this was discussed with Tim, he asked: 'Do you want to be rich?' This was an unexpected reply, as making money was never seen as the prime goal of such a business – but it would help!

Some more considerations were made, but ultimately (though not that same day) Tim convinced us that giving WWW away was the best option. But how? It was late 1992, and there was only a handful of servers, all in academia. Business was not really allowed to use the Internet since it was funded exclusively by academia – that is, indirectly by governments. Some universities were experimenting with royalties on software that they produced and maintained on their budgets, but that was actually used by other academic institutes. The University of Minnesota had discovered the amount of resources that went into maintaining

---

23) Practical Extraction and Reporting Language (PERL) is a family of high-level, general-purpose, interpreted, dynamic programming languages originally developed in 1987 as a general-purpose Unix scripting language to make report processing easier.
24) Hypertext Preprocessor (PHP), originally Personal Home Page created in 1995, is a server-side scripting language designed for Web development used as a general-purpose programming language installed on more than 244 million Websites and 2.1 million Web servers.

and distributing Gopher, and they started charging for it. However, while their charging model was well-defined and cheap, it still put a lot of users off, and Gopher's expansion stalled. No one wanted that to happen to WWW.

After agreeing that the best way forward was to put WWW in the public domain, a visit was made to see CERN's Legal Service. Some six months later, after many meetings and demonstrations to CERN's management up to the level of the Directors, a document of less than two pages was crafted, and this is now available on the Web.[25] The first page described what WWW was; the second page was only half-filled, but contained three very important statements: (1) CERN put the WWW software into the public domain; (2) it relinquished all intellectual property on that software; and (3) this did not constitute a precedent for any other past or future CERN software. (Some time later CERN made a blanket statement for its other software and even for future versions of the WWW software, keeping them out of the public domain, but those are details that do not concern us here.)

The document was published on 30th April 1993, a little more than 20 years ago. There is a certain pride in having brought that endeavour to a good end.

Although . . . exactly what that document now means is not clear. The technology itself was not put in the public domain, only the software as it was at the time. But what would it mean to put a technology in the public domain, if that technology has already been in use for a few years, and is mostly built on other existing, free technologies?

3.6.2
The Conferences

Towards the end of 1993 it became clear that a meeting of Web software developers was urgently needed. These guys (nearly all male, the women could be counted on the fingers of one hand) had never met each other, were frantically writing e-mails and downloading program code files, but had never been physically together over a coffee or a beer!

In December 1993, I placed a call on the net for the First International Web Conference (there had been small attempts at gatherings before, but locally and not very successfully). WWW1 took place at CERN in May 1994.[26] It was oversubscribed and filled the CERN auditorium to as much as it was legally allowed to hold. September 1994 saw the second conference, organized by the National Centre for Supercomputing Applications (NCSA) in Chicago, US, and there were also two meetings in 1995. After that, it became a yearly event, with some of the proceedings published on paper.[27]

---

25) http://cds.cern.ch/record/1164399.
26) See footnote 3.
27) R. Holzapfel (Ed.), *Proceedings of the 3rd International World-Wide Web Conference*, 1995.

3.6.3
The Consortium

The most unpleasant episode in the whole WWW history was the setting up of the WWW Consortium.[28] It is a tale of misunderstandings, conflicts of interest, friction between egos and personalities, cover pulling, occupying territory, jealousy and intrigue, and most of the unsavoury events occurred at higher managerial levels in the different organizations involved, with very little or no participation of the Web's developers. The story may one day be told in full by historians, though already at this point in time those aspects are no longer important because the final outcome is extremely positive.

Of course, there was also a certain naiveté on the part of the group, and Mike Sendall had well perceived this. In his opinion, as people managers, on a scale from 1 to 10, he gave them a zero; this was actually not surprising as the group was a collection of doers, not leaders.

The Consortium saved the Web; the main threat being from incompatibility. There were several different types of server, but worse, there were several different types of browser that treated the Web page source differently. There were also the extensions to the definitions, which perhaps is the most evil way of adulterating the Web and splitting it up into incompatible islands. It must be remembered that the Web is not based on physical laws; it can be made into whatever anyone desires. The progress is not slow because it is not regulated by the difficulties of understanding some physical phenomenon.

Making the Web grow was dependent on serving the users. There were two contradicting requirements: on the one hand all software had to be compatible, whatever machine or operating system it ran on, while on the other hand there had to be a sufficiently fast evolution in features beneficial to users. The first is a serious difficulty, the second is a motor of divergence.

The Consortium's goal was to bring all major players together so that they could agree on standards, split the work over their respective competencies, and cooperate rather than engage in cut-throat competition. It was necessary to make them see that there was no place for several world-wide Webs. The early members of the Consortium had many differences, and conflicts in the meetings, in the corridors and elsewhere, were not uncommon. Participants were simply not used to thinking in terms of a shared platform.

Progress was slow at first, but fortunately for the world it ultimately all got off the ground and worked because of the immense personal commitment of Jean-François Abramatic, Director of Development of INRIA,[29] the French Institute for Research in Computer Science and Automation, who was a capable and always positive manager, a *'herder of cats'*. 'Jeff', as he was known to the team, understood very well the psychology of making software; he cooled heated debates, saw to everyone's needs, and never lost sight of the ultimate goal – to

---

28) www.w3.org.
29) INRIA (Institut National de Recherche en Informatique et en Automatique).

bring standards to a business that did not want them but could not survive in any other way.

Indeed, without the work of Jean-François Abramatic there might not be any Web standards. He is one of the unsung heroes of the Web.

3.7
On the Nature of Computing

To understand better the past and current situations in the networked world, it is necessary to contemplate the deeply different nature of the digital world.

During the late 1970s, 'home computers' came to the market, and mainly catered for hobbyists and tech-savvy enthusiasts. During the 1980s, the Personal Computer (PC) began to attract more people, as the interface became more graphical and more software was available. However, PCs were still mainly used either by small companies, or at home by enthusiasts. It was the WWW that really made the market take off since, with access to the Internet, an increasing amount of information was available almost instantaneously.

In the large majority of connected homes the net was – and still is – used to consume information. Rare indeed were those who also published something.

Today, mobile devices are used much more to consume than to produce. What people put out there are images and videos, 'dumping' this information on social sites (of which there are only a handful) and perhaps adding a little text. Overwhelmingly, the information produced is of a very personal nature.

The net is used as a cheap and instantaneous alternative to writing letters and sending copies of photos to family and friends. It is also an alternative to the traditional mass-media.

However, there is a fundamental difference between the digital world of computing and the physical world of atoms.

3.7.1
Copy

Anything physical is difficult to copy, as it takes time and materials. Printed books can be photocopied, and vinyl music records can be copied to cassette tape, but it is not convenient. A Portable Document Format (PDF) file or an MP3 (short for MPEG3; Moving Picture Experts Group Audio Layer 3) song can be copied with almost no effort.

In the physical world the idea of copyright is, therefore, easy to enforce: in order to be profitable, the fraudster would have to set up a fairly large operation and/or hide in a foreign country.

Trying to enforce the same type of copyright in the digital world is almost impossible. Copy protections schemes have had some success as the average person has no knowledge to break the codes. Yet, as soon as some clever kid writes a program that does break such protection, the software to get around it can be downloaded and used by anyone. The operation itself becomes distributed.

3.7.2
See

Physical objects are visible. To hide a spy microphone is possible, but with some effort it can be detected and removed. Spy software is much more difficult to find, however, and most people have no understanding of how such software even works, as is demonstrated by the very confused and confusing articles that appear in the press.

The computer you own – whether in your mobile device, on your desk, or in your car – is not a machine; it is the infrastructure on which the 'machines' run, and those machines are software. Programs (applications) are the real machinery in the digital world; they are invisible except through their interaction with the physical device, but they are not obliged to effect any such visible interaction. They can work quietly in the background, totally hidden. In traditional computing jargon, such hidden programs are called 'daemons', and the connotation is relevant. Google's applications such as Google Earth now require their users to accept the installation (and probably constant running) of a daemon that will, behind your back, update Google's software, whether you like it or not. Digital objects are invisible.

3.7.3
Understand

Physical objects are quite simple. The item with the most complex interface that you own is probably your car; it is unlikely that you have anything else that has so many buttons and levers and pedals (unless you run a fairly realistic flight simulator). Your computer may have a screen, keyboard and mouse – just three controls – but the software you run is far more complex to handle than your car.

Most programs have about a hundred menu items, and you need many programs to run your digital life.

3.7.4
Remember

When you boil water for tea, you flip one switch on your kettle, and that switch comes back to its off position when the water is hot. Your kettle has only two states: 'on' and 'off'. Tomorrow, when you wake up, it is the same kettle as yesterday, and it is off.

By contrast, if you just use your mobile device to see the time you will have changed its state. If you do as little as make a single phone call, you will have added data to the call log, and if you look at your mail you may have changed the file system by several million bytes of data.

Computers have an enormous amount of state. Much of the evolution of software has been towards hiding this state or making it more understandable. It is highly questionable that any of these attempts have succeeded.

3.7.5
Interact

Your tea kettle has a fairly simple interface: a switch, a lid, a power cable and perhaps a gauge on the side to show how much water it contains. But no application is as simple as this: there are many controls, usually spread over a multitude of two-dimensional screens full of coloured buttons. To obtain any data or do anything with digital equipment, you need a computer – to run the app and to control it.

The only interoperability your tea kettle needs is with the electricity grid. This is usually not even something you think of, except perhaps when you move to a country with different electric sockets. The only time it becomes a real nuisance is when you move to a place that uses 120 V instead of 240 V.

But try to move any app to another device, or even to get the data across from one app to another inside the same device. The physical world may have interoperability problems, but they are trivial compared to those of the digital world.

3.7.6
Share

People expect variety and want to appear different from their peers, but in the digital domain it seems to be exactly the opposite: we want a single search engine, a single Cloud, a single social network, kids want the same computer and system as their friends have, and we all want to be able to phone each other seamlessly. For example, most people go directly to the now famous Google for any Internet search without visiting other search engines.

3.7.7
Think

As a conclusion, *'Robert Cailliau's first law of computing'* is: 'No paradigm of the physical world can be transposed to the digital world'. They are truly different worlds in a deep way. This difference is most likely based on two facts: (i) data and software are not electronic, they are completely abstract, above any substrate on which they are implemented; and (ii) computing is a mental activity.

A few recent happenings should illustrate how little understanding there is of the fundamental nature of computing.

During moments of inattention, some people publish intimate data about themselves on social sites, and then later regret it. A well-meaning group has hailed the proposed law in California, US, that will require social sites to give teenagers the ability to delete such content. That law ignores the fact that anyone could have made a copy (by screen copy or other means) of that content, that the site probably will keep a backup anyway and, worst of all, that such a law will probably stimulate pre-emptive copying by rivals: the same teenagers will now quickly copy any compromising content before the originators change their mind. One comment seen on a blog stated that a law in only one State would probably be unenforceable. As if a law applying to the entire US would help: Internet companies simply set up a copy of their server in another country when they get in trouble.

Some companies are introducing a model of renting their software instead of selling it, presumably to counteract the making of illegal copies. An installation of the software is still made on local storage, so how long will it take before those are illegally copied? In addition, the subscription model requires the regular use of identity (name and password), but at least one of the companies involved (Adobe) has been burgled and millions of account identities stolen.

Apple's latest version of their operating system (called 'Mavericks') is free, but it has removed the ability to synchronize content between devices unless this content passes via a Cloud server. This is extremely inconvenient for some users. It is also against the privacy laws of the European Union (EU) for businesses to store personal data outside the EU jurisdictions, therefore making it impossible for many to continue using their business devices and at the same time obeying the law, unless they build their own servers.

The identification of users by biometric data such as fingerprints has also been attempted. However, one recent fingerprint system was circumvented by hackers within three days of its appearance. The net is light speed fast, and information packets do not carry fingerprints.

Almost all of these problems do not originate with the manufacturer or the technology, but rather come from the obvious need for worldwide laws where there is a World Wide Web.

3.8
Science 'Un-human'

Some time ago, intellectuals introduced the split between the so-called 'soft' and 'hard' sciences. The 'soft sciences' or 'humanities' are essentially about human beings and involve studying the behaviour of a highly complex system. Whilst this is no doubt difficult, it is the very detailed study of the outward effects of only a single organism, the brain. This is not to look down upon the humanities, and neither should it be interpreted as arrogance.

By contrast, the 'hard sciences' such as physics, chemistry, biology, geology or astronomy, function through observation and experimentation, rigorously testing hypotheses against reality. It is just a simple fact: the physical sciences result in knowledge that could be defined as 'un-human'.

CERN studies really hard physics. It is concerned with a science that is related very little to human beings. That 'un-humanity' of physics is what makes it so difficult to grasp, so unappealing and weird to most people. Nonetheless, physics research on this planet is conducted by human beings, with all that implies: emotions, social organization, disappointments, and achievements. The humanity of physicists is also the root of the common misunderstanding that physics laws are the property of some persons, much as when we refer to 'Einstein's relativity', as if it were possible to go to a different restaurant to get a 'relativity' from a different cook.

An even more 'un-human' science is mathematics. Both mathematics and physics are at the foundations of computing, but computing is a technology and technologies are immensely linked to human nature. They apply knowledge to humans' desires as living, emotional, hormonal beings. Perhaps the term 'Computer Science' is inadequate: how can one have a science of something that is essentially constructed to specifications? Would it be better to talk about a duo of computer engineering and mathematics of computation?

3.9
Lessons to be Learned

With the help of INRIA and Georges Grunberg, touring Europe in 1995, trying to convince European industry to take a close look at WWW, I found a lukewarm response everywhere: the endeavour was largely unsuccessful.

Rather than pointing out that CERN should have asked for money for the Web, journalists should ask the painful, much more worrying question: 'Why did European industry not pick up WWW and run with it? What is wrong with our attitudes on the Old Continent?'

Industries of CERN's Member States should look at what is done at CERN and other research laboratories, and transform their results into industries that benefit humanity. WWW is an excellent example of a missed opportunity, but not by CERN.

3.10
Conclusions

CERN's role in the birth of WWW is significant: without the environment being free from commercial pressures, without the continued support of our direct superiors, and without the academic freedom that pervades the institute, we would never have succeeded.

The Web is also one of the first spin-offs of the LHC before that machine even became operational.

4
Grid and Cloud
Bob Jones

Today, the research community is facing a deluge of data produced by the latest generation of scientific instruments. This increased scale of scientific data is putting pressure on existing Information and Communications Technology (ICT) services which are struggling to store, distribute, process, analyse and preserve this precious commodity. At the same time, the need to demonstrate a return on investment is accelerating the scientific process itself, reducing the time from the acquisition of data to the publication of results. This demands that ICT services be more reliable and dynamic in their ability to serve the research community. As a consequence, the cost of ICT to the research community is continuously increasing and becoming a significant item in organisations' budgets.

The Large Hadron Collider (LHC) at CERN is a prime example of a research instrument with exceptional data processing needs. To address such needs, it has been necessary to revitalize the ICT models used by the research community in order to adapt to the changing needs of researchers, and to embrace new disruptive technologies and profit from the restructuring of the commercial sector which is introducing new products and services.

The LHC complex provides research facilities for several thousand High-Energy Physics (HEP) researchers from all over the globe. The LHC experiments are designed and constructed by large international collaborations and will collect data over a period of 15–20 years. These experiments run millions of computing tasks per day, and generate petabytes of data per month. The computing capacity required to analyse the data far exceeds the needs of any other scientific instruments today, and relies on the combined resources of some 160 computer centres worldwide. CERN and the particle physics community have chosen Grid technology to address the huge data storage and analysis challenge of the LHC. CERN leads the World-wide LHC Computing Grid (WLCG) project, which is a global collaboration linking Grid infrastructures and publicly funded computer centres worldwide. The WLCG was launched in 2001 to provide a global computing resource to store, distribute and analyse the data generated by the LHC. The infrastructure, built by integrating thousands of computers and storage systems in hundreds of data centres worldwide, enables a collaborative computing environment on a scale never seen before. WLCG serves a community of more than 10 000 physicists around the world with near real-time access to LHC data, and the power to process it.

4.1
Why a Grid?

Grids are an effective mechanism for bringing together computing and storage resources located in, owned, and operated by different organizations. By connecting through Internet networks, a Grid is a means for sharing computer power and data storage capacity, and providing secure access. It permits the creation of virtual research communities, making use of computers located all over the globe to become an interwoven computational resource for large-scale and compute- and data-intensive grand challenges.

The highlights of a Grid are:

- Federated yet separately administered resources, spanning multiple sites, countries and continents.
- Heterogeneous resources (e.g. hardware architectures, operating systems, storage back-ends, network set-ups).
- Distributed, multiple research user communities (including users accessing resources from varied administration domains) grouped in VOs (Virtual Organizations).
- Range of data models, ranging from massive data sources, difficult to replicate (e.g. medical data only accessible at hospital premises), to transient datasets composed of varied file sizes.

There were several important considerations in the design of the Large Hadron Collider (LHC) computing environment. Primarily, the volume of data generated by the experiments exceeds 20 petabytes per year of new data. This data is generated at a rate of gigabytes per second by the two largest experiments, ATLAS (A Toroidal LHC ApparatuS) and CMS (Compact Muon Solenoid). Thus, during most of first two years of LHC running, the total data rate is around 7 gigabytes per second into CERN's data centre. This data must be archived at CERN and a second copy distributed between regional (Tier 1)[1] centres in real time, together with an equal amount of data resulting from the initial processing. Thus, the distribution system must be capable of supporting these rates continuously.

The Tier 1 centres are large data processing centres capable of providing reliable data archiving for the lifetime of the accelerator, and they also have a role in distributing the data to a large number of Tier 2 centres (more than 150) once it has been processed in the Tier 1 centres. At the Tier 2 sites, physics analysis is performed on the processed data. Thus, another important consideration is the ability to rapidly process and distribute the data to the Tier 2 centres making it available to the thousands of scientists around the world.

The amount of raw computing capacity – both processing and storage – is driven by the volume of data, and a certain amount of simulation that is required by all experiments. These resources must be accessible by the many scientists using several different workflows from centrally organized and managed data processing for the bulk of the real and simulated data, to more widely varied flows during physics analysis, where the need for specific data sets is unpredictable and evolves as analysis progresses.

In considering how to build a system to satisfy all these needs, it was clear from the outset that locating all of the required computing and storage power at the CERN site would not be possible for a variety of reasons. First, the infrastructure of CERN's computing facilities would not scale to the required level without significant investments in a power and cooling infrastructure many times larger than existed at the time. Second, the CERN Member States have historically encouraged the provision of a significant fraction of the computing needs for an experiment to be located outside of CERN. The physics institutes and national laboratories participating in LHC experiments have local computing facilities, often of significant scale, and that with the expectation of

---

1) WLCG is made up of four layers, or Tiers; 0, 1, 2, and 3. Each Tier provides a specific set of services:

Tier 0: This is the CERN operated Computer Centre. All data from the LHC passes through this central hub, but it provides less than 20% of the total compute capacity. CERN is responsible for the safe-keeping of the raw data (first copy), first pass reconstruction, distribution of raw data and reconstruction output to the Tier 1s, and reprocessing of data during LHC down-times.

Tier 1: These are large computer centres with sufficient storage capacity and with round-the-clock support for the Grid. They are responsible for the safe-keeping of a proportional share of raw and reconstructed data, large-scale reprocessing and safe-keeping of corresponding output, distribution of data to Tier 2s and safe-keeping of a share of simulated data produced at these Tier 2s. There are currently 12 Tier 1 sites.

Tier 2: These are typically universities and other scientific institutes, which can store sufficient data and provide adequate computing power for specific analysis tasks. They handle analysis requirements and proportional share of simulated event production and reconstruction. There are currently around 140 Tier 2 sites covering most of the globe.

Tier 3: Individual scientists will access these facilities through local (also sometimes referred to as Tier 3) computing resources, which can consist of local clusters in a university department or even just an individual PC. There is no formal engagement between WLCG and Tier 3 resources.

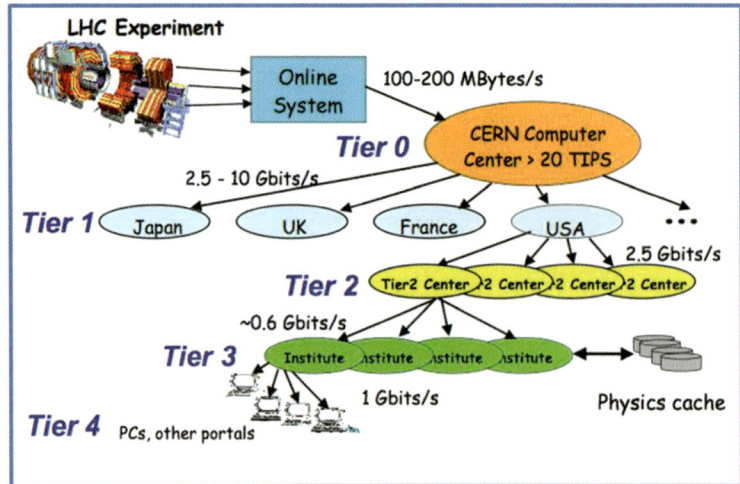

Figure 4.1 The original LHC computing model. (Reproduced with permission of CRC Press).[2]

improvements in wide area networking it was reasonable to expect that a distributed system that could couple these resources may achieve the scale needed. There are sociological advantages[3] in such a distributed system – investment is made locally and benefits the institute in several ways, providing training for students, and being able to leverage the resources for other local uses.

Figure 4.1 illustrates the key components of the model for LHC computing. At the time of developing this model (1999) it became clear that Grid computing technology could provide the practical implementation, permitting for example the integration of computing resources across multiple administrative domains; and the authentication and authorization infrastructure that provides a means for a user to have access to resources that may be distributed around the world.

This model was developed over the following 10 years by WLCG in conjunction with a series of Grid projects funded by the European Commission in Europe,[4] and by the Department of Energy (DoE) and National Science Foundation (NSF)[5] in the US. The infrastructure, as well as the tools to manage the exceptional data volumes and rates, were developed to provide a production environment that would be usable and manageable by the experiments and the computer centres involved. Tier 0 provides some 20% of the total computing resources, while Tier 1 centres and Tier 2 centres each contribute some 40%.

---

2) I. Bird *et al.*, *The Organization and Management of Grid Infrastructures*, Institute of Electrical and Electronics Engineers (IEEE) Computer, Volume 42, Issue 1, p. 36, 2009.
3) Ibidem.
4) E. Laure and B. Jones, *Enabling Grids for e-Science: The EGEE Project*, in Grid Computing: Infrastructure, Service, and Application, in L. Wang *et al.* (Eds), Grid Computing: Infrastructure, Service, and Application, p. 55, CRC (Chemical Rubber Company) Press, 2009.
5) R. Pordes, *The Open Science Grid (OSG)*, in A. Aimar *et al.* (Eds), Proceedings of the 14th Conference on CHEP (Computing in High Energy Physics and Nuclear Physics)'04, 2005.

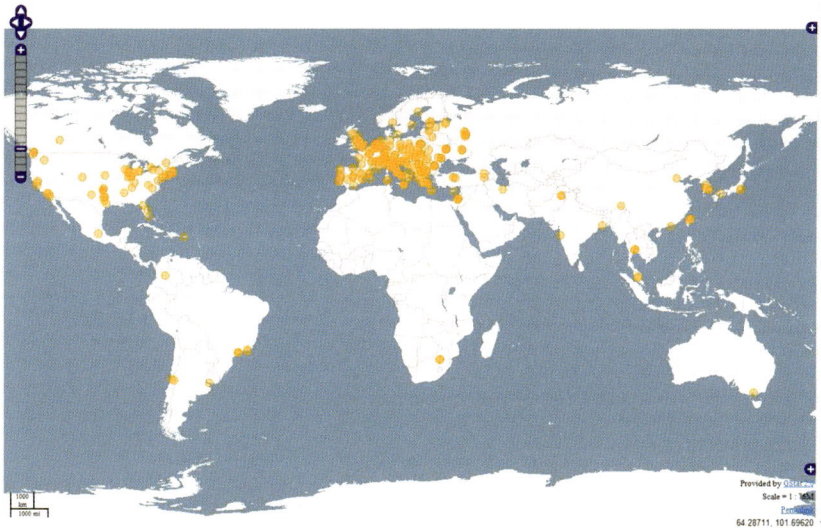

Figure 4.2 Worldwide distribution of WLCG Grid sites. (Reproduced with permission of CRC Press).[6]

In total today, there are more than 250 000 processing cores, and 150 petabytes of disk federated by WLCG. The worldwide distribution of the WLCG Grid infrastructure is illustrated in Figure 4.2.

4.2
A Production Infrastructure

In building the LHC computing environment, it was essential that this be an integral part of the production environment of the many computer centres participating, and that it be manageable by a realistically sized staff complement. Thus, in addition to the technical implementation of the Grid itself an entire ecosystem had to be built. This included processes and tools to support daily operation, problem solving and metrics reporting. As well as the service aspects, a set of policies was developed to govern the use of the infrastructure and to address privacy and security concerns. These ranged from acceptable use policies (for users and for virtual organizations), to policies of information retention and publication related to resource usage. A significant achievement was the development of a worldwide network of trust for both authentication and authorization. This unique trust framework[7] is what enables the many thousands of physicists to transparently access compute and storage resources anywhere in

---

6) Ibidem.
7) IGTF (International Grid Trust Federation): www.igtf.net.

the world. It has been necessary to strike a balance between defining infrastructure-wide policies and allowing freedom of choice to individual participants. The Grid infrastructure is under continuous development. Changes are considered that simplify the operation of the Grid service and reduce the level of operational effort needed. Technology in this area evolves very rapidly, and there is scope to benefit from recent advances. All of these points contribute to ensuring the long-term support and sustainability of the infrastructure.

4.3
Transferring Technology: Grids in Other Science Domains

Beyond HEP, the research communities making significant use of production Grid infrastructures include:

- **Life Sciences** research community uses Grid technology in the medical, biomedical and bioinformatics sectors in order to connect worldwide laboratories, share resources and ease the access to data in a secure and confidential way.
- **Astronomy and Astrophysics** institutes usually acquired the necessary resources on local computing facilities and quite often also contracted the access to a pool of resources at supercomputing centres. However, some of their applications grew in complexity and were considered very suitable to be run on Grid infrastructures. Large-scale projects currently under development such as the Square Kilometre Array (SKA) and the Cherenkov Telescope Array (CTA) are actively using the Grid infrastructure in preparation for their data processing tasks.
- **Earth Science** applications cover various disciplines such as seismology, atmospheric modelling, meteorological forecasting, flood forecasting and many others.
- **Fusion, Computational Chemistry and Materials Sciences and Technologies** make use of Grid's massive computational capacity for simulation purposes.

All of these research disciplines have been able to profit not only from the Grid technology developments driven by the needs of the LHC, but also from access to the production Grid infrastructure across Europe where the resources of the LHC community have been pooled by the WLCG project.

4.4
How CERN Openlab has Contributed to the WLCG Grid

CERN has been working closely with the Information and Communications Technology (ICT) industry to develop and evolve the underlying technology of WLCG.

The CERN openlab[8] is a structure for R&D collaboration with industry in the field of ICT. The primary mission of the CERN openlab is to develop new knowledge through the evaluation of advanced solutions or genuine joint research.

The CERN openlab has created a long-term link with industrial partners. Its framework has allowed industry to carry out large-scale ICT research and development in an open atmosphere – an 'Open Lab'. For CERN, openlab has contributed to giving the computing centre and, more broadly, the LHC community, the opportunity to ensure that the next generation of services and products is suitable to their needs.

At the start of the millennium CERN was in the process of developing the computing infrastructure for the LHC; significant research and development was needed; and advanced solutions and technologies had to be evaluated. The idea of Manuel Delfino, then head of the Information Technology (IT) department, was that, although CERN had substantial computing resources and a sound R&D tradition, collaborating with industry would make it possible to do more, and do it better.

4.5
Four Basic Principles

CERN was no stranger to collaboration with industry, and it had always conducted field tests on the latest systems in conjunction with their developers. However, what Manuel Delfino was proposing was not a random collection of short-term, independent tests governed by various different agreements. Instead, the four basic principles of openlab would be as follows:

1) Openlab should use a common framework for all partnerships, meaning that the same duration and the same level of contribution should apply to everyone.
2) Openlab should focus on long-term partnerships of up to three years.
3) Openlab should target the major market players, with the minimum contribution threshold set at a significant level.
4) In return, CERN would contribute its expertise, evaluation capacity and its unique requirements.

Industrial partners would contribute in kind – in the form of equipment and support – and in cash by funding young people working on joint projects. To this day, openlab is still governed by these same principles.

---

8) http://openlab.web.cern.ch.

4.6
Three-Year Phases

At the start of 2002, after a few months of existence, openlab had three partners: Enterasys, Intel and KPN QWest[9] (which later withdrew when it became a casualty of the bursting of the telecoms and dotcom bubbles). Subsequently, in July 2002, openlab was joined by Hewlett-Packard (HP), followed by IBM in March 2003 and by Oracle in October 2003.

The concept of three-year 'openlab phases' was adopted, the first covering the years 2003–2005. Management practices and the technical focus would be reviewed and adapted through the successive phases.

The status of 'contributor' was created in January 2004, aimed at tactical, shorter-term collaborations focusing on a specific technology. Voltaire[10] was the first company to acquire the new status on 2nd April, to provide CERN with the first high-speed network based on InfiniBand[11] technology. A further innovation followed in July with the creation of the openlab Student Programme, designed to bring students to CERN from around the world to work on openlab projects. The second phase, openlab-II, began in January 2006, with Intel, Oracle and HP as partners and the security-software companies Stonesoft[12] and F-Secure[13] as contributors. They were joined in March 2007 by Electronic Data Systems (EDS), a giant of the IT-services industry, which contributed to the monitoring tools needed for the Grid computing system being developed for the LHC.

The year 2007 also saw a technical development that was to prove crucial for the future of openlab, when CERN and HP ProCurve[14] pioneered a new joint-research partnership. Until this time, projects had essentially focused on the evaluation and integration of technologies proposed by the partners from industry. In this case, CERN and HP ProCurve were to undertake joint design and development work, and consequently openlab's hallmark motto, '*You make it, we break it*', was joined by a new slogan, '*We make it together*'. Another major event followed in September 2008 when Siemens become an openlab partner. Thus, by the end of phase II, openlab had entered the world of control systems.

This third phase was characterized by a marked increase in education and communication efforts. Increasing numbers of workshops were organized on specific themes – particularly in the framework of collaboration with Intel – and the communication structure was reorganized. The post of openlab

---

9) KPN Qwest was a telecommunications company equally owned by the Dutch national telecom operator KPN (Koninklijke PTT – Post, Telephone, and Telegraph operating entity – Nederland: www.knp.com) and the American Internet communications company Qwest Communications International Inc.: www.qwest.com.
10) www.voltaire.com.
11) InfiniBand is a switched fabric (a network topology) where network nodes connect with each other via one or more network switches (devices linking network segments) computer network communications link used in High-Performance Computing (HPC) and enterprise data centres.
12) www.stonesoft.com.
13) www.f-secure.com.
14) HP ProCurve was the name of the networking division of HP from 1998 to 2010.

communications officer, directly attached to the openlab manager, was created in the summer of 2008. A variety of communication channels are used for active dissemination, including the publication of technical reports and articles, workshops and seminars regularly organized at CERN, and the participation of CERN experts at off-site events and education activities. A specific programme was drawn up with each partner, and tools for monitoring spin-offs were implemented.

Everything was therefore in place for the fourth phase of openlab to begin in January 2012 with not only HP, Intel and Oracle as partners, but also with Chinese multinational Huawei, whose arrival extended openlab's technical scope to include storage technologies. More recently, Rackspace[15] and Yandex[16] have also joined openlab.

So, what is it that attracts companies to join the CERN openlab project? Being part of this R&D public private partnership offers commercial partners the following advantages:

- Exceptional testing environment in terms of complexity and scale of deployment.
- Engagement of multidisciplinary CERN experts capable of product design, testing, and deployment.
- Confidential feedback on faults found with products, and suggestions for improvements.
- Association with the world's leading global science brand name – CERN.
- Dissemination of positive results as success stories to a wide range of audience via conference presentations, whitepapers, Web site, and so on.
- Exposure of products to enlarged research community (CERN is at the centre of a network of several hundred research organizations). as well to student engineers and scientists as part of an extensive CERN-driven education and training programme.
- On-site visits at CERN for customers.
- Possibility for CERN openlab personnel to transfer to the company at the end of assignment, resulting in reduced recruitment costs and risks.

Below are a number of striking examples of the results of the CERN openlab project that demonstrate the return on investment for participating companies.

Through the deployment of Oracle database services for the LHC data processing tasks, Oracle was able to test and publicly demonstrate its data mining capabilities on a vast scale, on the basis of CERN's open information-sharing policy.

The deployment led to important performance improvements for Oracle streams and data replication services, based on feedback from CERN from its use in WLCG. Oracle 10 g[17] also includes support for the Institute of Electrical

---

15) www.rackspace.co.uk.
16) www.yandex.com.
17) Oracle 10g is version 10 of the Oracle Database, where g stands for Grid.

and Electronics Engineers (IEEE) floating point types, permitting extremely efficient floating point computations based on the collaboration with CERN in openlab.

Working with CERN permitted Siemens to expand its existing market for its SIMATIC[18] Windows Control Center (WinCC) Supervisory Controls and Data Acquisition (SCADA) system.[19] CERN became the largest deployment ever for SIMATIC WinCC, which has encouraged its uptake in more than a 120 institutes across Europe, Asia and North America.

The openlab framework also provided a collaborative environment where Siemens could work with Oracle to greatly improve the scalability of the SIMATIC WinCC system so that it could meet the challenges of the LHC. This development increased the market size for SIMATIC WinCC and the sale of licenses for Oracle.

When, for instance, the Many Integrated Core (MIC) co-processor was in its very early stages of development, Intel asked openlab to perform a thorough evaluation. Intel chose CERN openlab as one of the first partners worldwide for this challenging task, and in return Intel received constructive feedback on many of the technical specifications, ranging from the internal architecture to the co-processor card's entire software environment.

Such examples highlight the development cycle at the heart of the CERN openlab project. A joint programme of work is defined by CERN and the company that is compelling enough to justify them investing their own resources. New, co-developed prototypes are deployed at CERN in a gruelling environment that tests them to their limits, and the insights gained are then used to improve the prototypes. The company integrates the results obtained into their new products, while CERN publishes the advances made among the research community and beyond. As a consequence, CERN is sure of finding products on the market that satisfy its needs, while the company is certain that its products are production-ready and relevant to the research market.

4.7
EGEE to EGI Transition

The Enabling Grids for E-Science (EGEE)[20] series of the European Commission projects and its companions nurtured the goal of a Grid infrastructure for e-Science from the initial concept to the operation of a scalable, federated, distributed production system. The adoption of the production Grid infrastructure by a range of scientific disciplines and candidate industrial uses meant that it was necessary to find a new governance model that went beyond physics and beyond limited-duration projects. The conceptual and logistical framework for a

18) SIMATIC is an integrated automation family of products produced by Siemens.
19) http://www.automation.siemens.com/mcms/human-machine-interface/en/visualization-software/scada.
20) http://eu-egee-org.web.cern.ch.

permanent organization to oversee the operation and development of the Grid on a Europe-wide level was the subject of a dedicated design study, the European Grid Initiative Design Study (EGI-DS),[21] which was funded by the European Commission (EC) and took place between September 2007 and December 2009. As a result, a new organization, called 'EGI.eu',[22] was created in February 2010 to coordinate and evolve the European Grid Infrastructure (EGI) by building on the findings of previous European Union-funded projects. This transition was an important step in ensuring that the European research community has access to a distributed computing infrastructure. A four-year project called EGI-Integrated Sustainable Pan-European Infrastructure for Researchers in Europe (EGI-InSPIRE)[23] was co-funded by the EC and national grid infrastructures with the goal of creating a seamless system ready to serve the demands of present and future scientific endeavours.

4.8
Lessons Learned and Anticipated Evolution

The successful collection, distribution and analysis of data from the four LHC experiments represent a major achievement and demonstration of the Grid's success. The original concept of a Grid – being able to engender collaboration and the sharing of resources – is even more relevant today than it was 10 years ago. However, the environment has evolved and today's 'Grid' infrastructures must be integrated into an overall science e-infrastructure framework, capable of evolving and adapting to the rapidly changing needs of the science communities, especially those that are coming to realize the potential of such infrastructures in being able to help them deliver better science.

The production Grid infrastructure has been running continuously since 2009. The infrastructure serves not only the LHC community but also a range of scientific disciplines.

It is still important to recognize that WLCG uses a Grid for a very good reason: to integrate the resources that are distributed worldwide across tens of different management domains. Whilst this is the very rationale for a Grid in the first place, and is still a prime requirement, it is also apparent that there are many other communities which perceive Grids as being difficult to use, or not suitable for their needs, and are seeking other solutions such as Cloud computing[24] to achieve their objectives.[25] It is necessary to continuously reassess the

21) http://web.eu-egi.eu.
22) www.egi.eu.
23) http://www.egi.eu/about/egi-inspire.
24) Cloud computing is a phrase used to describe a variety of computing concepts that involve a large number of computers connected through a real-time communication network such as the Internet. In science, it is a synonym for distributed computing over a network, and means the ability to run a program or application on many connected computers at the same time.
25) M. Lengert and B. Jones, *Strategic Plan for a Scientific Cloud Computing infrastructure for Europe*, CERN-OPEN-2011-036, 2011.

implementation of the Grid infrastructures in the light of the changes in the environment noted above, but never forgetting the important successes and lessons, and ensuring that any evolution does not disrupt the continued operation for existing communities.

Although networking, Grid and supercomputing have developed independently, they are now becoming interwoven and should rather be viewed as building blocks or services within an e-infrastructure ecosystem that a science community can make use of in a coordinated manner.[26] Blurring the boundaries between these elements of e-infrastructure will encourage their interoperation and present a more consistent role to the users. These existing elements must be supplemented by, and integrated with, new components including scientific Cloud services, emerging data infrastructures managing significant slices of the data continuum, and other new services. It must also be agile enough to be able to support the additional services required by new science communities, or to help those communities integrate their own specific services, while taking benefit from the overall structure (AAA,[27] policies, support, etc.).

In such an environment, ease of use and access are vital considerations. The Cloud model has been readily adopted as it is a simple concept from the user's point of view, and a similar simplicity must be applied to the more traditional services. Enabling the use of existing trusted federated identities would help address some of the initial complexity in being able to access a Grid, for example.

4.9
Transferring Technology: Grids in Business

The EGEE embarked on a strategy to bring on board companies of different sizes and commercial sectors that have explored Grid technology, thus leveraging the compute and IT resources needed to shorten the time required to take new products and services to market, thereby gaining efficiencies and cost savings.

As a result, a number of companies such as CGG Veritas,[28] Maat-g,[29] NICE,[30] Imense[31] and Constellation Technologies[32] developed products and services that integrated Grid components. Other companies have been able to create new

---

26) H. Davies and B. Bressan (Eds), *A History of International Research Networking*, Wiley, 2010.
27) AAA, or Authentication, Authorization, Accounting, is generally used to refer to the underlying security framework.
28) CGG (Compagnie Générale de Géophysique) is a French-based geophysical services company founded in 1931. In 1996, Veritas DGC was formed from the merger of Veritas and Digicon Inc. (DGC). Following its merger with Veritas DGC Inc. in January 2007, CGC became known as CGG Veritas until January 2013 when the company changed its legal name back to CGG: www.cgg.com.
29) www.maat-g.com.
30) www.nice.com.
31) www.immensetech.net.
32) www.contech.com.

niches in IT consultancy and training around EGEE technologies, such as Excelian,[33] GridwiseTech[34] and Linalis.[35] Two of these companies – Imense and Constellation – have been formed thanks to the strategic alliances forged with EGEE and underpinned by the technologies developed over the past few years.

The gLite suite of middleware components serves as the 'glue' that holds EGEE's Grid infrastructure together. The middleware is released under an Open Source business-friendly licence, Apache version 2.[36] gLite provides a framework for building Grid applications tapping into the power of distributed computing and storage resources across the Internet. The gLite distribution is an integrated set of components designed to enable resource sharing. It encompasses a wide range of services, while ensuring ease of installation and configuration on chosen platforms such as Scientific Linux and Windows. It is currently deployed on hundreds of sites worldwide, enabling global science and industrial research and development.

The gLite Open Collaboration[37] has been established between the EGEE partners involved in middleware activities as a framework for the maintenance and future evolution of the middleware beyond the end of the EGEE series of projects, ensuring the availability of the middleware, support and developments. This is complemented by the European Middleware Initiative (EMI)[38] which brings together the major European middleware providers such as Advanced Resource Connector (ARC),[39] gLite[40] (Lightweight Middleware for Grid Computing), and Uniform Interface to Computing Resources (UNICORE)[41] to deliver a consolidated set of services.

---

33) www.excelian.com.
34) www.gridwisetech.pl.
35) www.linalis.com.
36) Apache is a free software license, written by the Apache Software Foundation (ASF), which allows the software user the freedom to use such software for any purpose, to distribute it, to modify it, and to distribute modified versions of it, under the terms of the license, without concern for royalties.
37) http://indico.cern.ch/getFile.py/access?contribId=2&resId=1&materialId=slides&confId=68337.
38) www.eu-emi.eu.
39) Advanced Resource Connector (ARC), or Heavy Expertise-Light-Weight Solutions, one of the three main middleware solutions offering standard-compliant, portable, platform independent, open source middleware. The most challenging community using ARC is the unique distributed Nordic Tier-1 (http://www.nordugrid.org/arc) for the CERN LHC experiments: ATLAS (A Toroidal LHC ApparatuS); and ALICE (A Large Ion Collider Experiment).
40) gLite (glite.web.cern.ch), or Lightweight Middleware for Grid Computing, the next generation middleware for distributed Grid computing, born as part of the Projects EGEE (Enabling Grids for E-Science) series, provides a robust framework of services on which to build applications for a Distributed Compute Infrastructure. Its components satisfy the requirements of many research communities, such as the High-Energy Physics (HEP) with its need for CERN Large Hadron Collider (LHC) experiments.
41) UNICORE (Uniform Interface to Computing Resources: www.unicore.eu), or Seamless Access to Computational and Data Resources, initiated in the supercomputing domain, today is a general-purpose Grid technology. It contributes to European Middleware Initiative (EMI) knowledge on High-Performance Computing (HPC) and nonintrusive integration into existing infrastructures. In its recent version, UNICORE 6 follows the latest standards from the Grid and Web services world.

While Grid computing today is used in many academic disciplines, from HEP to geosciences, health and the humanities, businesses can also benefit from Grid technologies. They can be adopted by one department or a single office, and there is potential for Grid use in Small and Medium Enterprises (SME), as well as in bigger, more complex organizations.

Grid technologies can help companies to use resources more efficiently – making better use of computer power already available and opening the door to large-scale processing that might otherwise be out of reach.

By using an internal Grid consisting of a couple of hundred processors, drug discovery company e-Therapeutics[42] was able to analyse a database of 15 million compounds and 2.6 million proteins. 'Using this kind of approach has led to more productivity with very little money. As a scalable process, the use of e-Science orientated computing could completely change the economics of drug discovery', said the company's CEO, Malcolm Young.

4.10
Sharing Resources Through Grids

Grids give users easy access to shared resources and data, no matter where they are located. Through VOs, Grids bring geographically dispersed communities together to achieve common goals – by sharing problems, discussing solutions, and forming close-knit collaborations.

GridwiseTech[43] has used this approach to help hospitals form better collaborations with medical research. The company's software makes setting up and managing a VO simple and intuitive.

4.11
What are the Hurdles?

Moving from a research to an industrial environment is never easy, and the commercialization of Grids is no exception. Grid services and applications are not always intuitive to use, and can be tricky to harmonize with systems and practices already in place, such as security policies. Grids must also adjust to a commercial environment, allowing providers to log usage and bill for services.

The EGEE business use cases described in Sections 4.12 to 4.17 showcase a variety of companies that have taken advantage of the services developed by the EGEE project for WLCG, such as adopting technology on a company's own infrastructure or integration into a new or existing solution, running proof-of-concept applications on the Grid infrastructure or working on areas of joint technical development.

---

42) www.etherapeutics.co.uk.
43) www.gridwise.pl.

4.12
Philips Research: Scientific Simulation, Modelling and Data Mining Supports Healthcare

Philips Research is one of the world's largest corporate research organizations that develops new technologies and investigates potential growth areas for research and development. Over the past few years, Philips Research ICT has been investigating the deployment of Grid technology with the aim of enabling research that effectively draws together knowledge, joins expert networks for knowledge exchange, and also accesses external High-Performance Computing (HPC) resources, such as in the medical field.

Developments in medical imaging and bioinformatics applications fields involve many high-resolution sensors, leading to significant increases in data generation, as well as data correlation from different sources. Open Innovation[44] is aimed at addressing the need to have resources and/or knowledge in the right place and at the right time, in order to support the technology chain. The infrastructure must be able to cope with both 'data explosion' and 'large-scale data analysis' and ensure secure collaboration with partners. While large computations are necessary, much can be parallelized or distributed. A key end objective is to support the shift from experiments that are model-driven to data-driven experiments, as well as ensuring involvement in the ecosystem by actively participating in e-Science projects and being able to both consume external resources and provide resources to externals.

Philips Research's adoption of gLite is connected with this Open Innovation concept, which has resulted in an open organization leading strategic cooperation with companies, universities, and other institutions.

4.13
Finance: Stock Analysis Application

Athens University of Economics and Business (AUEB) in Greece and Abdus Salam International Centre for Theoretical Physics (ICTP) in Italy, forming part of the European Union IndiaGrid project,[45] have developed a Stock Analysis Application[46] for the intensive statistical analysis of financial research data related to securities (stocks, bonds, options) with the ability to be run on different Grid infrastructures and using its gLite middleware. The application automatically analyses a large mass of financial data focusing on the latest trade volume, best-buy price, best-sell price, and so on, at

---

44) Open innovation is a term promoted by Henry Chesbrough, a professor and executive director at the Centre for Open Innovation at the University of California, Berkeley, US, though the idea and discussion about some consequences date as far back as the 1960s. The concept is related to user innovation, cumulative innovation, know-how trading, mass innovation and distributed innovation.
45) www.euindiagrid.eu.
46) http://euindia.ictp.it/stock-analysis-application.

the time that a particular event occurs, for example the arrival of a new order. The application facilitates processes that could also be achieved by Grid scripting.

So why Grid? Grid is well adapted for applications dealing with large amounts of data. This finance application handles more than 4 terabytes of input from 700 securities such as stocks, bonds and options, each equalling several years' worth of data. It is suitable for the primary objective of analysing massive financial databases on an instrument-by-instrument basis (one instrument's data is analysed at each node), but may have many other application domains.

4.14
Multimedia: GridVideo

GridVideo is a Grid-based multimedia application using gLite for the distributed tailoring and streaming of media files, and lays the foundations for a Video On-Demand (VOD) service that is widely used in both educational and commercial environments.

The development of GridVideo was carried out using the Grid test bed built within the PI2S2 project[47] running at the University of Messina, Italy. For any commitment on Grid environments, the middleware is always a crucial component. By combining components from the best middleware projects currently available with components developed for the WLCG project, the gLite middleware facilitated the implementation of Grid applications, offering an opportunity to use the basic services in an easy and efficient manner.

GridVideo[48] enables the adaptation of media files to the actual characteristics of client devices, with enormous benefits in terms of download and processing time and without overloading user devices. With the diffusion of Internet-based multimedia applications, the provision of tailored content has emerged as a key issue, in particular in environments, such as mobile computing, characterized by a lack of resources.

The main idea behind GridVideo is to use Grid resources both to store high-quality videos and to perform on-the-fly parallel tailoring to enable the provision of user-adapted multimedia contents. Thus, it represents a powerful way to implement a VOD Service overcoming challenges associated with tailoring techniques. In a nutshell, a Grid-based solution, thanks to its distributed nature, allows the realization of a robust and scalable system without overloading user devices.

---

47) PI2S2 (Progetto per l'Implementazione e lo Sviluppo di una e-Infrastruttura in Sicilia basata sulla Grid).
48) D. Bruneo et al., GridVideo: A Practical Example of Nonscientific Application on the Grid, IEEE, Transactions on Knowledge and Data Engineering, Volume 21, N. 5, p. 666, 2009.

4.15
Imense: From Laboratory to Market

Imense,[49] a small, high-tech UK company based in Cambridge, UK, developed the product called 'Imense®' to significantly improve the accuracy and depth of a search to attract new users. The company targeted image agencies; publishing groups; digital asset management providers; search engines; photo-sharing and social networking sites. Images and video comprise over 70% of the digital data available on the Internet (equivalent to billions of images); unfortunately, traditional search engines cannot index this information directly but rely instead on text descriptions entered by hand. Imense's key innovation was a new form of image retrieval that automatically analyses images in terms of their content, without the need for human-generated captions. The company has also developed a powerful query language that allows people to search for the images they need. It is thought that the technology developed by Imense was acquired by Apple. The creation of Imense is a fine example of how Grid technology helped a young, high value-added start-up transform an idea into a business.

4.16
Total, UK

The Total Group Geoscience Research Centre (GRC), located in Aberdeen, UK, plays an important role in research and development for Total's exploration and production worldwide. Their geologists, geophysicists and reservoir engineers work together to develop and test new techniques in areas such as history matching, fault and sedimentary modelling, and seismic studies for reservoir and pore network modelling.

The engineers and geoscientists use seismic, geo-models and reservoir simulation models to improve their knowledge of how petroleum reservoirs behave during production, and their potential to deliver. Dynamic flow simulation and seismic processing and modelling are essential for this, and are computationally demanding, generating large amounts of data. To carry out its research, the GRC needs to exploit to the maximum its available computing resources, and Grid computing provides many benefits for these applications.

The GRC became an early Grid adopter in order to experiment with Grid computing and to assess the potential for external Grid computing. The potential benefits included providing computational power and data storage on demand, by linking together distributed computing resources and implementing an open source Grid in the oil and gas industry. The GRC operated a hands-on approach, porting an existing in-house application to the Grid. Although the outcomes remain internal, the Total UK case study provides an excellent

---

49) www.imense.com.

4.17
Seismic Imaging and Reservoir Simulation: CGG Veritas Reaping Benefits from the Grid

Finding new oilfields, monitoring resource recovery and practising sustainable development are all complex tasks. Worldwide collaboration, huge computing power, fostering innovation and cutting-edge geosciences software all require major financial investment. Whilst building huge processing centres is an industrial challenge in itself, small laboratories face the most difficult challenges when the time comes to make the investments needed. The competitive edge and overall benefits for the geoscience community derive from access to state-of-the-art geosciences software, shorter processing times for seismic data, and the ability to collaborate with colleagues working on the same project, irrespective of the physical location of the research groups involved.

Modern seismic data processing and geophysical simulations require ever-greater amounts of computing power, data storage and increasingly sophisticated software. However, small and medium-sized research centres encounter difficulties in exploiting their innovative algorithms due to an inability, at research level, to keep pace with any enhancements to the current state of the art. Grid computing offers an opportunity to foster the sharing of computer resources and provide access to large computing power for a limited period of time at an affordable cost, as well as to share data and sophisticated software. The opportunity to solve new complex problems and to validate innovative algorithms on real scale problems is also a way of attracting and retaining the brightest research workers for the benefit of both academic and industrial R&D geosciences communities.

Back in 2007, CGG Veritas, a French company leading the development of natural resources through geophysics,[50] embarked on a mission to evaluate the benefits of Grid computing through the Expanding GEOsciences on DEmand (EGEODE) Virtual Organization. The approach adopted included access to EGEE Grid resources, coupled with a proposed solution to geophysicists working in research laboratories who needed access to a generic platform for seismic processing.

A year on, CGG Veritas developed software called 'Geocluster'[51] to process seismic data from a variety of sources, including trucks that mechanically send sound-waves into the Earth with the use of large, hydraulics devices, and also ship-borne 'airguns' that fire at regular intervals just below the water's surface as the vessel moves along predetermined survey lines. Regardless of the source of the sound-waves, or whether they take place on land or sea, the principle is the

---

50) See footnote 28.
51) http://www.cggveritas.com/default.aspx?cid=1925&lang=1.

same: the reflected compression wave created is detected by a network of sensors, forming a signal. The challenge for the developers of Geocluster was to process this signal, distinguish it from background noise, store and interpret the information, and put it all into a format that is easily understood by humans.

Geocluster is able to accomplish this by creating three-dimensional underground maps that outline the properties of the subsurface, along with the locations of oil and gas reserves. A sister application, known as Reservoir Simulation,[52] models how these reserves evolve during the drilling process, enabling a more efficient extraction. The data are interpreted by a geophysicist who instructs the exploratory company on where to drill. As with Geocluster, Reservoir Simulation was developed to operate in a Grid environment using EGEE as a research environment and test bed. As an outcome, CGG Veritas has marketed its Geocluster application, called 'Geovation'. This includes the new generation of massive high-density wide-azimuth datasets and the latest compute-intensive imaging algorithms with a full range of advanced applications for interactive processing and quality control.

4.18
Societal Impact

The public research community, the European Commission and governments, have all invested heavily in data-supporting infrastructures during the past decade. Distributed Computing Infrastructures such as Grids and Clouds are critical because science is fundamental to addressing the problems confronting our planet and our lives. Data is fundamental to science, and the scientific investigations that we now carry out require ever-increasing data sets. Flexible, powerful computing systems are required to support this. Whilst computing does not have the power to save our planet from global warming or energy shortages, it does have an underlying role to play in preventing such crises.

For example, it has never been more important to have powerful and accurate climate information. Indeed, elaborate computer models are the primary tool used by climate scientists and bodies such as the Intergovernmental Panel on Climate Change (IPCC) to report on the status and probable future of planet Earth. Models like those used by the IPCC need data from the atmosphere, land surface, ocean and sea ice, all originating from different communities, along with diverse accompanying metadata (data which describe data). The amount of data that climate scientists need to manage is enormous (on the petascale), yet a broad and global community needs to be able to access and analyse it. This is ideal for a Grid solution – where information stored around the world can be woven together without moving databases. The Earth System Grid Federation (ESGF)[53] is a project sponsored by the DoE, the NSF and the National Oceanic

---

52) http://www.cggveritas.com/default.aspx?cid=4.
53) www.esgf.org.

and Atmospheric Administration (NOAA) in the US that will support the data for the next generation of climate models used by the IPCC in their next assessment.

Recent developments in IT have made it possible to consolidate IT infrastructure in a way that delivers increased flexibility and responsiveness to business needs, while reducing costs. This change involves a move from IT being provided individually by organizations procuring their own separate IT infrastructure, to a new model in which IT is provided as a utility, which is known as 'Cloud computing'. The National Institute of Standards and Technology (NIST) defines Cloud computing as ' . . . a model for enabling convenient, on-demand network access to a shared pool of configurable computing resources (e.g. networks, servers, storage, applications, and services) that can be rapidly provisioned and released with minimal management effort or service provider interaction'.[54]

Since the late 1980s, all science domains have had the following processes in common: data collection, processing, and analysis. This is applicable to the entire domain of laboratories or research environments, and to the science community affiliated to them. Furthermore, for the data collection, extreme-scale laboratory instruments such as the LHC or Joint European Torus (JET), satellites and molecular biology each have their own large-scale IT systems to collect, process and analyse their data. Until today, every player has been creating an enormous amount of duplicated IT infrastructure and data assets, and this is introducing barriers to collaboration and avoiding the synergetic exploitation of data and models.

Over the past few years, consumer-facing firms delivering products in large volumes have actively adopted Cloud computing. Currently, commercial service providers are expanding their available Cloud offerings to include the entire stack of IT hardware and software infrastructure, middleware platforms, application system components, software services and turn-key applications. The private sector has taken advantage of these technologies to improve resource utilization, increase service responsiveness, and accrue meaningful benefits in efficiency, agility and innovation. Similarly, for research organizations, Cloud computing holds significant potential to deliver public value by increasing operational efficiency and responding faster to constituent needs.

By leveraging shared infrastructure and economies of scale, Cloud computing presents a compelling business model. Organizations will be able to measure and pay for only the IT resources they consume, to increase or decrease their usage to match requirements and budget constraints, and leverage the shared underlying capacity of IT resources via networks. The resources needed to support mission-critical capabilities can be provisioned more rapidly, and with minimal overhead and routine provider interaction.

---

54) http://www.nist.gov/itl/cloud/index.cfm.

The Helix Nebula initiative[55] combines the computing needs of CERN, the European Space Agency (ESA) and the European Molecular Biology Laboratory (EMBL) with the capabilities of several IT companies. Working together, these two groups are addressing the technical, legal and procedural issues that today make it difficult to move jobs seamlessly from one Cloud to another at scale.

Helix Nebula aims to pave the way for the development and exploitation of a Cloud computing infrastructure, initially based on the needs of European IT-intense scientific research organizations, while also allowing the inclusion of other stakeholders' needs (enterprises, governments and society). The scale and complexity of services needed to satisfy the foreseen needs of Europe's IT-intense scientific research organizations are beyond what can be provided by any single company, and hence will require the collaboration of a variety of service providers.

The initiative is a learning process to understand what a formal procurement of Cloud computing would involve. For publicly funded institutions, monetary transparency, accountability and value are increasingly important, and Helix Nebula offers an environment in which potentially complex questions around Service Level Agreements (SLAs), intercloud transfers, and more can be explored ahead of procuring a more lasting solution.

Not only are the questions that the initiative is addressing likely to be of value, but the relationships that are being built between the different research communities and the IT industry may well prove to be the industry's most valuable and long-lasting asset.

CERN and its scientific programme has been a driving force that has achieved the largest collaborative production Grid infrastructure in the world for e-science. By providing seamless access to a vast computing resource, 24 hours over 7 days, it has shown that such a production infrastructure can be used by a wide range of research disciplines, producing scientific results which, without Grid, would not have been possible to achieve. Through such a structure, scientists were able perform more scientific investigations on a larger scale, and obtain results in a shorter time frame. CERN has spawned collaborations within Europe, and allowed Europe to collaborate as a whole with other regions worldwide. Grid computing has acted as a good showcase of what is possible with distributed computing infrastructures, and laid an important foundation for future forms of utility computing, including Clouds, which are currently having an immense impact on many business sectors and the IT industry in particular.

Acknowledgements

This chapter includes material contributed by many colleagues at CERN, and in the WLCG, EGEE, EGI-InSPIRE, Helix Nebula and CERN openlab projects.

---

55) www.helix-nebula.eu.

5
The 'Touch Screen' Revolution

Bent Stumpe

It is probably not known that the first capacitive transparent touch screen was developed at CERN as early as 1973 and used at the CERN Super Proton Synchrotron (SPS) accelerator. The technical developments carried out at CERN and elsewhere, which have had a major impact on modern communication devices, are described in the following chapter.

Capacitive Touch Screens have provoked a big change in daily life. If the computer is an iPad or a similar device, the interaction process is performed directly on the touch screen. Therefore, you will most probably use a touch screen when you call someone using a smart phone, or very often when you make a payment in the supermarket or you use the Global Positioning System (GPS) in your car. Today, train stations, airports, safety systems, medical apparatus and so on are all using touch screens. Likewise, many products currently use touch screen technologies, and the numbers of these will continue to increase in the near future. In fact, the touch screen market forecasts a threefold increase by 2022.[1]

5.1
The Birth of a Touch Screen

A touch screen is an electronic device that is able to detect the direct presence of an object, for example touching a touch-sensitive screen of a computer or microcomputer with a finger. It can be either transparent or nontransparent, and can have many different shapes, many types of mechanical supports, as well as a number of different electronic circuits associated with the detection of the presence of the object. Normally, a touch screen is used as a separate overlay in front of a display device. The name 'touch screen' covers two combined technologies which include the screen technology and the electronics associated with the device.

---

1) C. Thiele, *Touch Screen Modules: Technologies, Markets, Forecasts 2010–2020: Full analysis of technologies, players & markets*, IDTechEx (www.idtechex.com), Report, 2013.

From Physics to Daily Life: Applications in Informatics, Energy, and Environment, First Edition.
Edited by Beatrice Bressan.
© 2014 Wiley-VCH Verlag GmbH & Co. KGaA. Published 2014 by Wiley-VCH Verlag GmbH & Co. KGaA.

The idea of using touch screens is very old. Original prototypes were known in 1971 when the CERN development started, but these were designed with a different technology from what we see today. It was at the time when the SPS accelerator had to be built and the traditional methods for its control were not sufficiently cost-efficient. A general-purpose console was needed, and from an idea based on my studies performed at a television factory in 1959, whereby coils of 80 micrometers width were etched directly onto the substrate of standard circuit boards, a new touch screen came to light at CERN in 1973; this was based on the principle of a change in the capacity of transparent capacitors printed onto a glass plate.[2]

The thinking went that, if it was possible to etch a coil onto a printed board substrate, it might also be possible to print a capacitor on a transparent substrate such as Mylar, Kapton or Glass. In fact, the proposed development described how such a capacitor could be a part of an electronic circuit able to detect the presence of a finger on a transparent touch screen. The requirement for the capacitor to be transparent to the user was achieved because the distances between the printed lines were so small that the human eye could barely observe the structure of the thin lines. Ion sputtering technologies and conventional methods to produce printed circuit boards were used at CERN to deposit a very thin layer of copper on a small sheet of transparent Mylar. It was then possible to make the first transparent capacitor for use in what is called the 'self-capacitance touch screen'. The first prototype of this touch screen had only nine capacitors, each with its own independent detection circuit connected to the computer, and was able to detect nine simultaneous touches. In other words it was a 'multi-touch' touch screen.

This novel device was incorporated in the system for interfacing the operator's control desk consoles with the computer. The touch buttons were drawn by computer on the face of a cathode-ray tube over which was placed a transparent glass touch screen (Figure 5.1). In order to protect the capacitor from being short-circuited when touched by a finger, a very thin, nonconductive, transparent lacquer was used for protection.[3]

The touch screen had a network of fine conductors deposited on it. When a finger was placed over one of the now 16 possible button positions, a change in capacitance was caused, which was then detected by the associated circuit and provided information to the computer about which button had been touched.

A click was associated, simulating the action of pushing the button. The panel enabled the operator to display all the actions they wanted to perform on the control consoles (Figure 5.2).[4] The touch button screen was a multitouch screen as it had the ability to register 16 touches at the same time. Later, in 1977, the CERN x/y multitouch screen was developed.

2) F. Beck and B. Stumpe, Two devices for operator interaction in the central control of the new CERN accelerator, CERN 73-6, 1973.
3) B. Bent Stumpe and C. Sutton, *The first capacitative touch screens at CERN*, Volume 50, N. 3, p. 13, CERN Courier, 2010.
4) J.F. Lowe, *Computer creates custom control panel*, Design IDEAS, 1974.

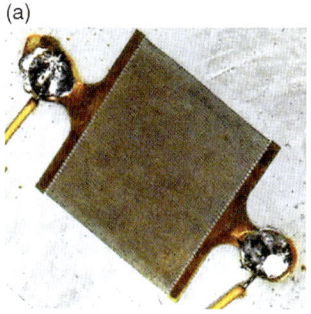 

Figure 5.1 (a) The first self-capacitive transparent touch-button capacitor, developed in 1972. (Courtesy of CERN); (b) The transparent multitouch-button screen used to form the interface between the computer and the system operator, in 1977. (Courtesy of CERN).

The early CERN touch screens had a relatively high resolution, and were found to be very reliable; however, they were relatively expensive to produce due to the small number needed. Only one screen in one console was needed for controlling the entire accelerator. The control room was equipped with four consoles. Subsequently, modern technologies have overcome these high-cost problems due to a high integration and mass production.

Only well-equipped laboratories could at that time afford the use of touch screens because of the high cost of the computers controlling them. At that time, very few people had ever heard about a touch screen, and the vision for what applications such a device could be useful for still had to be created!

During the evolution of touch screen technology, the early touch screens were programmed to have only a fixed number of buttons, with the text written under

Figure 5.2 (a) Bent Stumpe in front of the consoles of the CERN Proton Synchrotron (PS) control room; (b) A close-up of the console. (Courtesy of CERN).

computer control in black and white characters. Although the computer power at that time was limited, the computers behind the early touch devices could cover many square metres of surface. A console with one integrated touch screen was therefore very expensive.[5]

5.2
The Novelty for the Control Room of the CERN SPS Accelerator

A new concept was adopted for the control room of the Super Proton Synchrotron (SPS) accelerator, namely the use of general-purpose consoles with a built in multitouch screen, each attached to its own computer system. At the time, minicomputers were just emerging onto the market, and those from Norsk Data[6] (about 30 in total) were used in the SPS accelerator. These were partly placed near the control room itself, and partly in the strategic areas close to the equipment of the machine, which was 7 km in circumference. Despite the fact that the touch screen had only 16 fixed buttons, it was possible for a single person to control the entire accelerator from one such console (Figure 5.3).

The need to have the equipment installed in the 7 km-long accelerator hardwired to the control room was completely eliminated by the introduction of computer-based message transmission systems that connected all of the computers together. Hence, control of the accelerator itself was greatly simplified by the touch screen technology.

(a)

(b)

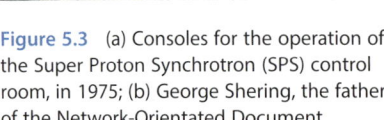

Figure 5.3 (a) Consoles for the operation of the Super Proton Synchrotron (SPS) control room, in 1975; (b) George Shering, the father of the Network-Orientated Document Abstraction Language (NODAL) used in the SPS, seated in front of a SPS console, in 1977. (Courtesy of CERN)

---

5) E.A. Johnson, *Touch Display – A novel input/output device for computers*, Electronics Letters, Volume 1, Issue 8, p. 219, 1965.
6) Norsk Data was a minicomputer manufacturer located in Oslo, Norway. Existing from 1967 to 1992, it had its most active period in the years from the early 1970s to the late 1980s. At the company's peak in 1987 it was the second largest company in Norway and employed over 4500 people.

The second important novel point was the development of the Network-Orientated Document Abstraction Language (NODAL) special software for controlling the accelerator. One part of NODAL was used to write the text for 'software buttons' and to identify which button had been touched for a very specific control operation to be performed at a given location of the accelerator.[7]

Another essential development for accelerator operation control was the new type of TV raster scan displays in colour that allowed visualization on larger screens and, in real time, the general running state of the SPS accelerator. In this way, the effect of any important interaction made by the accelerator operators would be immediately visible all over CERN.

As an additional support to facilitate operator interactivity, a pointing device (tracker ball) using mechanical and electric principles was developed in 1972 and added to the consoles. This was a precursor of the computer 'mouse', that would come onto the market during the 1980s.

To complete the control room equipment, a single computer-controlled knob was developed which had a mechanical and electronic construction that allowed it to perform any operation for which a knob might need to be used. The single computer-controlled knob could be used to perform an operation that required the use of any type of switch with a programmable number of positions, and it could also be a potentiometer having any electric value. The knob also had a computer-controlled brake that allowed a tactile feedback which could indicate to the operator that he or she was remotely controlling the movement of heavy equipment. The end stops could also be programmed.

At the start of 1973, the decision was taken that CERN should involve the industry in any further developments made at the unit. This related to all research and development in which CERN was involved, such as physics, electronics, vacuum technology, computer technology, communication via the Internet such as World Wide Web and software developments. During the 1970s, the development of microchips and their integration into the microcomputer permitted a further step in the evolution of the touch screen at CERN. At this point, the suggestion was made at CERN to switch from the 16 fixed button set-up, which was well tested and valuable, to the development of the 'mutual capacitance' multitouch screen. The consequent development was rapid; from the time of the initial proposal in 1977, an x/y touch screen prototype was built by the Polymerphysik GmbH[8] firm in Germany and delivered to CERN in 1978.

The self-capacitance transparent touch screen was produced and commercialized by Ferroperm[9] in Denmark as early as 1974. NESELCO (a subsidiary of

---

7) M.C. Crowley-Milling, *How CERN broke the software barrier*, New Scientist, Volume 75, N. 1071 p. 1, 1977.
8) Super Proton Synchrotron (SPS), *Experiments to find a manufacturing process for an x/y touch screen*, SPS/AOP/BS/jf/B-48, Report on a visit to Polymer-physik GmbH: www.polymerphysik.de, 1978.
9) Ferroperm: www.ferroperm.com, *Touch Screen brochure*, Vedbaek, Denmark, 1974.

LK-NES, now LKE Power)[10] was convinced about the future role that touch screen-based control systems might play, and was interested in a proposal to develop x/y mutual capacitance multitouch screens.[11] As regards this technology, a joint CERN/NESELCO proposal was sent in July 1979 to *Udviklingsfondet*,[12] the Danish development fund, for the sum of 11 444 000 million Danish kroner[13] to develop a new printing method which would be cheaper and more suited to the manufacture of x/y touch screens. This research resulted later in several publications and patents being taken out by Århus University in Denmark. New, simpler and cheaper touch detector circuits for the new x/y mutual capacitance technology ware also developed, eliminating at the same time the need to employ the more complicated phase-locked loop circuits previously used for the detection of a touch.[14]

The appearance of microcomputers on the market, and the ongoing effort by the computer industry to miniaturize essentially any electronic component in order to integrate them into smaller units that could each perform a different logic operation, pushed CERN to invest in redesigning the original consoles. Subsequently, smaller and more powerful, independent and intelligent units capable of having the same control of an accelerator as the large consoles appeared. An example of these was the CERN 'Intelligent Touch Terminal' commercialized by NESELCO and entered into operation for control of the Antiproton Accumulator (AA) at CERN in 1979 (Figure 5.4).

The Intelligent Touch Terminal had its own Central Processing Unit (CPU), and made extensive use of other microcomputers built into special electronic modules that partly drove the touch screen, the touch screen display, and the bigger colour display. Another microcomputer in the terminal controlled a serial link if the 'Stand Alone Terminal' had to communicate with any other apparatus or other computer systems. It should be noted that the Intelligent Touch Terminal had all of its functionalities built into a small table-top cabinet about the size of a Personal Computer (PC) that later came onto the market. In particular, it should be noted that this terminal already had a touch screen, whereas the later-

---

10) In 1968, Lauritz Knudsen (LK) merged with another switchgear manufacturer, NES, and through the next decade the company was named LK-NES. The early product range of generator control and alarm systems was marketed under the name NESELCO. Originally, LK-NES Ltd was established to service at regional level, and in the late 1980s moved its operations at international level. In 2001, the company Lauritz Knudsen Electric Ltd grouped the manufacturing LK-NES Ltd operations, consolidating its medium voltage manufacturing under LK-Electric Ltd with the trade name of LKE Power Distribution Systems: www.lke-power.com.

11) NESELCO, Advertisement for the 4 × 4 touch screen and the CAMAC (Computer Automated Measurement And Control) associated modules, Journal of the Atomic Energy Society of Japan, 1979.

12) *Udviklingsfondet* was a Danish Development Foundation that existed from 1970 to 1990 and was the main source of funding for industrially relevant research in those years.

13) The Danish krone is the official currency of Denmark introduced on 1 January 1875: one Danish krone is equivalent to €0.134 012.

14) C.P. Torp, *Neselco har udviklet verdens mest avancerede kontrolrums-udrustning*, LK-NES BLADET, 1980.

(a)  (b)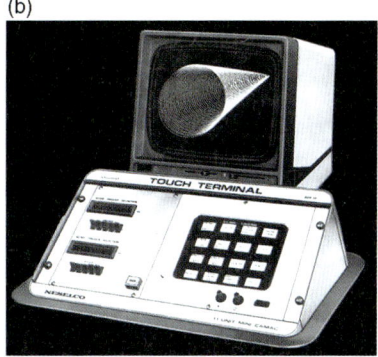

Figure 5.4 (a) The Intelligent Touch Terminal developed for the CERN Antiproton Accumulator (AA). (Courtesy of CERN); (b) The New Touch Terminal from NESELCO. (Courtesy of NESELCO).

developed PCs would have to wait many more years for this technology to arrive.[15]

The CERN touch screen systems were soon in use in many bigger control systems around the world, including: the Joint European Torus (JET), the Rutherford Appleton Laboratory (RAL) and the Daresbury Laboratory in the UK; HMI,[16] the Hahn Maitner Institute and DESY,[17] the National Research Centre in Germany; Fermilab, the Fermi National Accelerator Laboratory in the US; KEK,[18] the High-Energy Accelerator Research Organization, Mitsubishi and TOYO Corporation in Japan; and the Institute of the High-Energy Physics in China.

It is important to remember that most of the ongoing developments in industry on PCs, software and hardware were promoted by the needs of the CERN physicists and engineers.

Today, the CERN control centre uses a high number of PCs connected to many control centres located around the various stations of the Laboratory. The operators are also now equipped with smart phones, which makes it possible to remotely control the installations anywhere with a single push on the touch screen (Figure 5.5).

15) M.C. Crowley-Milling, *Recommendation for the Touch Screens' development and their knowledge transfer to the Danish Industry*, CERN SPS-DI/MCCM/tr, Enclosure N. 5, 1978.
16) HMI (Hahn Meitner Institut).
17) DESY (Deutsches Elektronen-Synchrotron).
18) KEK (Kō Enerugī Kasokuki Kenkyū Kikō).

Figure 5.5 General view of the CERN accelerator control console in 2013. (Courtesy of CERN).

5.3
A Touch Screen as Replacement for Mechanical Buttons

The features that make a modern touch screen such a useful device can be summarized as follows:

- Both, input and output information appears on the same screen.
- Audible and visual feedback to the user are provided in multiple forms.
- The hierarchy of a pyramid sequence of possible choices is defined in advance.
- Only one amongst the many possibilities can be selected at a given time.
- The display under the touch screen has high definition colour displays.
- X/y capacitive screens allow gestures and advanced software writing for most applications.
- Touch-sensitive devices are much easier to operate than keyboard systems.
- As it is software-controlled, visual presentation under the touch screen has no limits. Movies, graphics, character sets of any type can be presented, which provides endless possibilities for writing software to be used for any new application.
- No mechanical movement is involved in the 'touch' control process.
- Both, the touchscreens and the electronics behind them are extremely reliable.
- A few touches on the screen can replace thousands of fixed mechanical buttons.

The flexibility related to possible programmable applications is far superior compared to other systems. It is now possible to program the device with a choice between a few large buttons or a high number of much smaller buttons, or to use a mix of these. Due to the limited physical space available on small screens, some modern touch devices often present only a few buttons on the screen. As an example, the new Swatch touch watch presents four buttons on the screen.[19] As there can be different layers on the screen (i.e. three), there are $4 \times 4 \times 4 = 64$ possibilities; however, by having 16 fixed buttons on the screen one would have $16 \times 16 \times 16 = 4096$ possibilities. So, imagine how many possibilities a touch screen can offer having a still larger number of buttons! The software developer prepares the control sequence applications, allowing the user to react directly to what is displayed.

5.4
Attempts at Early Knowledge Transfer

As noted above, during the early 1970s very few people had any knowledge of touch screens, but major interest was subsequently trigged by an article entitled *Computer creates custom control panel*,[20] which described the CERN development and led to enquires being raised by 288 companies. Consequently, only a few years later several companies were manufacturing various types of touch screen on a commercial basis. For example, in August 1981 a 'Touch Sensitive Terminal' using a capacitance-sensing faceplate was advertised,[21] while in November 1983 Hewlett-Packard (HP) introduced a PC with touch screen control to the market.[22]

An attempt at knowledge transfer was made as early as 1977, with the proposal of an innovative watch called 'The Intelligent Oyster' by a well-known Swiss manufacturer of watches.[23] This new type of wristwatch was proposed to be an interactive miniterminal, having an integrated touch control with the possibility for almost unlimited applications. However, the original proposal was returned with the mention that only applications submitted on the original entry form could be taken into consideration by the selection committee.

Believe it or not, about three decades later 'Touch Watches' were being produced by the Swatch group[24] and other manufacturers.

Another example of knowledge transfer is in medical applications for developing countries. In 1980, it was envisaged by the World Health Organization (WHO) that touch screen technologies could be used in small handheld devices with incorporated medical software to allow paramedics to make medical

---

19) http://www.swatch.com/zz_en/watches/wi2011_swatchtouch.html.
20) J.F. Lowe, *Computer creates custom control panel*, Design IDEAS, 1974.
21) H.H. Ng and S.J. Puchkoff, *Touch-sensitive screens ensure a user friendly interface*, Computer Design, The Magazine of Computer Based Systems, p. 135, 1981.
22) *Hewlett-Packard introduces MS DOS personal computer with Touchscreen control*, HP Measurement Computation News, 1983.
23) B. Stumpe, *The Intelligent Oyster*, Proposal to Rolex Awards for Enterprise, 29 March 1977.
24) http://www.swatch.com/zz_en/watches/wi2011_swatchtouch.html.

diagnoses based on the symptoms presented on screen, and to prescribe treatments according to the materials at their disposal in the field.[25] Unfortunately, the cost and computer size to apply touch screen technology at that time were too prohibitive and inadequate, and instead another type of development based on semiconductor technology with the support of CERN was introduced for the early detection of leprosy in areas where access to medical care was limited.[26] Today, touch screen technologies are built into large numbers of medical equipment and used in many medical applications.

In 1977, CERN decided for the first time to participate in the Hanover fair in Germany in order to show to the public some of its latest inventions in collaboration with NESELCO, which in the meantime had managed to commercialize nearly all the components built into the central touch screen consoles for the control of the SPS accelerator.

In order to demonstrate that touch screen control technologies were not limited to an accelerator, but could also be used for control of various industrial processes, it was decided to build for the occasion an innovative drink distributor, called 'Drinkomat'. It used touch technologies, special software and hardware developed at CERN for control of the fluids in the chemical process for bubble chambers[27] film development at CERN. The visitors at the exhibition could use the touch screen to choose a drink (bloody Mary, whisky on the rocks, gin and lime, etc.), or compose a drink of their own. On shelves visible to the visitors, acting on the equipment like an operator in a control room, visitors could control the various basic components of the drinks stored in large bottles, each with a computer-controlled valve connected to a plastic tube passing from the valve to the mixer of the drink. Today, many manufacturers are producing different types of Drinkomats, and many of these are using touch screens.

CERN has always had a policy stimulating knowledge transfer and free access to its scientific and technological results. The economic impact and partnership with industry triggered many developments which have been beneficial not only to European industry. Only since the end of the twentieth century has the attitude of CERN towards patenting been formalized. The reason why many technologies are known only now to have been invented at CERN a long time ago is due to the fact that the formalization of the CERN patent policy was only made in the year 2000. Although the x/y multitouch technology was invented at CERN in 1977, it is interesting to observe that later it has been possible for different firms to grant patents on this invention.

---

25) B. Stumpe, *Touch screen applications on medical devices for development countries*, Presentation at the World Health Organization (WHO), 1980.
26) H. Srinivasan and B. Stumpe, *Leprosy diagnosis: a device for testing the thermal sensibility of skin lesions in the field*, WHO Bulletin, Volume 67, Issue 6, p. 635, 1989.
27) A bubble chamber is a vessel filled with a superheated transparent liquid (most often liquid hydrogen) used to detect electrically charged particles moving through it. It was invented in 1952 by Donald A. Glaser, for which he was awarded the 1960 Nobel Prize in Physics: http://www.nobelprize.org/nobel_prizes/physics/laureates/1960.

5.5
Evolution Turned Into Revolution

The real touch screen revolution took place only after the development and massive use of portable telephones. However, a constant evolution with parallel progress of many different technologies occurred over a long time until the maturity of the technologies allowed to combine them into portable computers, tablets, and finally smart phones.

Contrary to CERN, for industry the driving force behind these developments has always been purely commercial. The business and expected financial benefits are directly linked to the number of end users. In 2010, the total world population of both sexes combined without migration was 6 916 100 000, and for 2020 the estimated number without migration will be 7 718 083 000.[28] The number of worldwide mobile phones subscriptions up to 2011 and the mobile traffic forecast is reported in Figures 5.6 and 5.7, respectively. In 2011, when the total mobile-cellular subscriptions was 5.9 billion, penetration had reached 87% globally and 79% in the developing world. Between 2007 and 2011, mobile-broadband subscriptions grew by 45% annually, with numbers of mobile-broadband subscriptions being twice that of fixed-broadband subscriptions (see Figure 5.7).

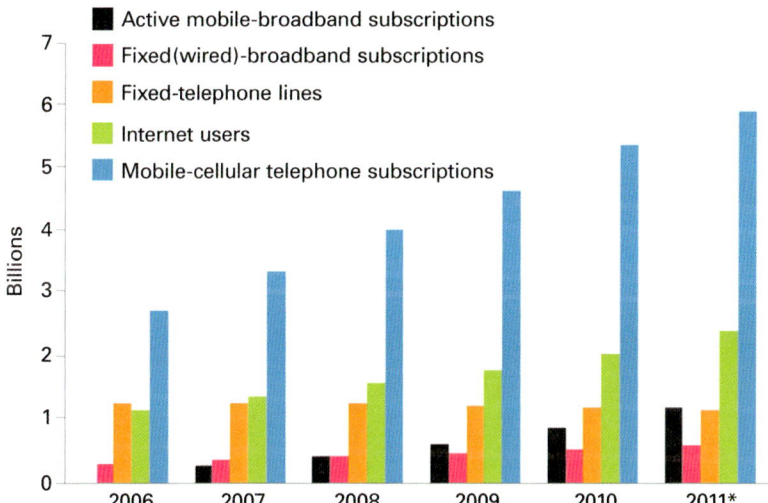

Figure 5.6 Evolution of the worldwide mobile phone subscriptions. (Source: ITU World Telecommunication/ICT Indicators database).[29]

---

28) UN (United Nations), *World Population Prospect: the 2012 Revision*, Department of Economic and Social Affairs, Population Division, 2013: http://esa.un.org/wpp/Documentation/publications.htm.
29) ITU (International Telecommunication Union), *The World in 2011: ICT facts and figures*, ITU Telecom World '11, 2011: http://www.itu.int/ITU-D/ict/facts/2011/material/ICTFactsFigures2011.pdf.

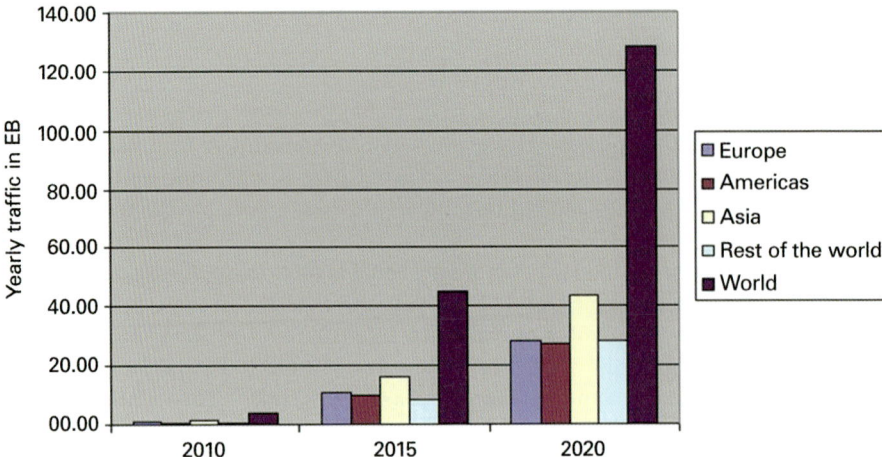

Figure 5.7 Mobile traffic forecasts 2010–2020. (Reproduced with permission of Universal Mobile Telecommunication System, UMTS).[30]

Today, compared to 2010, the total worldwide mobile traffic has increased 33-fold and is forecast to reach more than 127 exabytes by 2020 (see Figure 5.7).

The trend for global mobile subscriptions, including machine-to-machine per continent and summarized for the whole world, is reported in Table 5.1, and demonstrates the very rapid increased forecast for Asia in 2020. Clearly, these are the figures behind the business!

As with any revolution, the use of touch screen technology in mobile devices has created fundamental changes over a relatively short period. The miniaturized fixed buttons under the display and the very dense x/y grid pattern used in modern multitouch screens, which are completely transparent and can be incorporated under a glass cover, have been fundamental.[31]

Table 5.1 Global mobile subscription forecasts (in millions), including machine-to-machine[31]

Global Base (million)	2010	2015	2020
Europe	1033	1222	1427
Americas	915	1166	1437
Asia	2579	3825	4957
Rest of the world	801	1276	1863
World	5328	7490	9684

(Reproduced with permission of Universal Mobile Telecommunication System).

30) UMTS (Universal Mobile Telecommunications System), *Executive Presentation*, Forum Report 44, 2011.
31) Ibidem.

Modern software developments capable of detecting motions made with one or more fingers on the touch screen allows the use of what are known as 'gestures'. The speed of one or more finger movements on the surface, and the direction of these movements, can be detected without technical problems such that today, these gestures allow a smart phone or tablet to perform almost any operation.

The speed of modern computer systems with integrated databases, coupled to an extremely fast access to information through modern telecommunication techniques, has opened up an endless number of new real-time applications. Could it have been imagined in 1971, how a simple idea coming from another electronic application, would be at the origin of such technical evolution affecting the daily lives of most of the world's population?

5.6
Touch Screen and Human Behaviour

Today, it is obvious that major changes in human behaviour have taken place as a result of touch screen technology. Simply by observing the people around you, an extensive use of modern smart phones, tablets and other touch screen-based devices will be obvious that allows oral and visual communication with anybody else having a similar device. These devices also provide access to any database connected to the Internet as well as real-time commercial transactions, including the purchase of items or services and subsequent payments. The possibility of any easy participation in social networks of all types can have a major impact on the social life of many people. Currently, there is no sign that this revolution is about to stop; on the contrary, the major question to be answered in the future is whether this technology will be able to keep up with other technologies that exploit the possibility of misuse, whether of a criminal or political nature.

So, what is next in line? Perhaps the first major change will be smart phones where touch interaction is enhanced by voice control. Although the systems currently being developed will surely encounter technical problems, these will undoubtedly be overcome in a relatively short time.

A much more futuristic suggestion is that new devices will become available for which speech is no longer required; rather, the device will be activated simply by the transmission of brain thoughts.

As Jules Verne said: 'Anything one man can imagine, other men can make real'.[32]

Acknowledgements

The author would like to thank the following persons for their strong support and encouragement throughout the developments: Frank Bech, Michael C. Crawling-Milling, Vincent Hatton, and George Shering.

32) J. Verne, *Around the World in Eighty Days*, Pierre-Jules Hetzel Publisher, 1873.

Section 2
Developing Future

Availability of new tools, massive data handling and networks capability, originally developed for the needs of fundamental research in Physics, have paved the way and directly or indirectly impacted many different fields.

It is easy to see how the Internet and the Web have influenced our day-by-day reality. Other technologies have a significant though less visible impact. A great many technologies have been developed in the course of research activities at research laboratories such as CERN, and these technologies have subsequently led to industrial applications and not only.

6
Solar Thermal Electricity Plants
Cayetano Lopez

Since physics can be broadly defined as the general analysis of Nature, conducted in order to understand how the Universe behaves,[1] it is easy to understand why physics is present all around in our lives. It covers a wide range of subjects: electricity, heat transfer, thermodynamic, optics, and so on, which are essential in our daily activities. Energy is another crucial domain where physics technology can provide new solutions, in particular in the field of solar energy. Solar energy as such has appealing qualities: it is environmentally friendly; it is virtually infinite; and the energy source (solar irradiation) is free of charge.

For many years, the former CERN Director General and Nobel Prize Carlo Rubbia[2] advocated the need for developing new knowledge to improve the concentration solar technologies by means of several scientific projects. He has been advisor to CIEMAT,[3] the Spanish Centre for Research in Energy, Environment and Technology, that runs PSA,[4] the Almeria Solar Platform since 1980.

Amongst many other scientists and technology developers, Professor Rubbia stressed the need for basic science to attempt various alternative solutions in order to propose new forms of efficient and sustainable energy and to collaborate with the industry to provide cheap and abundant clean energy when the utilization of fossil fuels will end.

---

1) H.D Young and R.A. Freedman, *University Physics with Modern Physics*, 11th Edition, Pearson/Addison Wesley, 2004.
2) In 1984, the Nobel Prize in Physics was awarded jointly to Carlo Rubbia and Simon van der Meer ' . . . for their decisive contributions to the large project, which led to the discovery of the field particles W and Z, communicators of weak interaction': http://www.nobelprize.org/nobel_prizes/physics/laureates/1984.
3) Under the CIEMAT (Centro de Investigaciones Energéticas, Medioambientales y Tecnológicas) Directorate of Professor Juan Antonio Rubio.
4) PSA (Plataforma Solar de Almería).

From Physics to Daily Life: Applications in Informatics, Energy, and Environment, First Edition.
Edited by Beatrice Bressan.
© 2014 Wiley-VCH Verlag GmbH & Co. KGaA. Published 2014 by Wiley-VCH Verlag GmbH & Co. KGaA.

6.1
The Four STE Technologies

Solar Thermal Electricity (STE) plants have many links with physics. The solar concentrators, the power block, the antireflective and selective coatings used to increase the thermal efficiency of solar receivers, are just a few examples of STE plant components based on physical principles or laws. Since there are many links between STE plants and physics, only those having a more significant influence in the plant performance are described here, grouped into four categories: optical issues; thermodynamic issues; heat transfer mechanisms; and fluid mechanics.

The typical STE plant is a power plant using concentrated solar radiation to produce electricity. Direct solar radiation is concentrated and converted into thermal energy in the form of latent or sensible heat of a working fluid, and this thermal energy is then used to feed a thermodynamic cycle (mainly Rankine, Brayton or Stirling cycles).[5] Figure 6.1 shows the simplified scheme and subsystems of a typical STE plant. Although there are four different STE technologies at present, they all fulfil the simplified scheme shown in Figure 6.1.

The optical concentrator accomplishes the mission of increasing the flux density of the solar radiation beam, so that the solar flux density on the surface of the receiver may be up to several thousand times higher than the flux density of the direct solar radiation reaching the Earth's surface, which is usually of the order of, or lower than, 1000 W m^{-2}. This concentration is achieved by reflecting the collected solar radiation onto a receiver with a smaller surface. The concentrated solar radiation reaching the receiver outer surface is then converted into thermal energy by increasing the specific enthalpy of a working fluid as it circulates inside the receiver. Several working fluids are used nowadays, such as water, thermal oils, molten salts or air.

---

5) The Rankine cycle, named after the Scottish polymath and Glasgow University professor William John Macquorn Rankine, is a mathematical model used to predict the performance of steam engines. It is an idealized thermodynamic cycle of a heat engine that converts heat into mechanical work. The heat is supplied externally to a closed loop, which usually uses water as the working fluid. In the form of steam engines, it generates about 90% of all electric power used throughout the world, including all biomass, coal, solar thermal and nuclear power plants.

The Brayton cycle, named after the American engineer George Brayton, sometimes known as the Joule cycle, is a thermodynamic cycle describing the workings of a constant-pressure heat engine. Gas turbine engines and air-breathing jet engines use the Brayton cycle. Although the Brayton cycle is usually run as an open system (and indeed must be run as such if internal combustion is used), it is conventionally assumed for the purposes of thermodynamic analysis that the exhaust gases are reused in the intake, enabling analysis as a closed system.

The Stirling cycle, patented in 1816 by Reverend Dr Robert Stirling and his engineer brother, is a thermodynamic cycle describing the general class of Stirling devices which includes the original Stirling engine. It is reversible; that is, if supplied with mechanical power it can function as a heat pump for heating or cooling, and even for cryogenic cooling. It is defined as a closed regenerative cycle (i.e. containing the working fluid permanently within the thermodynamic system and using of an internal heat exchanger called a 'regenerator') with a gaseous working fluid.

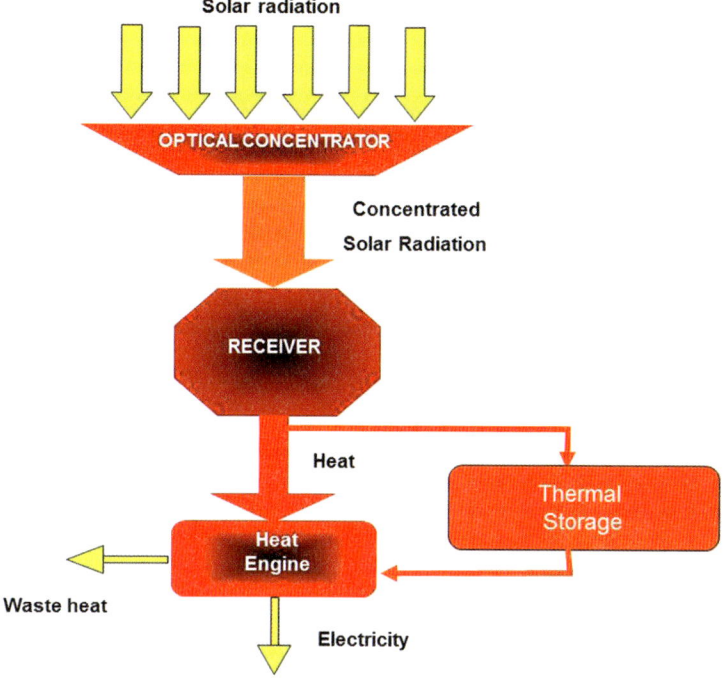

Figure 6.1 The main components and subsystems of a Solar Thermal Electricity (STE) plant, including storage.

The thermal energy gained by the working fluid at the receiver, in the form of sensible or latent heat, is then either converted into electricity by means of a heat engine using a thermodynamic cycle, or sent to a thermal storage system to be used later to feed the heat engine when solar radiation is not available. Some waste heat is usually available as a byproduct of the electricity generation.

At present, there are four different technologies for STE plants, termed 'Parabolic-trough collectors', 'Central receiver systems', 'Compact linear Fresnel concentrators', and 'Stirling dishes' (see Figure 6.2).

The main difference amongst these four technologies is the way in which they concentrate the direct solar radiation, as described in the following:

- **Parabolic-trough collectors.** These are solar concentrating devices used to concentrate the direct solar radiation onto a receiver pipe placed at the focal line of a parabolic-trough concentrator (Figure 6.2a). This concentrated solar radiation increases the specific enthalpy of a working fluid as it circulates inside the receiver from the collector inlet to its outlet. This technology can be used to achieve temperatures of up to 773 K.
- **Central receiver systems.** In this case the receiver is placed at the top of a tower, and many reflecting elements known as 'heliostats' reflect the direct solar radiation onto the receiver. The heliostat field may be composed of

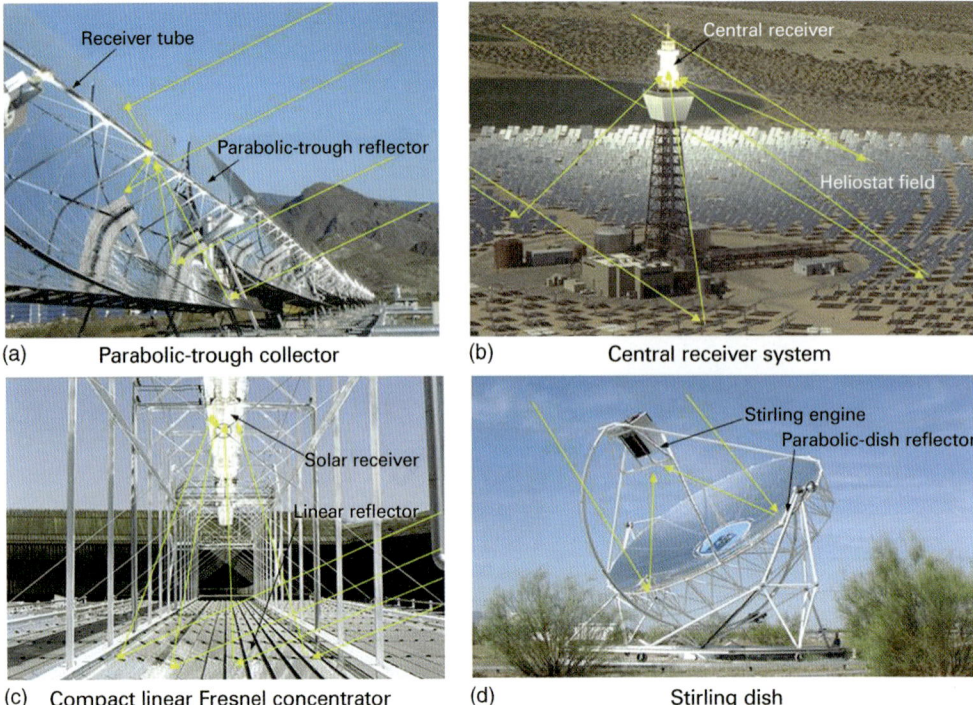

Figure 6.2 (a–d) The four Solar Thermal Electricity (STE) technologies currently available.

hundreds or even thousands of heliostats placed around the tower holding the receiver (Figure 6.2b). Temperatures of 1200 K and even higher can be achieved with this technology because the solar flux incident on the receiver is significantly higher than in parabolic-trough collectors.

- **Compact linear Fresnel concentrators.** These are somehow a hybrid design between parabolic-troughs and central receiver systems, because the solar concentrator is composed of many long parallel reflectors that reflect the direct solar radiation onto a common linear receiver, similar to that of a parabolic trough, placed on top of them at a fixed distance (i.e. the receiver does not change its position at any time; Figure 6.2c). The concentrated solar radiation incident on the receiver is converted into thermal energy in the form of sensible or latent heat of the working fluid as it circulates inside the receiver. This technology has the same temperature range as parabolic-trough collectors, because the solar flux onto the receiver is similar.
- **Stirling dishes.** In this case, the optical concentrator is a parabolic dish reflecting the solar radiation on the absorber of a solarized Stirling engine (Figure 6.2d). The crankshaft of the Stirling engine is mechanically connected to the shift of an electricity generator. Therefore, the concentrated solar radiation is converted into mechanical energy by the Stirling engine and then into electricity.

Andasol-I plant — Puerto Errado-I&II plant

Gemasolar plant (central receiver system)

Figure 6.3 Aerial views of three Solar Thermal Electricity (STE) plants in Spain, with parabolic troughs, linear Fresnel and central receiver. The Germasolar plant is owned by Torresol Energy[6] (© SENER).[7]

More detailed information about the four technologies used in STE plants is available on the Web site of the European Solar Thermal Electricity Association (ESTELA)[8] and in several handbooks.[9],[10]

In March 2012, three plants with compact linear Fresnel concentrators were in operation worldwide, with a total output of 36.4 MWe,[11] while three plants were in operation with central receiver systems and a total output of 49.9 MWe. At that time, the parabolic trough technology had 50 plants in operation with a total output of 2401.5 MWe, while only 100 kWe of parabolic dishes with Stirling engines were in operation. Figure 6.3 shows different STE plants in Spain, with parabolic trough collectors (Aandasol-I plant), compact linear Fresnel concentrators (Puerto Errado-I&II plants), and central receiver (Gemasolar plant).

6) www.torresolenergy.com.
7) www.sener.es.
8) www.estelasolar.eu.
9) K. Lovegrove *et al.* (Eds), *Concentrating Solar Thermal Technologies: Principles, Developments and Applications*, Woodhead Publishing Ltd., 2012.
10) M. Romero-Alvarez and E. Zarza, *Concentrating Solar Thermal Power*, in F. Kreith and D.Y. Goswami (Eds), Handbook of Energy Efficiency and Renewable Energy, p. 21, CRC (Chemical Rubber Company) Press, 2007.
11) In the electric power industry, the MWe (megawatt electrical) is a term that refers to electric power; that is, the rate at which electric energy is transferred by an electric circuit.

6.2
Optical Issues in the STE Plant

Optics, together with heat transfer mechanisms, is probably the field of physics with a more outstanding presence in the STE plants. Optical concentrators are key elements in these plants, because they accomplish the concentration of the direct solar radiation required to convert solar radiation into thermal energy at medium or high temperatures.

6.2.1
Solar Concentrators

The maximum temperature achievable in a cost-effective manner when nonconcentrated solar radiation is converted into thermal energy is about 100–150 °C. This conversion into thermal energy can be achieved by using flat plate solar collectors or evacuated tubes. If higher temperatures are desired, the solar radiation must be concentrated to increase its flux density per surface unit. The higher the concentration, the higher the temperature achieved in the solar receiver.

The solar concentrators used commercially nowadays in STE plants concentrate the primary solar radiation by means of reflection. There are also some prototypes of solar concentrators that use optical lenses to concentrate the solar radiation by refraction, so that the lens is placed between the receiver and the Sun; however, there is not yet a commercial plant using this type of concentrator.

Most solar concentrators are based on the *law of reflection*, which is one of the fundamental laws of optics. According to this law, when a ray of light strikes a mirror, the light ray reflects off the mirror at an angle (angle of reflection) equal to the angle of incidence. These angles are the angles between a normal line drawn to the surface of the mirror and the reflected and incoming rays, respectively. Figure 6.4 shows the *law of reflection* schematically, while Figure 6.5 shows the application of this law to a parabolic concentrator (which is the shape usually adopted for solar concentrators) in both versions: parabolic-trough and parabolic dish.

6.2.2
Selective and Anti-Reflective Coatings

Since the curvature of the parabola can be easily defined in accordance with the desired focal distance, F, most solar concentrators are designed with a parabolic shape because when the *law of reflection* is applied to a parabola, the result is that all the incident sunlight rays arriving at the parabola surface in a direction parallel to the parabola axis are reflected towards a common point (the parabola focal point) where the receiver is placed. Since the surface of the parabola aperture plane is much larger that the receiver surface, a high solar flux is achieved

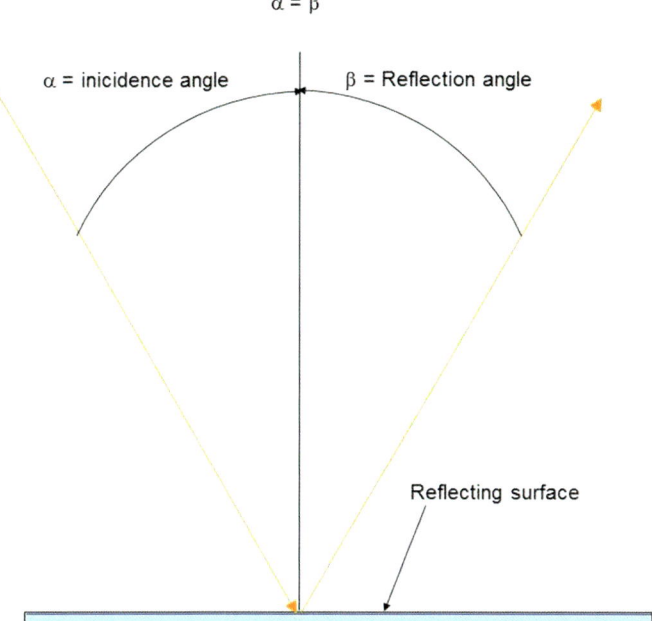

Figure 6.4 The law of reflection.

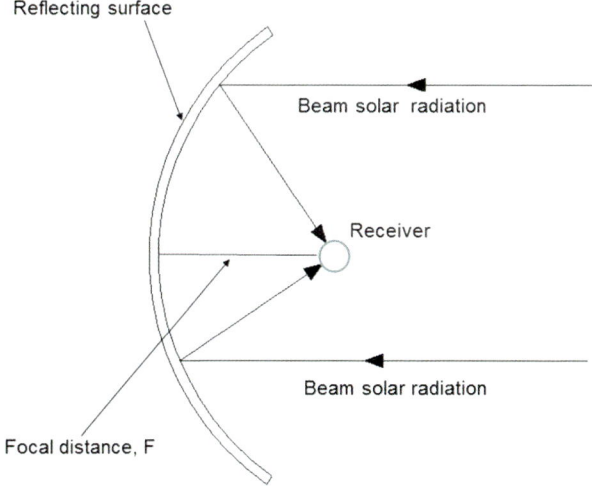

Figure 6.5 Parabolic concentrator.

on the receiver. Theoretically, a maximum concentration ratio (i.e. the ratio between the parabola aperture area and the receiver area) of about 220 could be achieved with parabolic-trough concentrators, but for practical reasons the concentration ratio is usually kept below 80. When higher concentration ratios are required, three-dimensional (3D)-concentrators (i.e. parabolic dishes) have to be used. In a central receiver plant, the sunlight rays reflected by hundreds or even thousands of heliostats are superimposed on the receiver outer surface to achieve concentration ratios of several hundreds or thousands. Solar flux densities of 700 kW m^{-2} are typical on central receivers.

One very interesting physical application of STE plants is the development of selective coatings for the outer surfaces of solar receivers, and anti-reflective coatings for the glass elements. A selective coating is a coating with a high absorption in the more energetic wavelength range of the solar spectrum (250–1250 nm), while its emissivity is very low in the infrared wavelength range (780 nm to 1 mm). With these properties, most of the concentrated radiation reaching the receiver is absorbed, while thermal loss by radiation from the receiver to the environment is very low. The fundamentals of these special coatings are explained below.

In physics, *interference* is a phenomenon in which two waves superimpose to form a resultant wave of greater or lower amplitude. Interference usually refers to the interaction of waves that are correlated or coherent with each other, either because they come from the same source or because they have the same (or nearly the same) wavelength. As interference effects can be observed with all types of wave, such as light, acoustic and waves on the surface of liquids, they also occur with the light waves coming from the Sun (solar radiation).

When a light wave with a wavelength λ falls on the outer surface of the coating of an opaque substrate (see Figure 6.6), it is partially reflected (r_{01}) with a lower intensity, and it is also refracted through the coating of thickness d and index of refraction n. The refracted ray is then reflected at the surface of the opaque substrate and finally refracted when passing from the material 1 to the air. The reflectance ρ is defined as the ratio of the energy flux reflected by a surface to the radiation incident on it. Reflectance is a dimensionless variable ranging from 0 to 1, which depends on wavelength λ, the direction of the incident radiation θ, the polish and the surface temperature. In Figure 6.6, the global reflectance depends on the intensity of the individual rays r_{01} and r_{02} resulting from the reflection and refraction phenomena of the incident light ray. The thickness and the refraction index of the coating material can be adjusted to maximize or minimize the reflectance. It is easy to proof that high or low absorption values can be obtained with a correct adjustment of the coating material and its thickness.

An ideal solar receiver should have the reflectance spectrum shown in blue in Figure 6.7, with a low reflectance in the more energetic range of the solar spectrum and a high reflectance at wavelengths above 1500 nm. However, the raw materials that are available and suitable for solar receivers usually have a very different spectral reflectance, with high reflectance at lower wavelengths.

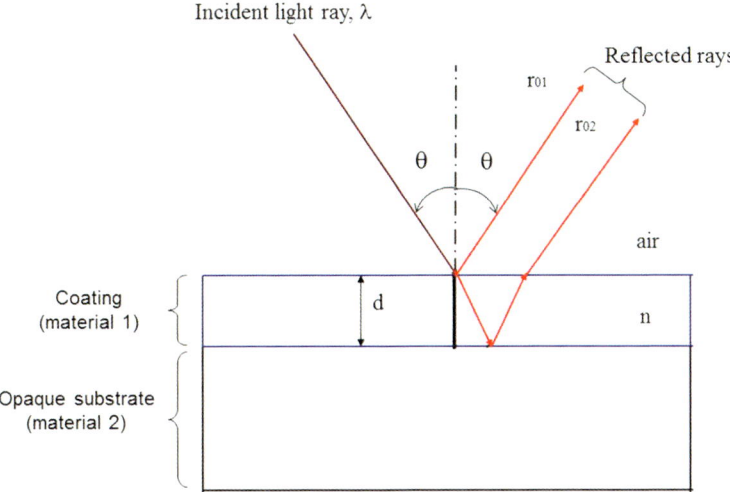

Figure 6.6 Reflection and refraction of a light wave incident on the coating of an opaque substrate.

If the optical parameters of the substrate are suitable, a spectral reflectance rather similar to that shown in Figure 6.7 can be obtained by making use of the interference phenomena and depositing several coating layers in a stack (the red spectrum in Figure 6.7).

The interference filters or multilayer solar absorbers are composed of several layers of absorbing material, anti-reflective layers in the solar spectrum and a

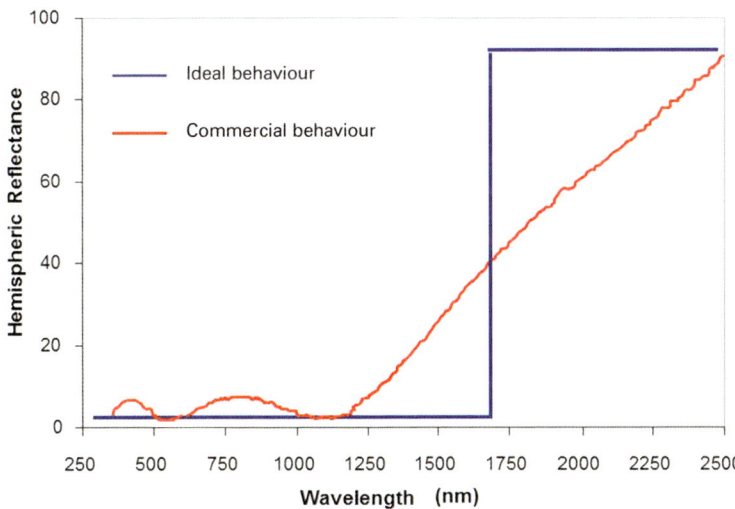

Figure 6.7 Ideal and commercial reflectance behaviour of a solar receiver.

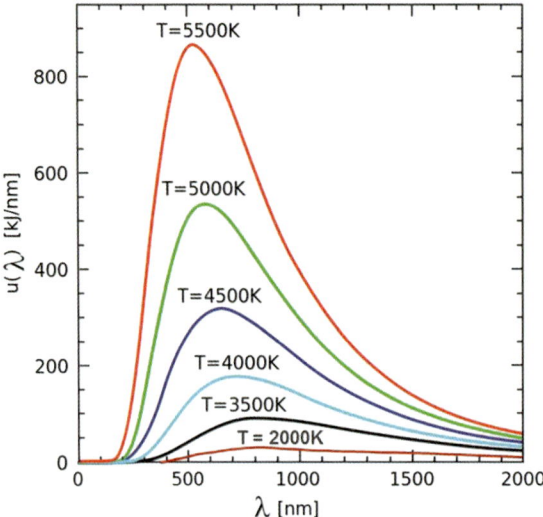

Figure 6.8 Radiant energy emitted by a black body at different temperatures.

reflective layer in the infrared region, by choosing the thickness of each layer in accordance with the desired final result. A physics criterion is followed for antireflective layers in the solar spectrum, and the thickness of solar absorbers is chosen to locate the absorption transition zone at the desired wavelength range.

For central receivers, the selective behaviour of the coating must be complemented with a good thermal stability at high temperatures, which makes it very difficult to develop coatings that fulfil these requirements simultaneously. It is for this reason that no commercial selective coatings are yet available for receivers operating at temperatures higher than 500 °C in contact with hot air.

6.2.3
Thermography

Any object emits radiant energy – so-called 'thermal radiation' – the spectral distribution of which depends on its temperature. In the black-body approximation, the spectral distribution of radiant energy emitted versus its temperature is represented in Figure 6.8 for the range of 3500 to 5500 K. The spectral distribution and the wavelength at the peak value are determined by the *Planck's law* and *Wien's displacement law*.[12] At the same time, the *Stefan–Boltzmann law* determines the total amount of radiant energy emitted by a black body per surface unit, stating that it is proportional to the fourth power of its temperature (in

---

12) F.P. Incropera and D.P. DeWitt, *Fundamentals of Heat and Mass Transfer*, John Willey & Sons, 1996.

Kelvin), the so-called 'Stefan–Boltzmann constant' (σ) being the constant of proportionality.[13] The wavelength range of thermal radiation extends from 100 nm to 10^5 nm, while most solar radiation (thermal radiation emitted by the Sun) is in the range of 300 nm to 3000 nm.[14] Visible radiation is in the range of 250 nm to 750 nm only, while infrared radiation covers the range from 780 nm to 1 mm.

As depicted in Figure 6.8, the higher the temperature of the black body, the smaller the wavelength of the peak of its thermal emission (*Wien's displacement law*). Consequently, there is a relationship between the temperature of a body and its thermal radiation. By making use of this property, very high temperatures can be measured using optical sensors (i.e. pyrometers or spectroradiometers).

In general, the use of thermocouples or any other contact-methods to measure very high temperatures is not feasible because of the lower melting points of the metallic materials currently available. In this case, optical and thermal devices based on the *Planck's law, Wien's displacement law* and the *Stefan–Boltzmann law* allow these temperatures to be measured. However, these three laws have other interesting applications in STE plants, as outlined in the following paragraphs.

Occasionally, the temperature of a body has to be measured without using sensors that are in contact with that body. A typical case is that of pipes or elements which are not easily accessible, but are visible. In this situation, the temperature measurements can be taken by using infrared sensors installed either individually or in an array composing a matrix. Single infrared sensors are used in pyrometers, while matrices of sensors are used for thermal-image cameras. The main advantage of a camera over a pyrometer is that the camera provides a complete image of the target, together with a thermal map of its surface temperature; in contrast, a pyrometer will only detect the thermal radiation emitted by a small target at which it is aimed.

Pyrometers are based on the *Stefan–Boltzmann law*, while thermal-image cameras are based on *Planck's law* and the *Stefan–Boltzmann law*. Figure 6.9 shows the thermal image obtained with a camera to detect failures in the thermal insulation in a piping. The temperature is displayed using a colour code, from cold (blue) to hot (yellow). Today, thermography is widely used not only in STE plants to reduce thermal losses by detecting failures in the thermal insulation, but also in most industries employing thermal equipment, because it provides a fast and reliable means of measuring temperatures, with reasonable accuracy.

The parameter that most affects the accuracy of the temperature measurements collected with thermography is the emissivity of the surface being measured. As emissivity is the ratio of the radiant energy emitted by a material from its surface to the radiant energy emitted by a black body under the same conditions, the emissivity value must be provided for the device (whether pyrometer or camera) in order to allow an accurate calculation of the temperature, using

---

13) $\sigma = 5.6705119 \times 10^{-8} \, W \cdot m^{-2} \cdot K^{-4}$.
14) D.Y. Goswami *et al.*, *Principles of Solar Engineering*, Taylor & Francis, 2000.

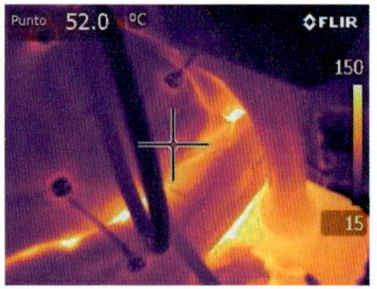

(a) Thermal-image camera (b) Thermal image camera obtained by a camera

Figure 6.9 (a) The thermal-image camera; (b) Thermal image obtained by an infrared camera.

the three physical laws (*Planck's law*, *Wien's displacement law*, and the *Stefan–Boltzmann law*).

One very peculiar application of thermography in STE plants is the detection of vacuum losses in the receiver tubes of parabolic trough collectors. As explained in Section 6.4, the annulus that exists between the glass cover and the inner steel tube is evacuated to eliminate conduction and convection losses between the steel tube and the glass cover. However, this vacuum can be lost partially or totally due to ageing or failure of the glass/metal welds (see Figure 6.13), and thermal losses are thus increased significantly. The sooner the vacuum loss is detected the better, as the defective receiver tube can be replaced by a new one. When the vacuum is lost, the temperature of the glass cover increases due to a convection phenomenon in the annulus, and a temperature difference of up to 30 °C can be measured between the bottom and top of the glass cover. As the receiver tubes are at a 3 metre height, the measurement of the glass cover temperature by thermography is very suitable because it allows a rapid detection of any vacuum loss without having to install thermocouples at the glass covers.

The transmissivity of the 3 mm-thick glass cover to infrared radiation is very important for this particular application, because it decreases very rapidly for wavelengths higher than 4500 nm, and is 0 for wavelengths (λ) $\geq$ 5000 nm (see Figure 6.10).

If a band-pass filter centred in $\lambda = 5500$ nm (approximately) or a little higher is installed in the thermal sensor, it will only detect the thermal radiation emitted by the glass cover, because the latter is opaque to the radiation emitted by the steel tube with $\lambda \geq 4500$ nm. Consequently, the thermal radiation emitted by the glass cover can be measured without being disturbed by the thermal radiation emitted by the steel tube, which will be at a higher temperature. It must be borne in mind that the temperature of the steel pipe will be about 650 K, while that of the glass cover will be in the range of 340–425 K.

6.3
Thermodynamic Issues in the STE Plant

Thermodynamics is another field of physics that has a significant influence on the performance of STE plants. If a simulation model of a generic STE plant is developed and introduces some simplifications (i.e. assuming a perfect optical concentrator, a receiver behaving as a black body and a heat engine with an efficiency equal to that of a Carnot cycle[15] operating between the temperature at the receiver and the ambient temperature), it will be found that the overall efficiency of the STE plant (the ratio between the electricity produced and the available solar radiation) depends on the working temperature at the receiver and the concentration ratio. Figure 6.11 shows the result of plotting the plant efficiency versus receiver temperature for several concentration ratios.

It can be seen from Figure 6.11 that higher concentration ratios lead to higher efficiencies for optimum receiver temperatures, the Carnot efficiency being the enveloping curve at the top. Therefore, the benefits of achieving higher temperatures at the receiver is clear from a thermodynamic standpoint, and this is the

---

15) The Carnot cycle, proposed by Nicolas Léonard Sadi Carnot in 1823, is a theoretical thermodynamic cycle that can be shown as the most efficient cycle for converting a given amount of thermal energy into work, or conversely, creating a temperature difference (e.g. refrigeration) by doing a given amount of work. Every single thermodynamic system exists in a particular state. When a system is taken through a series of different states and finally returned to its initial state, a thermodynamic cycle is said to have occurred. In the process of going through this cycle, the system may perform work on its surroundings, thereby acting as a heat engine. A system undergoing a Carnot cycle is called a 'Carnot heat engine', although such a perfect engine is only a theoretical limit and cannot be built in practice. When acting as a heat engine, it consists of the following four steps:
1) Reversible isothermal expansion of the gas at the hot temperature (isothermal heat addition or absorption). During this step the gas is allowed to expand and does work on the surroundings. The temperature of the gas does not change during the process, and thus the expansion is isothermal.
2) Isentropic (reversible adiabatic) expansion of the gas (isentropic work output). During this step, where the piston and cylinder are assumed to be thermally insulated (thus they neither gain nor lose heat), the gas continues to expand, doing work on the surroundings, and losing an equivalent amount of internal energy. The gas expansion causes it to cool to the cold temperature. The entropy remains unchanged.
3) Reversible isothermal compression of the gas at the cold temperature (isothermal heat rejection). Now, the surroundings do work on the gas, causing an amount of heat energy and of entropy to flow out of the gas to the low temperature reservoir. This is the same amount of entropy absorbed in step 1.
4) Isentropic compression of the gas (isentropic work input). During this step, where the piston and cylinder are assumed to be thermally insulated, the surroundings do work on the gas, increasing its internal energy and compressing it, causing the temperature to rise to the initial hot temperature. The entropy remains unchanged. At this point the gas is in the same state as at the start of step 1.

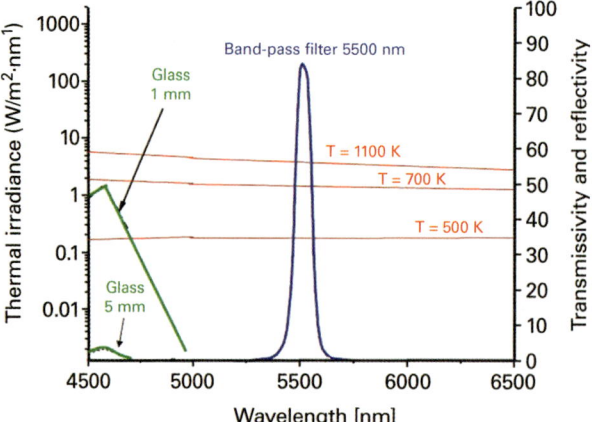

Figure 6.10 Glass transmissivity for two thicknesses (1 mm and 5 mm), thermal irradiance for blackbodies at several temperatures (600 K, 700 K, and 1100 K), and transmissivity of a band-pass filter centred at 5500 nm.

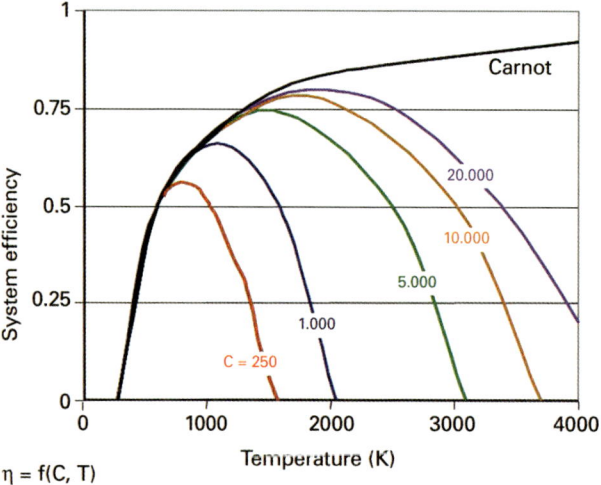

η = f(C, T)

Figure 6.11 Dependence of the Solar Thermal Electricity (STE) plant efficiency, η, as a function of the concentration ratio (C) and the receiver temperature (T).

reason why there is a tendency in the STE sector to develop new plant designs and technologies operating at higher and higher temperatures, because the Carnot efficiency increases with the working temperature. However, it must be taken into account that beyond the optimum working temperature, the efficiency decreases, due mainly to thermal losses by radiation; this becomes increasingly important as the temperature increases because the radiative thermal losses are proportional to the fourth power of the temperature (in K). So, for each concentration ratio there is an optimum working temperature beyond which the efficiency decreases because the radiative thermal losses become more significant than the efficiency increase due to a higher temperature.

The nominal conditions of the working fluid at the solar receiver outlet of the first commercial STE plant with central receiver technology (the Spanish PS10[16] plant installed by the Abengoa[17] company in 2008 in Seville) are 240 °C/24 bar,[18],[19],[20] while a commercial plant working at 525 °C is already in operation (the Spanish Gemasolar plant was inaugurated in 2011 by Torresol Energy),[21],[22] and several plants were either under construction or promotion in the US in 2012 to work at 550 °C/120 bar, such as the 'Ivanpah' plant promoted by Bright Source.[23]

However, higher temperatures also introduce new technical challenges, such as a greater stress in the raw material of the solar receiver and the need to develop very efficient and reliable operating procedures for the receiver during solar radiation transients (e.g. cloud passages during sunlight hours) in order to avoid overheating as the solar radiation quickly rises when the cloud is gone. This is the reason why metallurgy (another field of physics defined as a domain of materials science that studies the physical and chemical behaviour of metallic elements, their intermetallic compounds, and their mixtures, or alloys) is becoming a very important subject in STE plants, as stress-resistant raw materials are required in order to increase efficiency by operating at higher temperatures in the receiver. Clearly, metallurgy is becoming an increasingly important subject in relation to STE plants.

Another problem associated with higher working temperature is the existence of greater thermal losses in the plant, in spite of using selective coatings. The reason behind this is *the Stefan–Boltzmann law*, which states that the radiant power emitted by a body is directly proportional to the product of its emissivity and its absolute temperature rises to the power of 4. Therefore, the amount of radiant power emitted will be very high, even with low emissivity values. This is the technical incentive to develop advanced selective coatings with lower emissivities at higher working temperatures.

---

16) PS10 (Planta Solar 10).
17) www.abengoa.com.
18) The bar is a non-SI unit of pressure, defined by the International Union of Pure and Applied Chemistry (IUPAC) as exactly equal to 100 000 Pascal.
19) R. Osuna *et al.*, *PS10: Construction of a 11 mw solar thermal tower plant in Seville, Spain*, in M. Romero *et al.* (Eds), Proceedings of the 13th International SolarPACES (Solar Power And Chemical Energy Systems) Conference, 2006.
20) R. Osuna *et al.*, *Plataforma Solar Sanlúcar la Mayor: the largest European solar power site*, Proceedings of the 14th International SolarPACES Conference, 2008.
21) www.torresolenergy.com.
22) J.I. Burgaleta *et al.*, *Gemasolar, the first tower thermosolar commercial plant with molten salt storage*, Proceedings of the 17th International SolarPACES Conference, 2011.
23) www.brigthsourceenergy.com.

6.4
Issues in STE Plants Related to Heat Transfer

The importance of heat-transfer mechanisms in STE plants becomes evident from Figure 6.1. First, the *concentrated solar radiation* is converted into useful heat at the *receiver*, where a working fluid increases its specific enthalpy as it circulates through the receiver. This conversion of *solar radiation* into heat takes place by conduction through the walls of the receiver, and also by convection from the receiver inner surface to the working fluid circulating inside. At the same time, thermal losses from the receiver to the environment reduce the amount of useful heat delivered to either the *heat engine* or the *thermal storage system*.

Thermal losses at the receiver are due to convection and radiation to the environment, as well as to heat loss by conduction through the metallic elements supporting the weight of the receiver and keeping it in the correct position to receive the maximum amount of concentrated solar radiation.

At the same time, the hot working fluid delivered by the receiver is transported to either the thermal storage system or to the heat engine to produce *electricity*. The heat engine is usually included in a power block composed of heat exchangers, pumps and auxiliary systems that allow a correct operation of the heat engine, which is in turn mechanically connected to an electricity generator where the electricity is finally generated. The most common heat engine used in STE plants is a steam turbine operating with a water/steam Rankine cycle. Although small Stirling engines are used with parabolic dishes to form the so-called 'Stirling dishes', these systems are facing major problems in terms of their commercial benefits; hence, the number of Stirling dishes in operation is very small, with a total output power of less than 1 MWe in 2012. Innovative STE plant concepts using a gas turbine with a Brayton cycle are also under development.

Since a significant amount of heat is transferred from the solar receiver to either the thermal storage system or to the power block, and from the storage system to the power block, thermal losses are an important issue in any STE plant because the overall plant efficiency also depends on the thermal losses at the several subsystems. Hundreds, or even thousands, of meters of piping exist in a STE plant, depending on its size. For instance, a 50 MWe STE plant with parabolic trough collectors and a 1 GWh[24] thermal storage system requires about 80 km of receiver tubes and about 7 km of piping containing hot fluid. These numbers clearly show how important the thermal losses are in STE plants, and why thermal insulation is so important (see Figure 6.12).

Thermal insulation with a correct thickness is used to reduce thermal losses in vessels, heat exchangers and piping, while other more sophisticated means are used to reduce thermal losses at the solar receivers. As the solar receivers are the hottest parts in any STE plant, special attention is paid to reduce their

24) Gigawatt hour.

Figure 6.12 The pipes and components are thermally insulated in a Solar Thermal Electricity (STE) plant employing Parabolic-trough collectors.

thermal losses. The three basic mechanisms for heat transfer, namely conduction,[25] convection[26] and radiation,[27] are always taken into consideration by the plant designers, together with their respective laws: *Fourier's law*,[28] *Newton's law of cooling*[29] and the *Stefan–Boltzmann law*.[30] Whilst very efficient methods exist today to reduce thermal losses, they are not free of charge and a compromise between high extra costs and low thermal losses must be achieved during the design phase.

Currently, an increasing number of effective thermal insulation materials is becoming available, such that the amount of insulating material needed can be reduced, as can its installation costs, in order to achieve a low level of thermal losses.[31]

25) Conduction is the transfer of heat energy by the microscopic diffusion and collisions of particles or quasi-particles within a body due to a temperature gradient. They transfer disorganized kinetic and potential energy, jointly known as 'internal energy'. Conduction can only take place (in solids, liquids, gases and plasmas) within an object or material, or between two objects that are in direct or indirect contact with each other. Whether by conduction or thermal radiation, heat will flow spontaneously from a body at a higher temperature to a body at a lower temperature. Without external drivers, temperature differences decay over time, and the bodies approach thermal equilibrium.
26) Convection is the transfer of heat from one place to another by the movement of fluids. It is usually the dominant form of heat transfer in liquids and gases. It involves the combined processes of conduction (heat diffusion) and advection (heat transfer by bulk fluid flow).
27) Radiation is electromagnetic radiation generated by the thermal motion of charged particles in matter. When the temperature of the body is greater than absolute zero, interatomic collisions change the kinetic energy of the atoms or molecules, which results in charge-acceleration and/or dipole oscillation that produces electromagnetic radiation. Sunlight is part of the thermal radiation generated by the hot plasma of the Sun.
28) *Fourier's law* states that the rate of heat flow through a substance is proportional to the area normal to the direction of flow and to the negative of the rate of change of temperature with distance along the direction of flow.
29) *Newton's law of cooling* states that the rate of change of the temperature of an object is proportional to the difference between its own temperature and the ambient temperature (i.e. the temperature of its surroundings).
30) F.P. Incropera and D.P. DeWitt, *Fundamentals of Heat and Mass Transfer*, John Wiley & Sons, 1996.
31) www.microthermgroup.com.

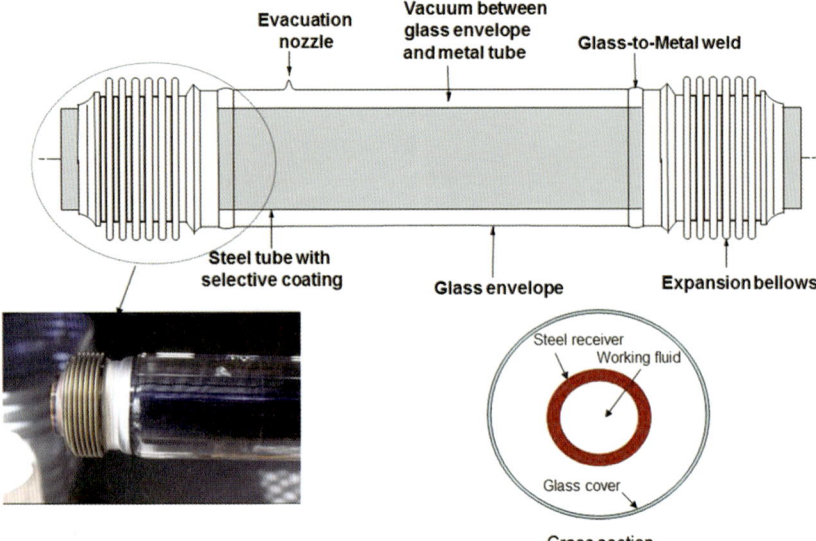

Figure 6.13 A typical receiver tube for a parabolic-trough collector.

Radiation heat losses are very important at the solar receivers of STE plants, because the receivers usually are at a temperature on the order of, or greater than, 600 K (and even at 875 K, depending on the technology used). The *Stefan–Boltzmann law* states that the energy radiated by a hot object per second and per unit area is proportional to the fourth power of its absolute temperature, T, and to its emissivity. As the working temperature of the absorber tube cannot be reduced without jeopardizing the overall plant efficiency (see Figure 6.11), and the Stefan–Boltzmann constant[32] is a constant by definition, the only parameter that can be modified to reduce thermal losses is the emissivity of the receiver. This is why the outer surfaces of solar absorbers are provided with the special coatings described in Section 6.2.2, with a low emissivity in the infrared wavelength range to reduce radiation heat losses. This is the first physical principle applied in solar receivers to reduce their thermal losses.

In addition to reducing radiation heat losses by using a special coating with low emissivity, convection losses can be avoided in linear receivers if a glass cover is used to envelop the steel solar receiver by leaving an annulus between them and creating a vacuum in this annulus. The glass envelope and the inner steel absorber tube are connected at both ends by glass/metal welds to avoid vacuum leaks over time. By using this approach, heat convection and conduction losses are significantly reduced in linear solar receivers like those used in parabolic trough collectors. Figure 6.13 shows a typical receiver tube for a parabolic trough collector. The use of a special coating (selective coating), and the existence of a annulus with vacuum around the hot steel absorber reduce thermal

---

32) $\sigma = 5.6705119 \times 10^{-8} \, \text{W} \cdot \text{m}^{-2} \cdot \text{K}^{-4}$.

losses to less than 10% of the solar radiant power absorbed at the receiver when it is at about 400 °C.

The use of special coatings with low emissivity, vacuum and a glass envelope (as shown in Figure 6.13) significantly reduces the heat losses in linear solar receivers, because these elements reduce or even eliminate the three heat-transfer mechanisms that provoke these heat losses, namely conduction, convection and radiation.

6.5
Thermal Storage of Energy

Dispatchability is one of the main benefits of STE plants when compared to other renewable energy systems. The possibility of storing thermal energy during sunlight hours to produce electricity when solar radiation is not available significantly enhances the integration of STE plants in the electrical grids, whilst at the same time allowing a greater integration of other renewable energy systems that are not dispatchable nowadays (e.g. wind farms and photovoltaic plants). Although this topic is very much related to heat transfer, and could have been included in Section 6.4, the importance of thermal storage for STE plants makes it worthy of further discussion at this point.

Today, thermal storage is affordable by using large tanks that are filled with a storage medium and provided with an effective thermal insulation to reduce thermal losses. As the storage medium is heated during the charging process, energy is stored in the form of sensible heat. During the discharging process, the storage medium is cooled in such a way that the stored thermal energy is recovered. Although most commercial thermal storage systems currently in operation employ a salt mixture of sodium and potassium nitrates, other storage mediums are available; details of these and their characteristics are provided in Table 6.1.[33]

The working fluid circulating in the solar receiver (see Figure 6.1) is usually different from the storage medium in the STE plants, and heat is transferred from/to the working fluid to/from the storage medium during the charging/discharging process by means of standard heat exchangers. This is the case for the parabolic trough STE plants, which use thermal oil in the solar field and molten salts in the storage system. Occasionally, however, the same working fluid is used for the storage medium as is circulating in the receiver, and this is the situation with the Spanish Gemasolar plant, where molten salts are pumped directly from a cold tank at 290 °C into the receiver. Once heated to 550 °C, the salts are then sent from the receiver to a hot tank at 550 °C. Due to their affordable price (ca.1 € per kg) and thermohydraulic properties, molten salts are currently the most widely used storage media in STE plants.

---

33) EPRI (Electric Power Research Institute), *Program on Technology Innovation: Evaluation of Concentrating Solar Thermal Energy Storage Systems*, Report N. 1018464, Technical Update, 2009.

Table 6.1 Thermal storage mediums and their characteristics.

Liquid media	Operating temperature (°C)		Average density (kg m^{-3})	Average thermal conductivity [W (m·C)$^{-1}$]	Average heat capacity [J (kg·C)$^{-1}$]	Volume-specific energy density[a] (kWh m^{-3})
	Cold	Hot				
Hitec XL	290	500	1913	0.519	1415	297
Binary nitrate salt	290	565	1818	0.524	1517	327
Therminol VP-1	290	390	768	0.089	2449	178
Caloria HT-43	200	290	715	0.090	2557	124
Solid media	Operating temperature (°C)		Average density (kg m^{-3})	Average thermal conductivity [W (m·C)$^{-1}$]	Average heat capacity [J (kg·C)$^{-1}$]	Volume-specific energy density[a] (kWh m^{-3})
Sand-rock-mineral oil	200	300	1700	1.0	1.30	60
Reinforced concrete	200	400	2.200	1.5	0.85	100
Cast iron	200	400	7200	37.0	0.56	160
Cast steel	200	700	7800	40.0	0.60	450
Silica fire bricks	200	700	1820	1.5	1.00	150
Magnesia fire bricks	200	1200	3000	5.0	1.15	600

a) Calculated at the hot operating temperature for each storage medium.

Thermal storage systems can be used to either extend the electricity delivery period or to displace the delivery period beyond sunset time to meet the energy demands from the consumers. Figure 6.14 shows an example of the use of a thermal storage system to extend the delivery period.[34]

6.6
Fluid Mechanics

It has been already explained that the conversion of solar radiation into heat takes place at the receiver by conduction through the walls of the steel receiver, and by convection from the receiver inner surface to the working fluid when the solar collectors are in operation. The plant efficiency increases not only when heat losses to the environment are reduced, but also when the heat transfer to the working fluid is enhanced by acting on the parameters of *Fourier's law* for

34) Ibidem.

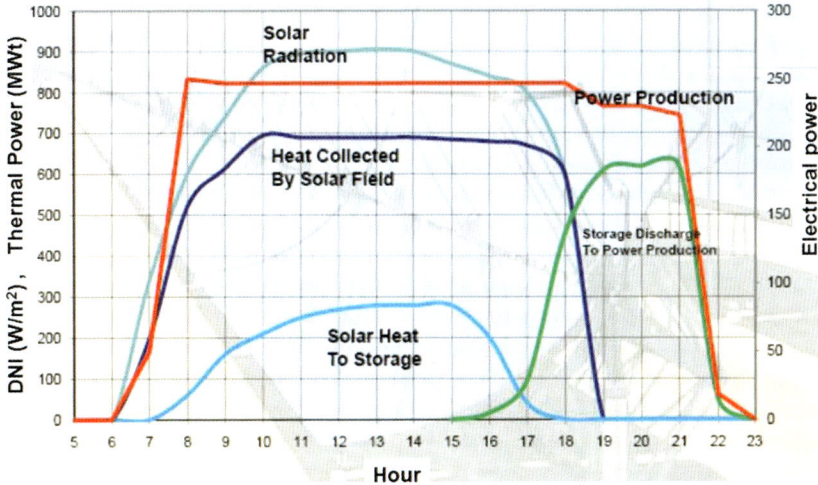

Figure 6.14 Extension of the electricity delivery period by using thermal storage.[35]

conduction losses in a cylindrical tube and *Newton's law of cooling* by convection losses.

For given temperatures, the heat convection can be enhanced by increasing h (the convection heat transfer coefficient at the inner wall of the steel receiver tube), and this can be accomplished if a turbulent flow regime is kept inside the steel absorber. This requirement will be fulfilled if a high working fluid flow is maintained while the solar collector is operating. Concerning heat conduction, it is clear that raw materials with a good thermal conductivity must be used for the solar absorber. In fact, the receivers in STE plants are made from carbon steel whenever the maximum working temperature and pressure make it feasible, because the thermal conductivity of carbon steel is greater than that of stainless steel. However, stainless steel or advanced alloys (e.g. Inconel) must be used for higher thermal loads in order to withstand the associated stress.

With regards to the heat transfer coefficient by convection (h) between the inner wall of the receiver and the fluid circulating inside, a fluid velocity which is sufficiently high to ensure a turbulent flow is required, not only to enhance the heat transfer from the receiver to the fluid but also to avoid too-high temperatures at the outer wall of the receiver. Many different correlations are available to calculate h as a function of the fluid properties and velocity, but all of them provide a value that is proportional to the fluid velocity.[36]

At the same time, the pressure losses in the piping are directly proportional to the length of the piping and the fluid velocity raised to the power of 2, which

---

35) Ibidem.
36) F.P. Incropera and D.P. DeWitt, *Fundamentals of Heat and Mass Transfer*, John Wiley & Sons, 1996.

demands an optimal velocity value between a high heat convection coefficient and high pressure losses in the piping, which would in turn significantly increase the pumping losses in the STE plant such that the overall plant efficiency would be penalized. The Reynolds number (Re) provides a measure of the ratio of inertial forces to viscous forces, and thus quantifies the relative importance of these two types of forces for given flow conditions.[37] Therefore, the Reynolds number represents the key parameter to be taken into account by the piping designer to find the correct fluid velocity that will ensure a good heat convection coefficient while maintaining the pressure losses at an affordable level.

Acknowledgements

The author is grateful to Eduardo Zarza, from PSA (Plataforma Solar de Almería), for his thorough collaboration and help all along the writing of this chapter.

---

37) G.K. Batchelor, *An Introduction to Fluid Dynamics*, Cambridge University Press, 2000.

7
Computers and Aviation

Antony Jameson

Although animal flight has a history of 300 million years, serious thought about human flight has a history of only a few hundred years, dating from Leonardo da Vinci,[1] with successful human flight having only been achieved during the last 110 years (see Figures 7.1–7.4). To some extent, this parallels the history of computing. Serious thought about computing dates back to Pascal and Leibnitz. While there was a notable attempt by Babbage to build a working computer in the nineteenth century, successful electronic computers were finally achieved in the 1940s, almost exactly contemporaneously with the development of the first successful jet aircraft. A visual presentation of the early history of computers is shown in Figures 7.5–7.8, while details of the more recent progress in the development of supercomputers and microprocessors are listed in Tables 7.1 and 7.2.

Although aeroplane design had reached quite an advanced level by the 1930s, as exemplified by aircraft such as the Douglas Commercial-3 (DC-3) and the Spitfire (see Figure 7.2), the design of high-speed aircraft requires an entirely new level of sophistication. This has led to a fusion of engineering, mathematics and computing, as indicated in Figure 7.9.

During the past five decades, computers have fundamentally transformed every aspect of aviation and aerospace. These impacts fall into three main classes. First, computing has completely transformed the design and manufacturing processes. Second, the advent of microprocessors with ever-increasing power has transformed the actual aircraft and spacecraft themselves, with computers taking over every aspect of the flight control and navigation systems. This parallels similar developments in automobiles, which are no longer directly controlled by their drivers, but instead use microprocessors to optimize engine performance and manage functions such as anti-skid breaking. The third way in which computers have transformed aviation is that the major aspects of aircraft operations are now controlled by computing systems such as electronic reservation and ticketing systems and automatic check-in. We shall discuss each of these aspects in more detail in the following sections.

1) L. da Vinci, *Notebooks*, Oxford University Press, 2008.

From Physics to Daily Life: Applications in Informatics, Energy, and Environment, First Edition.
Edited by Beatrice Bressan.
© 2014 Wiley-VCH Verlag GmbH & Co. KGaA. Published 2014 by Wiley-VCH Verlag GmbH & Co. KGaA.

7 Computers and Aviation

Figure 7.1 (a) Orville and Wilbur Wright, 1903. (U.S. Department of Defense photo. Use of military imagery does not imply or constitute endorsement of Wiley, its products, or services by the U.S. Department of Defense.); (b) The Wright Flyer, 1903. (Courtesy of John T. Daniels, U.S. Library of Congress)[2]

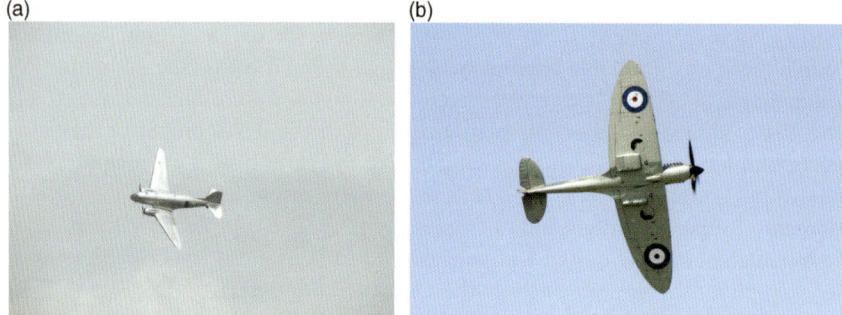

Figure 7.2 (a) Douglas DC-3, 1935. (© LAURENT MARDON – fotolia.com); (b) Supermarine Spitfire, 1936. (© jelwolf – fotolia.com).

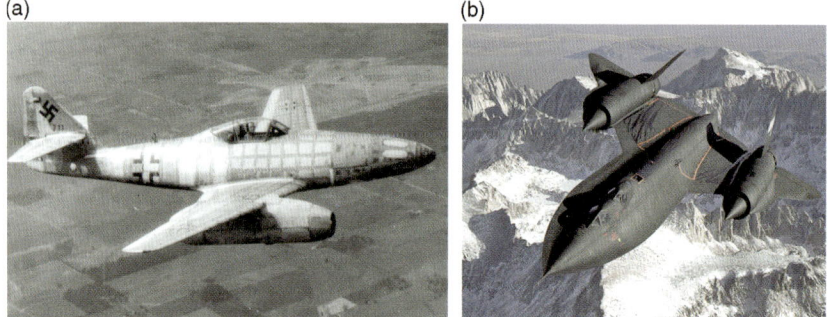

Figure 7.3 (a) Messerschmitt ME-262, 1941; (b) Lockheed SR-71, 1964. (a and b: U.S. Department of Defense photos. Use of military imagery does not imply or constitute endorsement of Wiley, its products, or services by the U.S. Department of Defense).

2) The Wright Flyer is the first successful powered aircraft, designed and built by the Wright brothers. They flew it four times near Kill Devil Hills, about four miles south of Kitty Hawk, North Carolina, US.

7 Computers and Aviation | 143

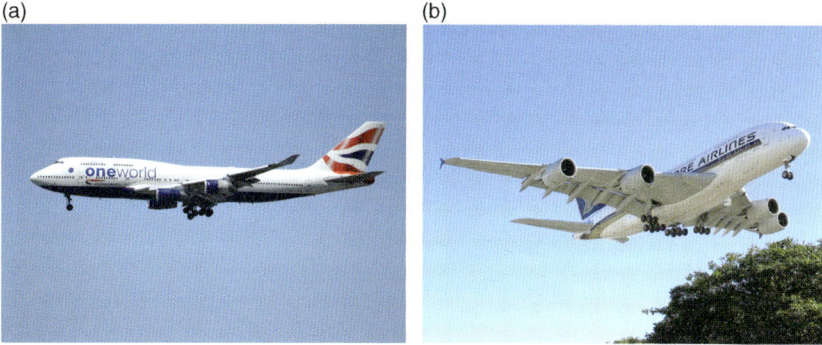

Figure 7.4 (a) Boeing 747, 1969. (Courtesy of Andre Chan, Stanford University); (b) Airbus 380, 2005. (Courtesy of Andre Chan, Stanford University).

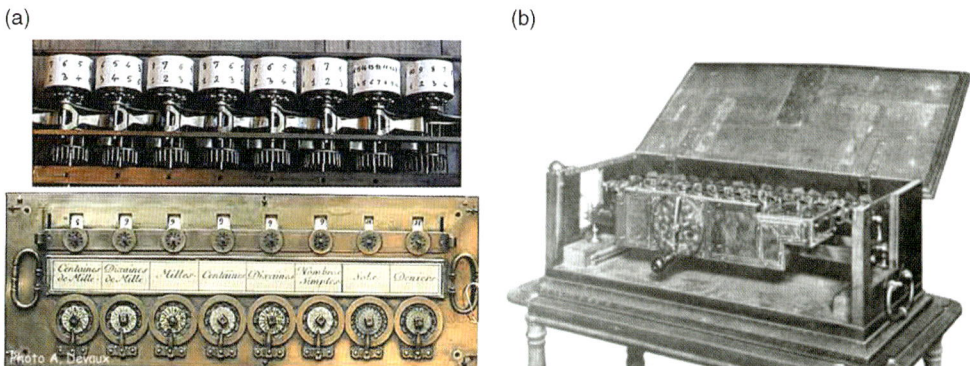

Figure 7.5 (a) Pascal's Pascaline, 1642. (Courtesy of André Devaux, Calmeca)[3]; (b) Leibniz's stepped reckoner, 1672.[4]

Figure 7.6 Babbage's difference engine, 1822. (Courtesy of Jitze Couperus, Flickr).[5]

3) http://calmeca.free.fr.
4) J. A. V. Turck, *Origin of Modern Calculating Machines*, The Western Society of Engineers, p.133, 1921.
5) www.flickr.com.

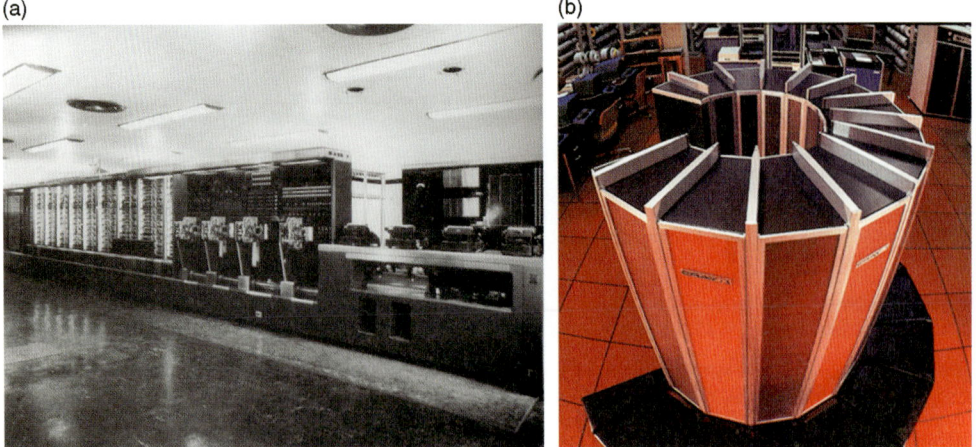

Figure 7.7 (a) Mark I, 1944. (Courtesy of John Kopplin and Michael Rothstein, Kent State University);[6] (b) Cray-1, 1976. (Courtesy of Cray Research).[7]

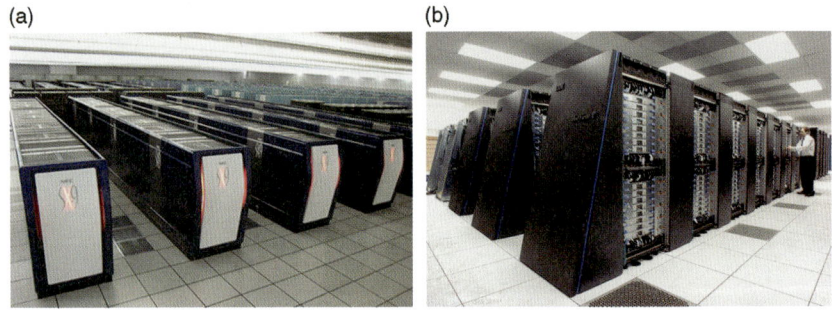

Figure 7.8 (a) NEC Earth Simulator, 2002. (Courtesy of Japan Agency for Marine-Earth Science and Technology, JAMSTEC); (b) IBM Blue Gene, 2005. (Courtesy of Argonne National Laboratory).[8]

6) Mark I, a computer which was built as a partnership between Harvard and IBM in the US, was the first programmable digital computer made in the US, but it was not a purely electronic computer.
7) The Cray-1 was a supercomputer designed, manufactured and marketed by Cray Research founded in 1972 by computer designer Seymour Cray in Seattle, Washington, US. After the Cray Research purchase in 2000, Cray was formed: www.cray.com.
8) Blue Gene is an IBM project designing supercomputers that can reach operating speeds in the petaflops range, with low power consumption. The project created three generations of supercomputers: Blue Gene/L, Blue Gene/P, and Blue Gene/Q.

Table 7.1 The supercomputers' timeline.

Year	Model	Performance
1964	CDC 6600	3 MFLOPS[9]
1976	Cray-1	250 MFLOPS
1993	Fujitsu Numerical Wind Tunnel	124.5 GFLOPS
2002	NEC Earth Simulator	35.86 TFLOPS
2007	IBM Blue Gene/L	478.2 TFLOPS
2009	Cray Jaguar	1.759 PFLOPS
2012	IBM Sequoia	20 PFLOPS

Table 7.2 Microprocessor timeline.

Year	Model	Manufacturing process	Transistor	Clock	Bits	Core
1971	Intel 4004	10 µm	2250	108 kHz	4	1
1978	Intel 8086	3 µm	29 000	4.77 MHz	16	1
2000	Intel Pentium 4	0.18 µm	42 M	1.5 GHz	32	1
2008	Intel Core i7	45 nm	774 M	2.993 GHz	64	4

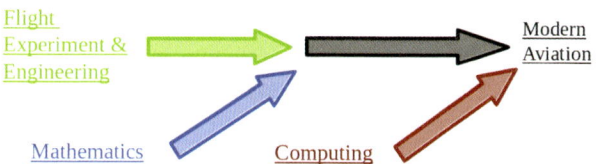

Figure 7.9 Fusion of flight experiments, mathematics and computing.

7.1
Computing in Structural and Aerodynamic Analysis

The first inroads of computing in the aerospace industry were in the design process, beginning with structural analysis based on the finite element method. In fact, the origins of the finite method may be found in the aerospace industry, in The Boeing Company, where it was developed under the leadership of Turner during the period 1950–1962.[10] Important early contributions were made by

---

9) In computing, FLOPS (FLoating-point Operations Per Second) is a measure of computer performance, useful in fields of scientific calculations that make heavy use of floating-point calculations. For such cases it is a more accurate measure than the generic instructions per second.

10) M.J. Turner et al., Stiffness and deflection analysis of complex structures, Journal of the Aeronautical Sciences, Volume 23, N. 9, p. 805, 1956.

Argyris, who was a consultant to Boeing.[11),12),13),14)] The NASA STRucture ANalysis (NASTRAN) software for structural analysis was developed under National Aeronautics and Space Administration (NASA) sponsorship between 1964 and 1968, and became a standard tool.

Computing methods for aerodynamic analysis followed soon afterwards, giving birth to the new discipline of Computational Fluid Dynamics (CFD). The Aerodynamic Research Group of the Douglas Aircraft Company,[15)] led by Smith, developed the first panel method for three-dimensional, linear, potential flows in 1964.[16)] Nonlinear methods were needed to enable the prediction of high-speed transonic and supersonic flows, and a major breakthrough was accomplished by Murman and Cole in 1970, who demonstrated for the first time that steady transonic flows could be computed economically. The first computer program that could accurately predict transonic flow over swept wings, FLO22, was developed by Jameson and Caughey in 1975, using an extension of the method of Murman and Cole, and this rapidly came into widespread use. At this time (1976), sweptwing calculations challenged the limit of the available computing resources. The most powerful computer available, the Control Data Corporation (CCD) 6600, had 131 000 words of memory, but this was not enough to store a full 3D solution, which had to be read back and forth from the disk drives. The CCD 6600 had a peak computational speed of about 3 megaflops, and a complete sweptwing calculation took about 3 hours with a cost of about US$ 3000. Nevertheless, Douglas found it worthwhile to run six or more calculations using FLO22 every day. The first major application was the wing design of the C17 (Cargo aircraft model 17). FLO22 was also used for the wing design of the Canadair Challenger; this was the first application of CFD to the wing design of a commercial aircraft. Today, FLO22 is still used for preliminary design studies and is very useful in this role as the calculations can now be performed in 10 seconds with a laptop computer.

11) J.H. Argyris, *The open tube: A study of thin-walled structures such as interspar wing cut-outs and open-section stringers*, Aircraft Engineering and Aerospace Technology, Volume 26, Issue 4, p. 102, 1954.
12) J.H. Argyris, *Flexure-torsion failure of panels: A study of instability and failure of stiffened panels under compression when buckling in long wavelengths*, Aircraft Engineering and Aerospace Technology, Volume 26, Issue 6, p. 174, 1954.
13) J.H. Argyris, *Energy theorems and structural analysis: A generalized discourse with applications on energy principles of structural analysis including the effects of temperature and non-linear stress-strain relations*, Aircraft Engineering and Aerospace Technology, Volume 26, Issue 11, p. 383, 1954.
14) J.H. Argyris and S. Kelsey, *Energy theorems and structural analysis: A generalized discourse with applications on energy principles of structural analysis including the effects of temperature and nonlinear stress-strain relations*, Butterworth, 1960.
15) The Douglas Aircraft Company was an American aerospace manufacturer based in Southern California. It was founded in 1921 by Sir Donald Wills Douglas, and later merged with McDonnell Aircraft in 1967 to form McDonnell Douglas: www.mdc.com.
16) J.L. Hess and A.M.O. Smith, *Calculation of the non-lifting potential flow about arbitrary three dimensional bodies*, Douglas Aircraft Report, N. E.S. 4062, 1962.

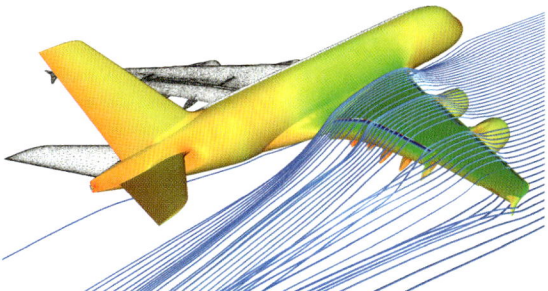

Figure 7.10 CFD simulation of Airbus A380. (Courtesy of DLR, the German Aerospace Centre, CC-BY 3.0).[17]

With the advent of the first supercomputers during the early 1980s, exemplified by the Cray-1, which achieved sustained computational speeds of around 100 megaflops, it became feasible to solve the full fluid flow equations (the Euler equations for inviscid flow and the Navier–Stokes equations for viscous flow) for complex configurations. The first Euler solution for a complete aircraft was accomplished by Jameson, Baker and Weatherill in late 1985, who were provided remote access to a Cray-1 by the Cray company, US.[18] By the 1990s, computer performance had advanced to the point where Navier–Stokes simulations could be performed routinely using meshes containing several million cells. This period saw the emergence of NASA-developed codes such as OVERFLOW (OVERset grid FLOW solver), CFL3D (Computational Fluids Laboratory Three-Dimensional), USM3D (Unstructured Mesh Three-Dimensional) and FUN3D (Fully Unstructured Navier–Stokes Three-Dimensional). During the 1980s and 1990s, there was a parallel development of commercial CFD software targeted at a wide range of industrial applications. The first commercial CFD software was Spalding's PHOENICS (Parabolic Hyperbolic Or Elliptic Numerical Integrated Code Series) code. Subsequently, Fluent, CFX[19] and STAR-CD (Simulation of Turbulent flow in Arbitrary Regions Computational Dynamics) emerged as the most widely used commercial software packages, though most aerospace companies still prefer to use codes that have been developed specifically for high-speed flow simulations.

The current use of CFD in aircraft design is illustrated in Figures 7.10 and 7.11. Figure 7.10 shows a simulation of the compressible viscous flow over an Airbus A380 wing, while Figure 7.11a and b illustrate the extent of CFD use in the designs of the A380 and the Boeing 787, respectively.

---

17) DLR (Deutsches Zentrum für Luft- und Raumfahrt e.V.): www.dlr.de.
18) www.cray.com.
19) Fluent and CXF are computational fluid dynamics software marketed by Ansys Corporation: www.ansys.com.

(a)

(b)
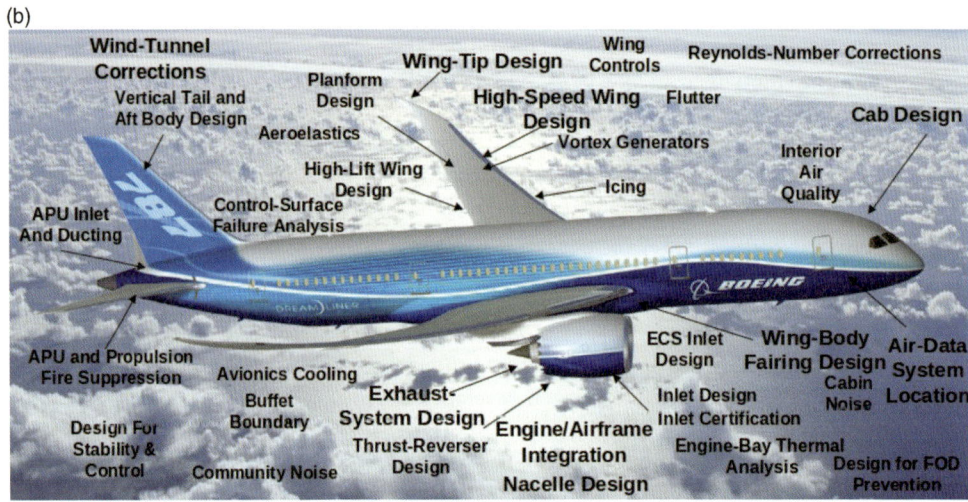

Figure 7.11 (a) CFD contributions to the Airbus A380. (Courtesy of DLR, the German Aerospace Centre);[20] (b) CFD contributions to the Boeing B787. (Courtesy of Ed Tinoco. Reproduced with permission of Boeing).[21]

20) DLR (Deutsches Zentrum für Luft- und Raumfahrt e.V.): www.dlr.de.
21) www.boeing.ch.

7.2
Computer-Aided Design and Manufacturing

Historically, engineering parts have been defined by engineering drawings and 'blueprints'. These required meticulous preparation by large teams of draughtsman working at drawing boards. By the 1960s, it was apparent that there was an opportunity for significant cost reductions if this process could be computerized, but this required the development of a new set of mathematical tools which could provide the foundations of modern computational geometry, and have since enabled the development of Computer-Aided Design (CAD) and Computer-Aided Manufacturing (CAM) systems.

The early development of geometric modelling technology was driven by the automotive and aircraft industries due to their unique engineering requirements for a wide range of curves and surfaces for their parts. Manually defining and manufacturing these components was becoming increasingly time-consuming and costly. However, by the early 1960s numerically controlled machine tools became more readily available and there was the need to generate the digital information to drive these machines. CAD systems began to emerge in this era, with some of the first developments taking place at Citroën, where de Casteljau developed CAD methods and introduced the de Casteljau algorithm in reports that were not published outside Citroën.[22],[23] The system was also introduced at Renault, where Bézier led the development of the UNISURF system[24] and introduced the concept of the Bézier curve.[25],[26],[27]

As in the case of CFD, CAD system development also experienced rapid changes as computer hardware became more capable. During the 1960s, CAD software was run on mainframe computers, and the earliest CAD systems were used primarily for replacing the traditional draughting practice. Although limited at that time to handling only two-dimensional data, the use of CAD for engineering drawing helped to reduce drawing errors and also allowed the drawings to be modified and reused. Large aerospace and automotive companies with the resources to cover the high costs of the early computers became the earliest users of CAD software, and most CAD development during that period was conducted

---

22) P. de Casteljau, *Outillages méthodes calcul*, Technical report, André Citroën Automobiles SA, 1959.
23) P. de Casteljau, *Courbes et surfaces à pôles*, Technical report, André Citroën Automobiles SA, 1963.
24) UNISURF was a pioneering surface system, designed to assist with car body design and tooling, developed in 1968, and fully in use at the company by 1975. By 1999, around 1500 Renault employees made use of it.
25) A. Bézier curve is a parametric curve that is frequently used in computer graphics and related fields.
26) P. Bézier, *Définition numérique des courbes et surfaces I*, Automatisme, Volume 11, N. 12, p. 625, 1966.
27) P. Bézier, *The mathematical basis of the UNISURF CAD system*, Butterworth-Heinemann, 1986.

internally in those companies. An example was the Computer-Augmented Design And Manufacturing (CADAM) system developed by the Lockheed[28] aircraft company. This system, which automated the production of 2D drawings was marketed by Lockheed after 1972, while Dassault purchased a license for its use in 1974 and also acquired UNISURF from Renault in 1976. Subsequently, this evolved into the 3D modelling system CATIA (Computer Aided Three-dimensional Interactive Application), which was originally used in conjunction with CADAM. Dassault began marketing CATIA in 1981, and it has become the most widely used CAD tool in the aerospace industry. During the 1970s, the emergence of powerful minicomputers made CAD software more affordable and accessible, and helped to create the commercial CAD software market. The very rapid growth of commercial CAD changed the way in which CAD was used and developed in the major automotive and aerospace companies as they began to use commercial software in conjunction with their internally developed CAD systems. Simultaneously, there were significant advances in the geometric algorithms on which CAD software was based, including B-Spline (Basis Spline)[29],[30] and NURBS (Non-Uniform Rational B-Spline).[31] During the 1980s, low-cost, low-maintenance and high-performance workstations using Unix[32] operating system were introduced. This again revolutionized the CAD software market, and effectively replaced the mainframe and mid-range computers as the preferred hardware for CAD systems. At the same time, 3D CAD software and solid modelling techniques matured and became a commercial reality. Subsequently, as the computer hardware and maintenance costs continued to fall and CAD software became more available and powerful, commercial CAD systems spread throughout industry.

In 1988, Boeing made the decision to use the commercially available CATIA to design and draft the new B777 aeroplane, which became the first CAD-based 'paperless' design of a commercial aircraft. This decision proved to be very successful, leading to reduced product development time and costs. From the 1990s to the present time, the same trend repeated itself, with more cost-effective and powerful personal computers replacing the less cost-effective workstations, and with a corresponding migration of CAD software from the Unix system to the mainstream Windows and Linux operating systems. The function of CAD systems also evolved from pure geometric modelling tools into a system of computer-aided engineering solutions that consists of CAM, digital assembly, and virtual production management.

---

28) Lockheed, the Lockheed Corporation (originally Loughead Aircraft Manufacturing Company) was an American aerospace company founded in 1912 and later merged with Martin Marietta to form Lockheed Martin in 1995.
29) In mathematics, a Basis Spline is a sufficiently polynomial function with derivatives of all orders that is defined by multiple subfunctions.
30) R. Risenfeld, *Applications of B-Spline Approximation to Geometric Problems of CAD*, PhD thesis, Syracuse University, US, 1973.
31) K.J. Versprille, *Computer-Aided Design Applications of the Rational B-Spline Approximation Form*, PhD thesis, Syracuse University, US, 1975.
32) Originally UNICS, UNiplexed Information and Computing System.

Using information technology such as CAM and production can effectively restore close interaction and communication amongst a large number of people in the design process. In a computer-assisted environment, the aeroplane designer has access to manufacturing processes and tools in the form of virtual environments, and these will allow the designer to virtually manufacture the product while designing it. A more optimal design trade and resource allocation between production and aeroplane performance can be achieved early in the design stage.

To conclude this section, some statistics are presented from the study of the digitally designed Boeing 777, which demonstrate the great benefits from design automation achieved through CAD systems. Boeing used CAD systems that combined geometric modelling using CATIA, finite element analysis using ELFINI (Finite Element Analysis System) and digital assembly using EPIC (Electronic Preassembly Integration on CATIA). The CAD systems allowed Boeing engineers to simulate the geometry of an aeroplane design on the computer without the costly and time-consuming investment of using physical mock-ups. More than three million parts were represented in an integrated database. Subsequently, a complete 3D virtual mock-up of the aeroplane was created that allowed the designers to investigate part interferences, assembly interfaces and maintainability using spatial visualizations of the aircraft components. The consequences were dramatic. In comparison with the earlier aircraft design and manufacturing processes, Boeing eliminated more than 3000 assembly interfaces without any physical prototyping, and achieved a 90% reduction in engineering change requests, a 50% reduction in cycle time for engineering change request, a 90% reduction in material rework, and a 50-fold improvement in assembly tolerances for the fuselage. Overall, CAD/CAM systems and digital preassembly greatly improve the quality of aeroplane designs and reduce the time required to introduce new airplanes into the marketplace. The application of CAD in the design of the Boeing 777 is illustrated in Figure 7.12.

7.3
Fly-By-Wire and Other On-Board Systems

Early high-performance computers were far too bulky and heavy to be carried on-board an aircraft, and consequently the role of computers was limited to functions that could be performed on the ground, such as design and manufacturing. The advent of the modern microprocessor has completely changed the situation, however. Today, a processor such as an Intel Core i7 with four cores clocked at 2.7 GHz is just as powerful as the supercomputers of the 1980s, and hence it is now possible to computerize critical on-board functions such as control, guidance, navigation and collision avoidance. In particular, the development of digital Fly-By-Wire (FBW) systems has revolutionized the operation of both military and commercial aircraft.

3D Fly-Thru  Full-Motion Human Modelling

Digital Pre-Assembly of a Boeing Airplane

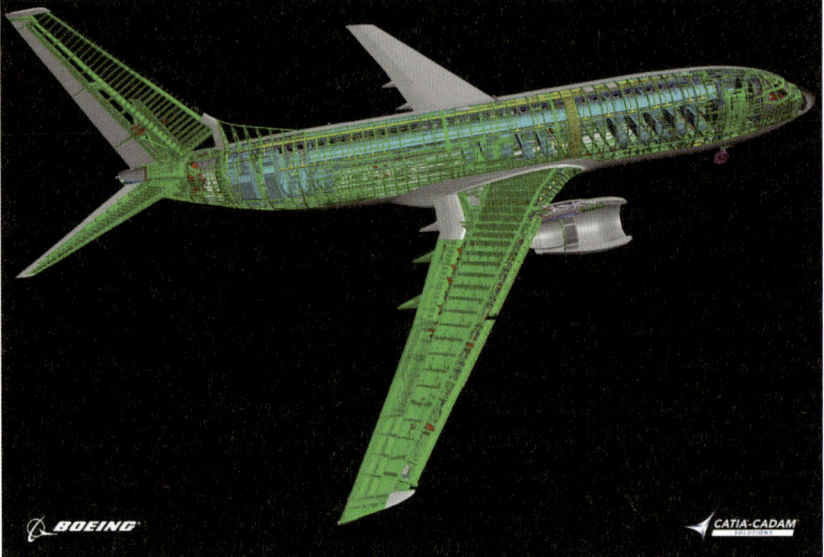

Figure 7.12 CAD applications in aircraft design and manufacturing. (Courtesy of Computer Aided Three-dimensional Interactive Application, CATIA).

The General Dynamics F16 was the first Fighter military aircraft with a full digital FBW control system while, led by Ziegler (a former fighter pilot), Airbus was the first company to use FBW for a civil aircraft, namely the Airbus A320. Shortly afterwards, FBW control systems were adopted for the Airbus 330 and 340, and the Boeing 777. The FBW control system has been credited with a key role in the successful descent onto the Hudson River of an Airbus 320 with both engines out of action after a bird strike.

In the FBW system, digital controls replace the conventional mechanically operated flight controls. The elimination of mechanical components in the new digital system is shown schematically in Figure 7.13. The pilot no longer physically moves the control surfaces through mechanical linkages; instead, the pilot's commands, or the orders from the autopilot computers (when in autopilot

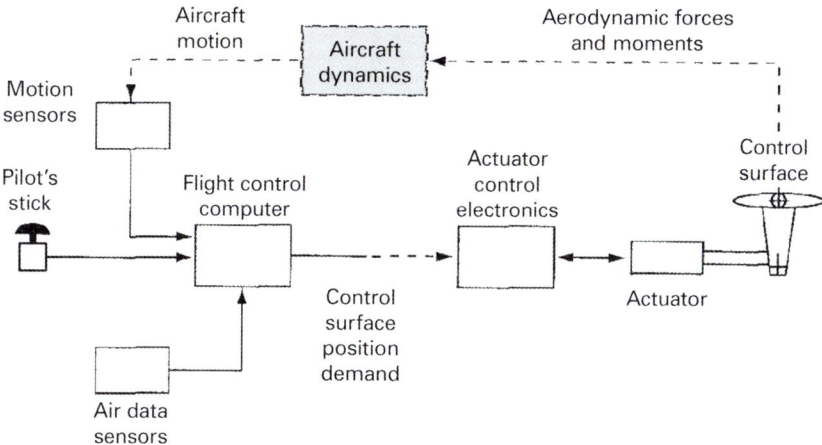

Figure 7.13   Schematic of a digital Fly-By-Wire (FBW) system.[33]

mode), are transmitted digitally to a group of flight computers which instantly interpret and analyse the control inputs and evaluate the aircraft's speed, weight, atmospheric conditions, and other variables to arrive at the optimum control deflections. The flight control surfaces are then moved by actuators which are controlled by the electrical signals. The replacement of the conventional mechanical components with electrically transmitted signals along wires led to the name FBW. Of course, realization of the FBW system is not possible without the development of digital flight computers and microprocessors that enable a fail-safe flight control system to be implemented economically, safely and reliably.

The flight computers take in all the information including pilot's order, the aircraft's current state and its external environment, and move the control surfaces to follow the desired flight path while at the same time achieving a good handling quality and ensuring that the aeroplane is not over-stressed beyond its flight envelope. There are multiple computers for redundancy. Sophisticated voting and consolidation algorithms help to detect and isolate failures in the event of faults occurring in any of the actuators. Another advantage of digital FBW actuation is the faster control surface position feedback that significantly increases the actuation response speed. A fast FBW system response is crucial for keeping an aerodynamically unstable aeroplane from divergence. An extreme example of this is the Lockheed F117 Stealth Fighter, which could fly for only about one-tenth of a second if it were not for its FBW system. In commercial aircraft, FBW systems allow the use of smaller tail surfaces, with a consequent reduction in both weight and drag.

---

33) R.P.G. Collinson, *Introduction to Avionics Systems*, 2nd Edition, Kluwer Academic Publishers, 2003.

7.4
Airborne Software

While FBW systems are one of the most visible uses of on-board digital systems, airborne software is now used to control almost every function of both military and commercial aircraft. The Block 3 software for the Lockheed F35 Fighter is planned to have 8.6 million lines of code written in C and c++.[34],[35] This will be used to provide a complete fusion of the flight control systems with the battlefield awareness systems. The first use of airborne software in a commercial aircraft was the Litton LTN-51[36] Inertial Navigation System on the Boeing 707 in 1968.[37] Since then, its use has been growing rapidly with each new generation of commercial aircraft, as microprocessors with ever-increasing power have become available. Recently, two-and-a-half-million lines of code were newly developed for the Boeing 777 in the ADA language.[38],[39] Including commercial-off-the-shelf software, the B777 has more than four million lines of airborne software. The US Federal Aviation Administration (FAA) has developed standards for the certification of airborne software. Modern commercial aircraft feature loadable systems that can easily be replaced or upgraded. On the Boeing 777, these include systems such as the Electronic Engine Control (EEC), the Air Data Monitor (ADM), the Cargo Smoke Detector System (CSDS), the Primary Flight Computer (PFC), the Global Positioning System Sensor Unit (GPSSU) and the Satellite Communications (SATCOM) system.

Airborne collision avoidance systems are a particularly important example of airborne software. The current standard is the Traffic Alert and Collision Avoidance System (TCAS), although the Lincoln Laboratory at the Massachusetts Institute of Technology (MIT) in the US has been developing advanced algorithms during the past few years which have been incorporated in the new Airborne Collision Avoidance System X (ACAS X),[40] and the FAA is undertaking trials with the aim of making this the new standard.

---

34) c++ is a programming language that is general purpose, developed by Bjarne Stroustrup starting in 1979 at Bell Laboratories (formerly known as American Telephone & Telegraph, AT&T, Bell Laboratories). It was originally named C with Classes, adding object-orientated features, such as other enhancements to the general-purpose programming language C programming language developed by Dennis Ritchie between 1969 and 1973 at AT&T Bell Laboratories.
35) G. Warwick, *Flight Tests Of Next F-35 Block Underway*, Aviation Week and Space Technology, 2010.
36) Litton LTN-51 was an inertial navigation system developed by Litton Industries, now part of the Northrup Grumman Corporation: www.northropgrumman.com.
37) J.P. Potocki de Montalk, *Computer software in civil aircraft*, Microprocessors and Microsystems, Volume 17, Issue 1, p. 17, 1993.
38) ADA is a structured and object-orientated high-level computer programming language originally designed by a team led by Jean Ichbiah of CII Honeywell Bull (French company resulting from the merge between CII, Compagnie Internationale pour l'Informatique, and Honeywell Bull, now Bull: www.bull.com) under contract to the US DoD (Department of Defense) from 1977 to 1983. It was named after Ada Lovelace (1815–1852), who is credited as being the first computer programmer.
39) R.J. Pehrson, *Software development for the Boeing 777*, The Boeing Company, Technical Report, 1996.
40) M.J. Kochenderfer et al., *Next-Generation Airborne Collision Avoidance System*, Lincoln Laboratory Journal, Volume 19, N. 1, p. 17, 2012.

7.5
Ground-Based Computer Systems

Ground-based computers of very large capacity are now used to control every aspect of commercial aviation. The Air Traffic Control (ATC) system is heavily dependent on computers, as there are around 7000 aircraft in the air over the United States at times of peak traffic (see Figure 7.14). Today, computers are needed for safety, efficiency, and to enable increased capacity.

Computer systems are equally crucial to airline management and operations. As passengers, we have all experienced electronic reservation systems. The first such system, named 'SABRE' (Semi-Automated Business Research Environment), was a joint development of American Airlines and IBM. Following its introduction in 1964, other airlines soon followed suit such that today, each

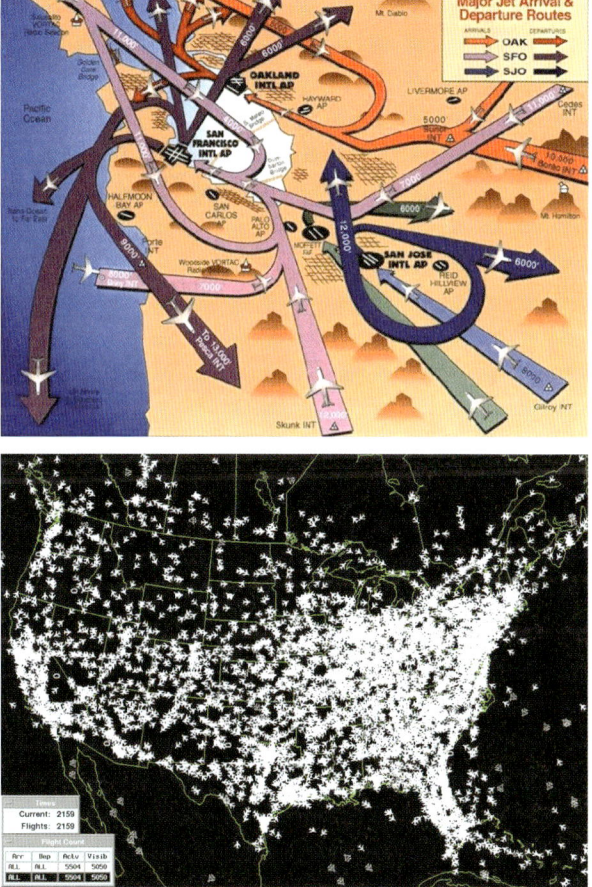

Figure 7.14 (a) Air traffic control chart; (b) Aircraft movement at peak traffic. (Courtesy of Federal Aviation Administration, FAA).

airline's computer reservation system interfaces with one of several Global Distribution Systems (GDSs). The major GDS providers are Amadeus,[41] Travelport,[42] and Sabre.[43] Computer systems are also used for online flight tracking. In addition, the airlines use yield management systems which adjust ticket prices from minute to minute, taking account of factors such as the number of unsold seats and the time to departure, with the aim of maximizing the revenue yield of each flight.

7.6
Conclusions

The external appearance of long-range commercial aircraft has changed very little during the past 50 years since the introduction of the first jet transports around 1960, and this reflects the qualitative understanding of the swept-wing design that had been achieved by aerospace engineers. The design process, however, has been completely revolutionized during the same period by the systematic use of computational simulation. Moreover, the role of information technology now extends well beyond the design and manufacturing processes to actual flight operations and management, through technologies such as digital FBW. Looking to the future, these trends will inevitably continue. According to the forecasts of Boeing and Airbus, air traffic is likely to continue growing at close to 5% per year for the next 20 years to more than double its current levels, with about twice as many aircraft in service. This will lead to increasingly severe environmental impacts in both emissions and community noise. Consequently, the European Union has announced an Aeronautics 2020 Vision which calls for:

- A 50% cut in CO_2 emission per passenger kilometre.
- A 80% cut in nitrogen oxide emission.
- A 50% cut in aeroplane drag.
- A 50% cut in perceived noise.

These targets are not likely to be realized without the pervasive use of advanced computational simulations. One major challenge is in aero-acoustics, which is paced by the demand to reduce the noise signature of both take-off and landing operations. The prediction of airframe noise due to high-lift systems and landing gear remains intractable with current computational methods, and will probably require a combination of high-order numerical algorithms with massively parallel computation at the exascale.

On the operational side, there is tremendous interest in Unmanned Air Vehicles (UVAs) for both military and civil applications. To date, the majority of

41) www.amadeus.com.
42) www.travelport.com.
43) www.sabre.com.

UAVs, such as the Predator drone, are remotely piloted by human operators based in ground stations, but in the future we will see an increasing use of autonomous UAVs capable of flying completely pre-programmed missions without human intervention. Autonomous UAVs can greatly reduce the cost of surveillance and remote sensing operations, and actually enable them to travel in inhospitable environments such as thunderstorms. Whilst it is not clear how soon passengers may be willing to fly in aircraft without pilots on board, the technology already exists for autonomous unmanned cargo operations, if the issues of the integration of UAVs into the air traffic control system can be satisfactorily resolved. In fact, unmanned operations may actually prove to be safer, given that pilot errors are one of the main causes of aeroplane crashes. The use of autonomous UAVs for customer deliveries is already being envisaged by companies such as Amazon. Overall, we can anticipate that the future will see an increasing penetration of autonomous UAVs into all aspects of aviation, including novel surveillance and transportation systems. The emergence of autonomous UAVs represents the ultimate fusion of the technologies of computing and flight. Such machines may ultimately be able to match the capabilities already achieved in Nature by insects and birds.

8
Antimatter Pushing Boundaries

Niels Madsen

Antimatter, normally the realm of science fiction, is one of the frontiers that CERN is pushing. The quote 'There is nothing new to be discovered in physics now', which has often been attributed to Lord Kelvin around the year 1900[1] – that is, before the discovery of quantum mechanics and relativity – well illustrates the dangers of underestimating apparently simple discrepancies. The Standard Model of physics, the great masterpiece of the twentieth century physics, is facing a number of small discrepancies that could have significant consequences. Antimatter is one of a number of promising means by which these issues are being tackled, the small problem with antimatter being that about half the Universe should be made of it – but effectively none is observed. Pushing the small problems in the early twentieth century led to a technological revolution driven by quantum mechanics that could not possibly have been predicted by contemporaries. Pushing today's frontiers is what CERN is all about, and antimatter plays a key role.

8.1
Science and the Unknown

While it is often purported that science is about the search for truth, it is more correct to say that science is the search for and the elimination of untruths. Along the way, temporary or partial truths are built up, such as *Newton's laws of gravitation*,[2] but it is in the nature of science to continuously question assumed truths and not to sweep anything under the carpet. The quote above, which is perhaps more correctly attributed to A.A. Michelson, who in 1894 remarked that in physics there were no more fundamental discoveries to be made, illustrates how wrong one can be when not sticking to the basic principle of scientific thought – which is precisely not about it being all over.

---

1) There is no evidence Lord Kelvin said this, but A.A. Michelson said something similar and seemed to allude to Lord Kelvin in 1894. See: L. Badash, *The completeness of nineteenth-century science*, Isis, Volume 63, p. 48, 1972.
2) I. Newton, *Philosophiae Naturalis Principia Mathematica*, London, 1687.

The current, possibly temporary, truth in our physical understanding of the Universe has led us to four fundamental forces of Nature to which (almost) all observed interactions can be attributed. One may split these fundamental forces into two categories. The three strongest forces, in descending order of strength, are the strong nuclear force, the electromagnetic force, and the weak nuclear force, and these are incorporated in what is referred to as the Standard Model.[3] The weak force and the strong force are those that dominate at the subatomic level, whereas the electromagnetic force governs the interaction between electrically charged particles. Beyond the Standard Model there is gravity, which regulates for example the movement of the planets. Gravity is the weakest of the known forces of Nature, something that remains a puzzle, as the current understanding is that the other forces, at high energies (i.e. in the early Universe), converge towards the same strength. Both our understanding of gravity, which is described by the *General Theory of Relativity*, and the Standard Model have celebrated great successes, though Gravity remains far less tested than the Standard Model due to its weakness. The Standard Model's latest success was the confirmation of the existence of a Higgs Boson at CERN,[4] announced on 4th July 2012.[5]

The success of our current understanding does not mean that we are done, to use the words of nineteenth century physicists. Significant issues remain, and as

3) The Standard Model includes all the known particles present in Nature, differentiating them by statistical properties and physical laws which they obey, into two families: *bosons* (which govern the interactions); and *fermions* (which make up matter). The latter are divided into two groups, *quarks* and *leptons*, with their respective antiparticles. The six types of quark are coupled in three generations: the lighter more-stable (*quark up, quark down*), the heavier less-stable (*quark charm, quark strange*), followed by the *quark top, quark bottom*. Quarks are electrically charged, and therefore are subjected to electromagnetic interactions; in Nature they are not isolated but are held together within the nucleus by strong interactions. There are six leptons, subdivided into three generations: the *electron*, the *muon* and the *tau*. To each of these three particles with electric charge and mass, is associated a neutral lepton called 'the *neutrino*': the *electron neutrino*, the *muon neutrino* and the *tau neutrino*. The leptons are elementary particles which are subject to weak interactions, electromagnetic interactions (with the exception of the neutrinos which, being electrically neutral, do not interact via electromagnetic interactions) and, like all objects having a mass, to gravitational interactions. In the Standard Model each fundamental interaction is described by a boson field and the boson carriers are the *quanta* of this field; however, in this theoretical framework massless particles are introduced, because otherwise the symmetry of the system would no longer be respected. To avoid this, the physicist Peter W. Higgs, together with François Englert and Robert Broût, speculated in 1964 that all space was permeated by a field (the *Higgs field*) that, interacting with the fields associated to the particles, would give them the right mass, thus creating a spontaneous breaking of the symmetry without altering the original one of the system.

4) ATLAS Collaboration, *Observation of a new particle in the search for the Standard Model Higgs boson with the ATLAS detector at the LHC*, Physics Letters B, Volume 716, Issue 1, p. 1, 2012; CMS Collaboration, *Observation of a new boson at a mass of 125 GeV with the CMS experiment at the LHC*, Physics Letters B, Volume 716, Issue 1, p. 30, 2012.

5) In 2013, the Nobel Prize in Physics was awarded to F. Englert and P.W. Higgs ' . . . for the theoretical discovery of a mechanism that contributes to our understanding of the origin of mass of subatomic particles, and which recently was confirmed through the discovery of the predicted fundamental particle, by the ATLAS and CMS experiments at CERN's Large Hadron Collider.': http://www.nobelprize.org/nobel_prizes/physics/laureates/2013.

Figure 8.1 The energy budget of the Universe. It is estimated that ordinary matter makes up only 4% of the Universe, while antimatter makes up about 0%.

with the issues that remained at the end of the nineteenth century, it is essentially impossible to predict what will be found and therefore what impact this may have on both our understanding of the Universe and on our everyday lives. To mention a few, we can start by highlighting the simple diagram in Figure 8.1 that shows how we believe the Universe is made up today.

Figure 8.1 highlights two main points. First, that more than 90% of the energy in the Universe is in the form of dark matter and dark energy, both of which are place holders for unknown particles, fields, or something else. They are both called 'dark' as they are invisible, unlike stars, and as we have yet to detect their interaction with anything but gravity. Furthermore, their interaction with gravity has thus far only been inferred through indirect means; that is, to make the movements of galaxies be self-consistent with our understanding of gravity. Thus, neither the Standard Model nor the *General Theory of Relativity* incorporates dark matter and energy at this point, and there is not enough information yet to say how this might happen nor anything about the practical impact from such understanding. Second, no bulk antimatter in the Universe is observed. The Standard Model predicts, with considerable success in the laboratory, that matter and antimatter are – to a large extent – mirror images of each other (in a metaphorical sense; see Section 8.2). This has been tested with some precision, and thus far holds sufficiently well to lead to the expectation that rather than the Universe all being made of ordinary matter, about half of the Universe should have consisted of antimatter, but it does not.

As was the case in late nineteenth century, we have a number of outstanding issues to deal with in physics today, of which two examples are given above. The issues that had to be addressed at the end of the nineteenth century eventually led to the discovery/invention of general relativity and quantum mechanics. Quantum mechanics is a cornerstone of materials science and has given us almost everything that we take for granted in our modern society today, from computers and lasers to advanced materials and chemicals, to technological breakthroughs that have brought us great leaps ahead in medicine.[6] General

---

6) B. Bressan (Ed.), *From Physics to Daily Life: Applications in Biology, Medicine, and Healthcare*, Wiley, 2014.

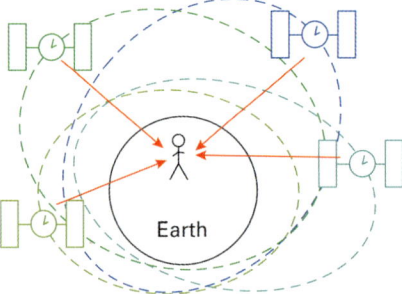

Figure 8.2 Schematic of the Global Positioning System (GPS). The GPS relies on very precise atomic clocks on satellites that transmit the time to a receiver on Earth. By knowing the satellites' orbits, receiving the times from four satellites will give the position. Atomic clocks build on decades of ever better measurements of atomic transitions, like those that are pursued on antihydrogen. Understanding General Relativity is necessary, as it explains e.g. how the clocks on satellites tick slightly faster than those at the surface.

relativity, while perhaps more abstract, has for example allowed us to have a functioning and relatively precise Global Positioning System (GPS), which would not have been possible by relying on the nineteenth century understanding of gravity (Figure 8.2). All of these developments have been a long time coming since the days of Niels Bohr and others in the early twentieth century, and it would be premature to say what developments may result from resolving the current outstanding issues. How antimatter fits into this program of understanding the basic nature of the Universe, and how CERN is pushing this edge, will be discussed in the following.

8.2
Antimatter and CERN

The visible Universe – the part we deal with directly in our everyday lives – is made up of relatively few fundamental building blocks. Protons, neutrons and electrons make up the bulk of what we see. For every such particle there exists – or one may say the possibility exists – of a so-called 'antiparticle' with the same mass but opposite charge. This curious fact was first realized by Paul Dirac during the late 1920s,[7] when he derived a version of quantum mechanics that took Einstein's theory of relativity into account. This was a key problem that needed solving at the time, and it is easy to see why; Rutherford, Bohr and others had realized that atoms are made up of a nucleus surrounded by orbiting electrons. However, these same electrons could easily find themselves orbiting with speeds approaching that of the speed of light, which was known to be the ultimate limit. Thus, in order to be able to describe these systems the effect of the speed limit

---

7) H.S. Kragh, *Dirac: A Scientific Biography*, Cambridge University Press, 1990.

had to be taken into account. It is in fact the relativistic effect that is the reason why gold is golden and not greyish, like most metals.[8]

After the postulated existence of such 'opposite' particles to the already known particles, the first observation came quickly by Anderson in 1933.[9],[10] Anderson discovered a so-called 'positive electron', that has been named the 'positron' and that is the antiparticle of the electron or antielectron. However, other antiparticles took a while to discover. The delay in discovering more types of antiparticles was due to a number of reasons, the first reason being that they are rare (as discussed above) and the Universe contains essentially none in stable conditions. The second reason was that even if they do appear, they will quickly disintegrate. This stems from a feature that gave these particles their name. When an antiparticle meets its particle 'twin' – that is, when a positron meets an electron – the two particles may disintegrate in a so-called 'annihilation'. Energy is conserved, so the energy they represent will be released, in this case in its purest form, namely that of electromagnetic waves or light. The energy can be calculated with Einstein's famous equation $E = mc^2$, where m is the mass of the particles and c^2 is the speed of light squared – a very large number. Thus, Einstein's equation gives us the exchange rate between energy and mass. Conversion between energy and mass is by no means a phenomenon unique to antimatter. In a nuclear reactor, the nuclei are split into lighter nuclei in a process called 'fission', and the energy released corresponds to the difference in mass of the fission products relative to the initial nuclei. In nuclear fission less than 1% of the mass is converted to energy whereas, in an annihilation, all of the mass is converted to energy. This brings us to a final reason for the length of time it took to discover more massive antiparticles, such as the antiproton that was discovered in 1955 at the Bevatron at Berkeley in California, US.[11] Antiparticles as well as particles may be created by converting energy to mass, thus the opposite process to the one just described. More massive particles will require more energy to be produced. The antiproton, which has a mass almost 2000-fold larger than the positron, therefore had to await the advent of accelerators, such as the Bevatron, to be created in the laboratory.

Thus, antimatter and matter can be created by converting energy to mass. One way of doing this is by accelerating particles to very high energies and colliding them with each other, or with a target material at rest. This is where CERN

---

8) N. Bartlett, *Relativistic Effects and the Chemistry of Gold*, Gold Bulletin, Volume 31, Issue 1, p. 22, 1998.

9) C.D. Anderson, *The Positive Electron*, Physical Review, Volume 43, Issue 6, p. 491, 1933.

10) The time from discovery to practical application has been much longer. The first mouse image acquired with Positron Emission Tomography (PET), using a small High-Density Avalanche Chamber (HIDAC), dates from 1977, while the prototype of the Advanced Rotating Tomograph (ART) scanner, the Partial Ring Tomograph (PRT), was also developed at CERN between 1989 and 1990. See: D.W. Townsend, *Detection and Imaging*, in B. Bressan (Ed.), From Physics to Daily Life: Applications in Biology, Medicine, and Healthcare, Chapter 4, p. 85, Wiley, 2014.

11) O. Chamberlain *et al.*, Physical Review, *Observation of Antiprotons*, Volume 100, Issue 3, p. 947, 1955.

comes into the picture. In fact, a key feature of the Large Hadron Collider (LHC) that accelerates particles to the highest energies is that it converts some of this energy to mass, creating for example the Higgs boson. Less will do to make antiprotons, but in the collisions of the LHC a large number of antiparticles are also created for brief moments until they decay or annihilate with surrounding matter. CERN serves a number of experiments that examine antimatter.

8.2.1
Antimatter at the LHC

At the LHC, protons are collided on protons with a nominal energy of each beam of 7 TeV. In such collisions a host of different particles and antiparticles are created, and the LHC was also constructed to create new more massive particles that could not be seen in previous machines, such as the Higgs boson.

However, as the available energy is increased in order to induce more massive particles or energy-demanding processes, lower energy, previously rare processes may also become more common. Some of these rare processes can shed light on small asymmetries between matter and antimatter. The Large Hadron Collider beauty (LHCb) experiment at the LHC is one example of an experiment that seeks such asymmetries specifically by examining how small fundamental particles called 'quarks' – some of which make up protons and neutrons – actually decay.

8.2.2
The CERN Antimatter Facility

There is another approach to studying antimatter that is also being pursued at CERN, that of precision measurements. Today, the heart of this effort is the Antiproton Decelerator (AD),[12] a unique machine that decelerates antiprotons to a low enough energy to be usable for a host of specialized experiments (Figure 8.3). The AD is the last in a long line of machines that has served to provide antiprotons to the CERN physics community. Initially, in the 1970s, antiprotons at CERN were collided with protons to create new particles, and this led to the discovery in 1981 of the Z and W bosons that carry the weak nuclear force.[13],[14] However, in 1982 CERN started the Low-Energy Antiproton Ring (LEAR),[15] a precursor to the AD, to decelerate antiprotons. It was at the LEAR facility that

---

12) S. Maury, *The Antiproton Decelerator: AD*, Hyperfine Interactions, Volume 109, Issue 1, p. 43, 1997.
13) The Nobel Prize in Physics 1984 was awarded jointly to Carlo Rubbia and Simon van der Meer ' . . . for their decisive contributions to the large project, which led to the discovery of the field particles W and Z, communicators of weak interaction': http://www.nobelprize.org/nobel_prizes/physics/laureates/1984.
14) P. Watkins, *Story of the W and Z*, Cambridge University Press, 1986.
15) R. Klapisch, *The LEAR Project and Physics with Low Energy Antiprotons at CERN (A Summary)*, Physica Scripta, Volume 5, N. T5, p. 140, 1983.

Figure 8.3 Part of the Antiproton Decelerator (AD) at CERN (Courtesy of CERN).

the first antihydrogen – the bound state of an antiproton and a positron – was made in 1995.[16] The low-energy antiproton facility involved three accelerators (LEAR plus the Antiproton Accumulator and the Antiproton Collector), and as CERN entered the LHC era and no longer used antiprotons at high energy, the antiproton complex was replaced by a single machine – the AD – that started delivering antiprotons for our physics community in 2000.

Precision measurements of atomic structure have for more than a century driven our understanding of atoms and quantum mechanics. This drive has led to the advent of atomic clocks, which now serve as the standards for timekeeping and that has given us the GPS. The best atomic clocks now have a precision of better than 1 part in 10^{18}, a precision which would be equivalent to measuring the distance to the Sun with a one-tenth of a micron precision.[17] As more precise measurements regularly lead to new discoveries, it has long been a dream to use these atomic physics tools on antimatter, and this dream is slowly coming true at CERN.

Precision measurements on antimatter to detect small differences between matter and antimatter take several forms at the CERN AD. The Antihydrogen Trap (ATRAP)[18] and Baryon Antibaryon Symmetry Experiment (BASE)[19] Collaborations are working with single antiprotons to detect minute variations in the magnetic moment (the small magnetic field) of the antiproton from that of the proton. Another group, Atomic Spectroscopy And Collisions Using Slow Antiprotons (ASACUSA)[20] is creating a bound state of a helium nucleus, an electron and an antiproton, which is also a sensitive probe of the antiproton magnetic moment as well as other antiproton parameters.

16) G. Baur *et al.*, *Production of Antihydrogen*, Physics Letters B, Volume 368, Issue 3, p. 251, 1996.
17) N. Hinkley *et al.*, *An Atomic Clock with 10^{-18} Instability*, Science, Volume 341, N. 1215, p. 1215, 2013.
18) http://home.web.cern.ch/about/experiments/atrap.
19) http://base.web.cern.ch.
20) http://asacusa.web.cern.ch.

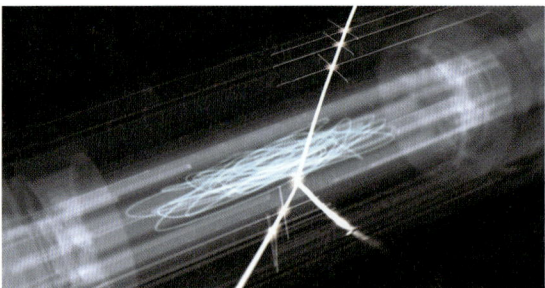

Figure 8.4 An artist's impression of an antiproton annihilation in the Antihydrogen Laser Physics Apparatus (ALPHA) experiment. The bright lines are the reconstructed tracks of the pions that result from an antiproton the annihilation. (Courtesy of Chukman So, ALPHA Collaboration, 2011).

The activity that has caught the most headlines, and which holds promise for the highest precision comparisons is from several groups, including the Antihydrogen Laser Physics Apparatus (ALPHA),[21] Antihydrogen Experiment: Gravity, Interferometry, Spectroscopy (AEGIS),[22] ASACUSA, ATRAP and Gravitational Behaviour of Antihydrogen at Rest (GBAR)[23] that are – or will be – making antihydrogen atoms; that is, atoms composed of an antiproton and a positron.

Some recent breakthroughs were made by the ALPHA Collaboration, which was the first to observe a resonant quantum transition in an antihydrogen atom, albeit without great precision at this early stage.[24] A separate (but potentially equally fruitful) endeavour is to compare the gravitational influence on hydrogen and antihydrogen. Due to the relative weakness of gravity, such measurements are essentially impossible with charged particles, and antihydrogen is therefore a perfect, pure and neutral antimatter candidate for such measurements. While it is not expected that antimatter behaves differently, the Standard Model does not incorporate gravity, and there is therefore only indirect evidence for how antimatter would behave. ALPHA recently made a first, very crude but direct, effort at detecting the gravitational influence on antihydrogen (Figure 8.4).[25]

The many successes of the antimatter facility and the very promising prospects for more precise measurements have led CERN to increase its investment in the facility. Currently, the facility is being upgraded with an additional small decelerator called 'ELENA' (Extra Low-Energy Antiprotons).[26]

21) http://alpha.web.cern.ch.
22) http://aegis.web.cern.ch.
23) gbar.web.cern.ch.
24) ALPHA Collaboration, *Resonant Quantum Transitions In Trapped Antihydrogen Atoms*, Nature, Volume 483, Issue 7390, p. 439, 2012.
25) The ALPHA Collaboration and A.E. Charman, *Description and first application of a new technique to measure the gravitational mass of antihydrogen*, Nature Communications, Volume 4, N. 1785, 2013.
26) W. Oelert et al., *AD performance and its extension towards ELENA*, Hyperfine Interactions, Volume 213, Issue 1–3, p. 227, 2012.

8.3
The Anti-World in Everyday Life

While antimatter research is an exciting field that is trying to answer some of our most basic questions about the Universe, antimatter has for a long time made its mark on science fiction. However, antimatter is not just food for science fiction authors; it also holds the key to solving some of today's – and perhaps some of the future's – problems. Without being exhaustive, a few of the current uses of antimatter in everyday life will be highlighted here.

The energy that would be released if we managed to completely annihilate 1 g of antimatter on 1 g of matter – that is, in total 2 grams of mass – is approximately 1.8×10^{14} Joules, or the equivalent of 42 kilotons of TNT (trinitrotoluene). This enormous amount of energy has inspired science fiction authors to use antimatter as an energy storage medium for interstellar travel. However, as there is no bulk antimatter in the Universe, it would all have to be made by hand, so for every Joule of energy we wish to take out we must supply at least the same amount of energy. For comparison, the total world energy supply in 2011 was 5×10^{20} Joule;[27] thus, with 100% conversion efficiency and doing nothing else, the entire human race could manage to make about 8 kg of antimatter per day. In practice there are many competing processes, so the loss in converting energy to mass is enormous. For example, when making antiprotons about one million high-energy protons are needed for each antiproton created. Taken together, these processes and the difficulty of storing antimatter render antimatter economically unviable and practically irrelevant as a bulk energy storage medium.

Yet, while larger amounts of antimatter are beyond our reach, small amounts already play an important role for some practical applications. When a positron annihilates on an electron the energy is typically released as two gamma-ray photons[28] (high-energy light). These gamma-rays will pass through the human body unhindered and can be identified by detectors that can reconstruct the path they took. By injecting a positron source into the human body that attaches itself to, for example, the red blood cells that transport oxygen, one may track where the red blood cells accumulate and, thus, identify the locations in the body with the largest oxygen consumption. This technique is known as Positron Emission Tomography (PET),[29] and is widely used in hospitals worldwide. The relative ease with which a positron-emitting radioactive source can be produced

---

27) IEA (International Energy Agency), *Key World Energy Statistics*: www.iea.org, 2013.
28) As unstable atoms decay, they release radiation in the form of electromagnetic waves and subatomic particles. Some forms of this radiation can detach electrons from, or ionize, other atoms as they pass through matter; this is referred to as ionizing radiation. Alpha- and beta-particles, X-rays and gamma-rays, are all forms of ionizing radiation. Gamma-ray photons refer to electromagnetic radiation of extremely high frequency and therefore high energy per photon.
29) D.W. Townsend *et al.*, *Positron Emission Tomography: Basic Sciences*, Springer-Verlag, 2003.

has consequently led to the positron becoming a workhorse of modern medical diagnostics.[30]

More recently, the technique has also been expanded to study live flows in particulate systems as an engineering aid; this procedure is referred to as Positron Emission Particle Tracking (PEPT) and was pioneered at the University of Birmingham, UK, during the 1990s.[31]

The relative ease of obtaining positrons and monitoring their annihilation has also driven the development of low-energy positron sources (often based on moderating positrons from a radioactive source) that are being used for studies of both inert and biological materials using Positron Annihilation Lifetime Spectroscopy (PALS).[32] PALS works by injecting low-energy positrons into a material and observing their lifetime. A positron may form a short-lived bound state with an electron, termed a 'Positronium' (Ps); the latter tend to seek out pores in a material where it will be bouncing until it eventually annihilates. The lifetime of a positron since its creation can therefore be used to elucidate information regarding the porosity of a material. More exotic uses of positrons and Ps have also been proposed, such as a gamma-ray laser.[33]

While positrons are relatively easy to obtain from radioactive sources, the larger mass of antiprotons requires higher energies and accelerator installations for their creation. Currently, CERN is one of the very few places in the world that provide a steady flow of antiprotons for experiments, and it is unique in providing low-energy antiprotons. Beyond the antiproton-based research discussed previously, antiprotons could also serve as a potential treatment for cancer. The irradiation of tumours is a standard component of the cancer therapy 'toolbox', where the most common types of radiation are X-rays and electron beams. However, heavier particles may be advantageous due to their loss profile when passing through biological material.[34] Heavier particles, such as protons or carbon nuclei that are used in more expensive (and therefore more scarce) facilities, will tend to have a limited penetration depth, close to which they deposit most of their kinetic energy, thus avoiding exposure of any living healthy tissue 'behind' the tumour.[35] Antiprotons will, until they annihilate, deposit energy almost like protons. However, the annihilation causes an added energy deposition that is fairly localized and therefore, potentially, may only influence the tumour.

---

30) D.W. Townsend, *Detection and Imaging*, in B. Bressan (Ed.), From Physics to Daily Life: Applications in Biology, Medicine, and Healthcare, Chapter 4, p. 85, Wiley, 2014.
31) D. J. Parker, et al., *Positron emission particle tracking - a technique for studying flow within engineering equipment*, Nuclear Instruments and Methods in Physics Research, Section A: Accelerators, Spectrometers, Detectors and Associated Equipment, Volume 326, Issue 3, p. 592, 1993.
32) J. Calloo, *Characterizing Defects in Metals Using PALS*, Lambert Academic Publishing, 2012.
33) D. Shiga, *How to build a gamma-ray laser with antimatter hybrid*, New Scientist, Volume 212, Issue 2844, p. 6, 2011.
34) W.R. Hendee et al., *Radiation Therapy Physics*, Wiley-Liss, 2004.
35) U. Amaldi, *Particle Beams for Cancer*, in B. Bressan (Ed.), From Physics to Daily Life: Applications in Biology, Medicine, and Healthcare, Chapter 3, p. 61, Wiley, 2014.

Measurements to characterize the influence of antiprotons on biological materials are also part of the CERN antimatter program, with a group called 'ACE' (Antiproton Cell Experiment).[36] When the first biological samples were irradiated in 2003 they showed interesting results; however, until cheaper and smaller antiproton facilities can be produced, the limited benefit from using antiprotons for cancer therapy will be swamped by the increased costs of creating antiproton facilities relative to the cost of carbon ion and proton facilities.

8.4 Beyond the Present Day

While antimatter is unlikely to become a household item in the foreseeable future, the applications discussed here demonstrate how fundamental research also generates wealth for society through serendipity. The discovery of the positron and other antimatter particles was motivated solely by the desire to understand and describe the world around us to the best of our ability. It is – and always has been – very difficult, if not impossible, to predict the future, and we will not fall into the trap here and try to do so. Science is no exception to this rule, as has been amply demonstrated.[37]

However, we have brought up a number of parallels – perhaps exaggerated – to the late nineteenth century, where seemingly good descriptions existed of most known physical phenomena, with only a few remaining to be clarified. This should serve as an inspiration for the present day where we have an almost complete Standard Model, but also a number of unexplained phenomena. The precision comparisons of matter and antimatter that are being pursued at CERN may help to elucidate some of these remaining unknowns. There is currently no reason to believe that matter and antimatter should be different in such a way that antihydrogen would 'fall' up, or that the spectrum of antihydrogen should be different from that of hydrogen. However, the effort involved in investigating these assumptions is outweighed by the momentous impact on physics that would result from any such difference being observed.

Seeing how quantum mechanics has had a finger in an uncountable number of technological breakthroughs since its discovery, probing the foundations of the theory using antimatter could potentially lead to truly groundbreaking discoveries and breakthroughs would follow thereof. Antimatter research is, therefore, a potential new vehicle by which the pursuit of ever more precise measurements will pay off to both science and society. This was perhaps the original point of A.A. Michelson's Kelvin quotation that said ' . . .our future discoveries must be looked for in the sixth place of decimals'.[38]

---

36) http://home.web.cern.ch/about/experiments/ace.
37) M. Irvine and B.R. Martin, *Foresight in science: picking the winners*, Frances Pinter, 1984.
38) A.A. Michelson, *Speech at the dedication of Ryerson Physics Laboratory*, University of Chicago, US, 1894.

Section 3
Sustainability and Learning

There are no limits to the infinite possibilities offered by the 'wonderful land' of technology. On all levels, technology represents the injection of science into daily reality.

Thanks to the technologies developed for the purpose of their research activities, research laboratories such as CERN have not only reduced the existing technological differences amongst the main developed countries, but have also benefited humanity, thanks to their scientists finding better solutions to technical and scientific problems.

Through technology, research has not only increased our knowledge and learning but also succeeded in making our daily environment more functional, practical and comfortable. Scientists know what Winston Churchill meant when he said: 'Success is never final, and failure is never fatal. It is the courage to continue that counts.' And so science goes on . . .

9
Towards a Globally Focussed Earth Simulation Centre
Robert Bishop

In the twenty-first century, humanity must grapple with the greatest convergence of global challenges in our history – the difficulty of living in balance with our fast-changing planet. Now, more than ever, as society comes face-to-face with the multiple situations of climate change, resource depletion, economic development and human security, a leadership capability is needed that tackles head-on the full complexity of the Earth and its many coupled systems.

Policy makers, community planners, emergency responders and scientists need to have at their disposal the expert assistance of an advanced capability Earth Simulation Centre that is able to operate at a global level. Such a Centre would assist in the analysis of global climate and environmental developments, and enable policy guidance in the case of future large-scale shocks and emergencies. It would complement local and national planning and prediction capabilities by offering an integrated global picture. As such, the Centre would be an invaluable tool for decision support while improving science-based policy making around the world.

The proposed Centre would include a fully dedicated high-performance computing capability and an appropriate complement of expert technical and scientific staff. The Centre would be capable of assimilating all available world data relating to the dynamic state of the planet, and providing risk analysis for human security. It would be heavily networked to relevant local, national and regional facilities, and be capable of accessing various Citizen Science networks that have recently entered the scene.

A centre of this nature would best be established for public good and created as a public private partnership that works closely with agencies of the United Nations, national governments, Non-Governmental Organizations and private corporations, so as to provide comprehensive insights into planetary change. It would be reasonable to allow the proposed Centre to generate operational income from fee-based services, while also receiving in-kind contributions from its partner organizations and from philanthropic sources.

From Physics to Daily Life: Applications in Informatics, Energy, and Environment, First Edition.
Edited by Beatrice Bressan.
© 2014 Wiley-VCH Verlag GmbH & Co. KGaA. Published 2014 by Wiley-VCH Verlag GmbH & Co. KGaA.

9.1
A String of Disasters

Open a newspaper today and you are likely to discover that another earthquake has struck somewhere in the world, that threatening wildfires have consumed precious forests and spread beyond control, or that a new '100-Year Storm' is inundating some region, causing flash floods and mudslides. Events like these are more likely in the present century, due to both natural variations of climate as well as the accumulated effects of our encroaching civilization, altered landscapes, burgeoning industry and a corresponding global warming.

Indeed, we can expect these events to be more impactful in future times than they were in the past – it was once the case that smaller and more resilient communities could more easily avoid harmful impacts when disaster struck – but not anymore, for in a highly urbanized world of multiple mega-cities, and with the world's population mostly clustered along coastlines, waterways and fault-lines, the geography of risk has profoundly changed.

A brief survey of the last decade is enough to cause alarm – earthquakes in Haiti, Italy, Turkey, Chile, New Zealand and Japan; tsunamis in both the Indian and Pacific Oceans; the Fukushima nuclear reactor meltdown; tornadoes across the Central and Southern United States; wildfires in Australia, California, Chile and Europe; multiyear droughts in the US Southwest, Mexico and the Horn of Africa; heat waves in the US, Russia and Europe; massive flooding in Asia, Africa and Latin America; disruption of international travel by Icelandic, Russian, Mexican and Chilean volcanoes; the list goes on.

In fact, we have come through a string of recent disasters that were neither accurately predicted, nor properly assessed for their severity – mostly because of the limited forecasting tools available, and our embryonic understanding of how Nature's various systems work in unison. Yet, the frequency of such harmful events will most likely increase in the decades ahead, not decrease. And any rise of sea level around our coastlines will most certainly exacerbate this situation.

The challenges associated with the management of natural disasters are compounded by the intensification of human activities in general, and their largely unforeseen impacts on the planet. Groundwater supplies are drawing short as demand for food production, sanitation and manufacturing increases. Deforestation and urban development continue to alter the global cycling of water, nutrients, and energy – disrupting natural balances that were in place for millennia before the age of industrialization began.

The need for detailed knowledge of planetary change has never been greater than it is today. An international centre would address this critical need for better and more integrated information, analysis and simulation, and for dealing with the difficult task of decision-making in these complex and increasingly turbulent times. Rivers, oceans, weather systems and climates cross national boundaries effortlessly, and so must our thinking (Figure 9.1).

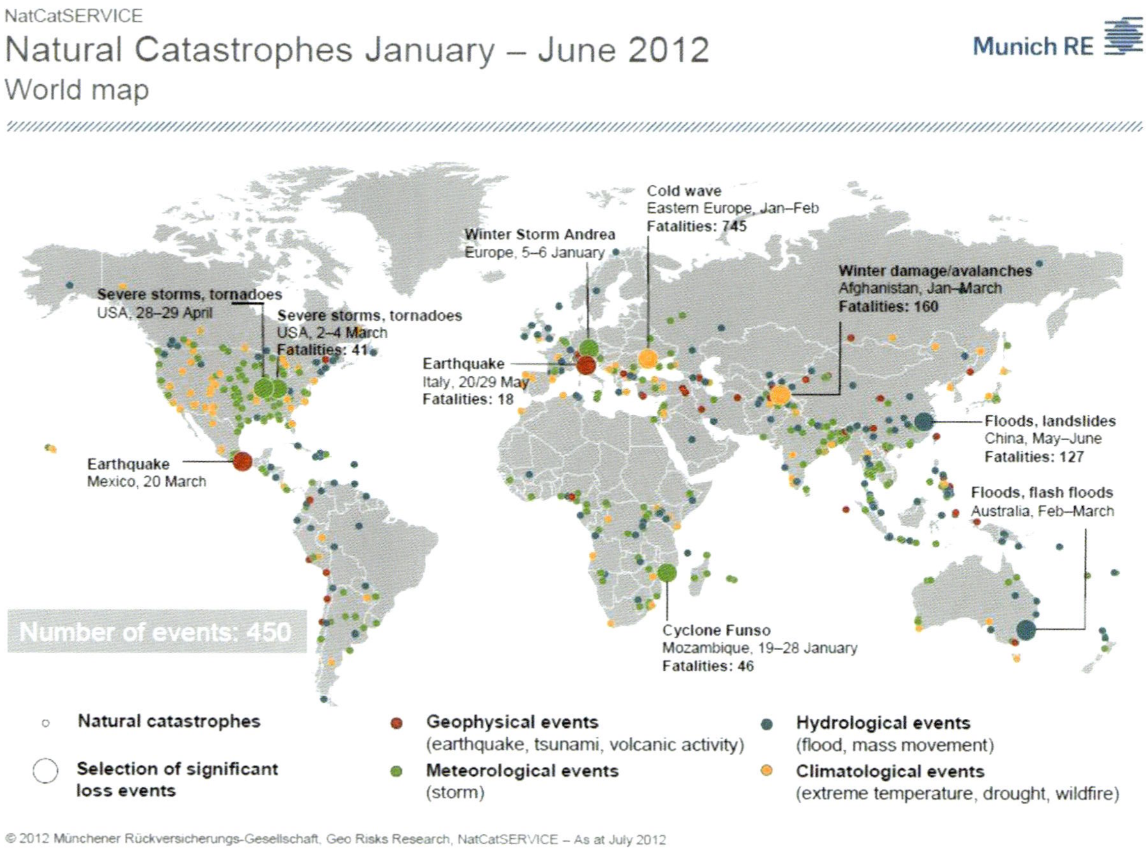

9.1 A String of Disasters | 175

Figure 9.1 World map: Natural Catastrophes, January–June 2012. (Munich Reinsurance Company, Geo Risks Research, NatCatSERVICE, 2014)[1]

1) http://www.munichre.com/app_pages/www/@res/pdf/media_relations/press_releases/2012/2012_07_13_worldmap_natcat_en.pdf.

9.2
Now is the Time

We stand at the nexus of possibilities. Just when the need for synthesis is greatest, the essential elements that can lead to powerful new insights and breakthroughs are within reach. Recent decades have seen the emergence of innovative new tools and sensor networks that weave together the strengths of specialized disciplinary knowledge – climate science, environmental studies, geophysics and public health, for example. Furthermore, giant leaps in the power of computing, mathematical modelling capabilities, data collection and analysis methods have all enabled diverse research communities of the world to run numerical simulations of the most complex processes in their respective domains, with higher and higher accuracy.

And yet the pinnacle of these technologies is not currently applied to the systemic threats of planetary change because top-tier supercomputers have thus far been dedicated primarily to the industrial–military complex, while the greatest threats to civilization of all – namely, climate change, natural disasters, environmental contamination, ecological collapse (together with the likelihood of social chaos) – have not received the dedication of resources that such a predicament demands. A quick glance at the Top500 Supercomputing Sites[2] worldwide shows that none of the ten most powerful machines is dedicated to the Earth Sciences.

Without doubt, our present-day world is in immediate need of a continuously updated, fully dedicated, leading-edge supercomputer meta-hub facility that can coordinate the global view and bring together the complex simulations and sciences of the dynamic planet on which we live.

Indeed, the power of simulation-based research within the industrial–aerospace–military complex has been well demonstrated over the past decades, and also proven itself as a potent tool within the scientific research community in general. Now is surely the time to stand up such a facility for the world at large, and tackle the most difficult simulation of them all – that of modelling the entire Earth System.

9.3
A Global Synthesis of Knowledge

The last two centuries are a story of knowledge specialization and splintering. Today, nearly one hundred separate research fields grapple with different aspects of our Earth System – each with their own academic departments, peer-review processes, professional journals, and research conferences.

Researchers, scholars and government officials must overcome many obstacles to keep abreast of developments in neighbouring fields. Insights are quite often

---

2) www.top500.org.

funnelled through parallel stovepipes and silos, with very few major institutions focusing on the Earth as a whole; knowledge integration is not an agenda item! As a result, many great advances of the last century have come at the expense of the Whole.

Today, no institutional centre exists that has the capability to model the entire Earth System. We have numerical models describing everything from solar winds in outer space down to models describing the Earth's inner core, with plate tectonics, land usage, atmospheric circulation, ice ages and ocean currents in between. But there is no robust Earth Simulation Centre that brings all of the many fragments together on a global scale.

There was indeed an early attempt by Japan to fill this void with its Yokohama Earth Simulator – launched in 2002 and still active today. It was a great first attempt, making history by using the most powerful supercomputer of the day (an NEC SX-6),[3] with its entire capacity dedicated to planetary dynamics. Sadly however, even though this machine was upgraded in performance by a factor of 3 in 2009, it has nevertheless slipped from first to the 472nd position in the November 2013 edition of Top500 Supercomputing Sites of the world, and therefore struggles to remain competitive at a time of severe Japan governmental funding challenges.

And here we should point out that knowledge synthesis delivers more than the sum of existing fragments of knowledge. So-called 'edge effects' that are found at the boundaries of knowledge domains are fertile ground for both breakthrough insights and new discoveries. There is much more to learn about the complexities of the total Earth System than we can imagine today, all lying in wait for us, but which will only be revealed through the application of advanced modelling methods, enhanced compute power, and extensive efforts to integrate the ever-expanding list of speciality sciences.

9.4
Modelling and Simulation as a Platform for Collaboration

For 50 years or more, computer-based simulation has proven to be a robust approach for the study of complex dynamic systems in industrial and scientific research. It offers a powerful extension beyond what theoretical and experimental methods alone are able to deliver. Simulation is now commonly used to explore the time-evolution of complex systems where experimentation is not feasible or, in many cases, where it is simply unsafe.

The formation of galaxies, solar systems, stars and planets, for example, cannot be studied experimentally – because such systems form on spatial and temporal scales that are far too large for experimentation.

---

[3] The SX-6 is a supercomputer built by the Japanese multinational NEC Corporation (known as Nippon Electric Company, Limited until 1983).

Likewise, geoengineering is an area of speculative science and engineering, with unfortunately, great potential for serious unintended consequences. Numerical simulations however, would enable such ideas to be thoroughly and safely investigated in minute detail, and with multiple experts looking on, either at their local desktop level or in a large-scale auditorium.

Today's advanced simulation methods also provide an effective platform for collaboration between specialists in differing areas of expertise – in fact, it is one of the few methodologies with a track record of success for bringing together the deep background knowledge from different fields of specialization.

Numerical models could certainly be used to combine global warming, flood patterns, soil erosion, landslides and load shifting, for example, to examine the combined effect they may potentially have for triggering earthquakes – or conversely, for earthquakes to trigger landslides and soil erosion. Normally intractable scenarios such as these would indeed benefit from the comprehensive integration of multiple complex processes into a single analysis model. The point being that modelling, simulation and visualization can provide the methodological platform to achieve major advances in very difficult areas.

Obviously, the proposed Centre will need to provide compute cycles in the case of extra-territorial disaster management and response, as is often needed in the case of global shocks and emergencies – the demand for international collaboration being considerably greater when such cross-border disaster strikes. It is during these times that the nations of the world would benefit most from a dedicated analysis and prediction centre with global capabilities.

9.5
Advances in High-Performance Computing

Because of *Moore's law*,[4] technological advances in the semiconductor industry produce an approximate 1000-fold improvement every decade in the price-to-performance ratio for computing hardware. Consequently, the highest performing machines today are a million times more powerful than the best machines built just 20 years ago. And for the same reason, machines that will be built 20 years hence most likely will have a performance level one million times faster than today's best machines, assuming that *Moore's law* holds up for the future 20-year period.

Today's top computers achieve multi petaflops[5] performance – they can perform trillions of calculations every second – allowing such computer systems to

---

4) *Moore's law* (named after Intel co-founder Gordon E. Moore) is the observation that, over the history of computing hardware, the number of transistors on integrated circuits doubles approximately every two years.

5) In computing, FLOPS (FLoating-point Operations Per Second) is a measure of computer performance, useful in fields of scientific calculations that make heavy use of floating-point calculations. For such cases it is a more accurate measure than the generic instructions per second.

conquer a number of previously intractable problems by deploying numerical modelling and simulation techniques.

The complex dynamics of ecosystems, atmospheric and oceanic circulation, as well as many of the complex aggregation processes involved in the formation of galaxies, stars and planets for example, have been successfully simulated on present day high-performance computers.

High-performance computing facilities are very expensive to maintain, however, and only a few governments can afford to invest in the infrastructure and expertise necessary to attempt simulations of the Earth on a global scale – with US, Japan, China, Germany, France, UK, Italy, Spain, Switzerland, Russia, Australia and India being exceptions to the rule. But the complexity of global simulation is so great that even these governmental efforts are very limited in scope.

It is clear that the only way to move forward so as to understand and successfully predict global-scale Earth System and socioecological threats is to pool resources globally and to thoroughly collaborate our research efforts across national borders. Just as CERN was expensive and above the budget and knowledge limits of any single national government to build and maintain, the same *raison d'être* applies for having a single large-scale globally chartered Earth System Simulation Centre capable of staying at the crest of all the component technology developments and acting as the meta-node[6] of a global network of collaborating institutes. In addition, just as CERN hosts an expert community of technical talent, so must the proposed Centre provide a resource pool available for assisting the various national scientific communities to tackle the most complex and difficult aspects of their specific local situations.

In summary, the main goal of the proposed Centre would be to build a supercomputing facility, annually ranked in the top ten machines of the world, and dedicated to combining all sciences of the Earth System: a global centre that complements the limited capabilities currently available at the local, national and regional levels, and a facility operating under open access, not-for-profit principles. The simultaneous threats of climate destabilization, frequent natural disasters and impending resource depletion surely demand the very best response from our prosperous civilization that can be mustered.

9.6
Creating Value from Massive Data Pools

We live today in an information age, but we can easily be overwhelmed by the deluge of data around us. Smart devices such as mobile phones, laptops and tablets proliferate and have rapidly become ubiquitous. Indeed, Internet access is essentially available to all and, as a consequence, many data streams have become truly astronomical in scale to feed this community of users – storing and forwarding petabytes and even exabytes of data per year.

---

6) Meta-nodes are nodes that contain subworkflows.

Emerging from this data-rich environment is the science of Big Data analytics – that is, the practice of mining huge volumes of data to reveal important patterns and insights. Of particular interest to this proposal are the pools of data recorded minute-by-minute by sensor networks[7] that monitor weather, ocean currents, seismic activity and many other aspects of our planet's behaviour, since every nation gathers information about the Earth in some fashion, either from *in-situ*, ocean-based or space-based devices.

The existence of such data pools is critically important to the proposed Centre, as the Centre will require access to such data pools for use as input to its modelling and simulation efforts. Toward this end, the efforts of all national governments to build portals that make their data archives open access and freely available, is strongly supported.

By using data of global scope we will already achieve substantial improvements in the forecasting of local extreme weather. Severe weather is known for wreaking havoc in many areas of the world, at very high cost to life and limb. The cost of building more accurate weather modelling and forecasting capability pales in comparison to the damages caused by unforeseen weather events. Improvements in storm track forecasting have a rapid return-on-investment, but they will also require access to the full panoply of world data, not just local data.

For example, the cost of hurricane evacuation on the US Atlantic seaboard exceeds US$ 1 million per coastal mile, or US$ 100 million in the case of a hurricane that cannot be predicted to come ashore within 100 miles. A 50% improvement in forecast accuracy would lower this cost by US$ 50 million, provided that it could be accomplished in a timely manner; this would be enough savings to provide for the cost of High-Performance Computing (HPC) equipment in a single event and, more importantly, saving lives along the way. The economic impact of only a few such events would make the proposed Centre a highly desirable investment.

In addition to Big Data analytics, the proposed Centre would also be provisioned to supply HPC consultation and compute cycles to developing nations and regions which do not have the necessary resources themselves – devoting as much as 25% of its total computing capacity to less-developed nations that cannot afford research facilities of their own.

9.7
Interactive and Immersive 4D Visualizations

Critical to Big Data analytics is the ability to extract, manipulate and visualize key information under investigation. Policy makers, elected leaders, journalists and, indeed, the general public will need compelling visual narratives that convey

---

[7] Sensor networks are dense wireless networks of small, low-cost sensors, which collect and disseminate environmental data. Wireless sensor networks facilitate monitoring and controlling of physical environments from remote locations with better accuracy.

important insights about our changing planet in order to convince themselves to take action at the local level. Researchers will also need increasingly sophisticated visual representations of the physical processes they are studying, especially as model complexity increases and more processes of Mother Nature are introduced.

One important advantage of visualization-based analysis is that computer simulation output can be presented as crisp pictures for every time-step in a key process. Although the brute force of numerical computation for validation purposes of the underlying physics and chemistry will always prevail, visual output enables the human mind to understand Nature's fundamentals intuitively and more thoroughly.

Furthermore, the concept of 'visual computing' allows any number of viewers to explore virtual worlds with precision and rigour, either locally or remotely, either individually or in an auditorium setting with others. In this respect, we have all witnessed the remarkable comprehension that Google Earth,[8] IMAX[9] theatre and daily-televised weather reporting has brought to the public at large.

Even in the arena of children's entertainment, modern video consoles demonstrate just how powerful immersive and interactive visualization can be. The image realism rendered within game worlds such as SimCity[10] and Second Life[11] is an essential ingredient for meaningful engagement today – and cultivates millions of young players, as a result. Similarly, we believe that when scientific simulation is combined with photo-realistic imagery in Earth Sciences, it will encourage a wider community of individuals to explore the dynamic Earth in ways that go beyond what is possible today. We all understand that seeing is believing and, in this case, we believe it will specifically open the gate to broad and constructive civic engagement.

For all of these reasons, visual simulation is now routinely employed as a professional training tool, from flight simulation for commercial pilots to surgical simulation for medical specialists.

Virtual worlds allow people to learn from their mistakes in a low-impact and safe environment before moving on to the real world. In a similar way, the proposed Centre would emphasize the importance and training value of intuitive, interactive and immersive visualization for understanding complex socio-economic systems, and exploring dangerous scenarios therein.

This 'practice' capability could easily be delivered over the Internet in the case of emergencies. Decision-makers and first-responders alike would benefit from interactive simulations of global or localized scenarios, providing them with the

---

8) www.google.com/earth.
9) IMAX (Image Maximum) is a motion picture film format and a set of cinema projection standards created by the Canadian company IMAX Corporation: www.imax.com.
10) SimCity is an open-ended city-building computer and console video game series originally designed by developer Will Wright.
11) Second Life is an online virtual world developed by the American company Linden Research, Inc., doing business as Linden Lab: www.lindenlab.com.

ability to minimize risks and avoid potentially catastrophic harm at effectively zero cost in a virtual world.

9.8
Leveraging the Many Layers of Computing

Supercomputing is regarded as the pinnacle of digital computing capability, and clearly the proposed Centre would make extensive use of it to accomplish its mission. Yet, there is a hierarchy of machine power leading up to this level, all of which may find relevance in the Centre's total operations.

The next layer below supercomputing is Grid computing, where several machines of lesser performance are networked together, and where an overlaying software architecture allows users to break their work into multiple segments, and then scatter the segments over the distributed machine network for execution.

If the machine network can be metered for usage and the various costs charged out to individual users, this is often labelled 'Cloud computing'. Recent advances in 'virtualization' and storage access have led to the widespread adoption of Cloud computing in the business world, and this same method is now finding its way into the world of science.

Grid computing or Cloud computing will work nicely if the calculations at hand do not require intensive communications between the scattered software segments; however, such methods will experience serious bottlenecks when references across the network become intense and frequent.

Attempting to calculate the vast array of parallel and contemporaneous physical processes of a full Earth System simulation running on a distributed computing network will obviously suffer such bottlenecks, and the corresponding latency effects will negatively impact the availability of final results. As such, Grid and Cloud computing are less capable of delivering on-time critical simulations compared to a single consolidated system of equal horsepower – which is the hallmark strength of supercomputing.

At a level below such Grid and Cloud computing we have the Internet itself, and below that, individual desk-tops, laptops, tablets and smart phones, most of which today can be connected to the Internet, and which can be viewed as access devices for downloading data from 'the Cloud', or sending data up to it.

It would be ideal for the proposed Centre to use whatever level of the computing hierarchy works best, decided case-by-case, for each of its component software developments. However, when it comes to modelling, simulating and visualizing the most complex of its global challenges, there is no doubt that communication considerations between each of the multiple job segments will force the work onto a centralized supercomputer that is fully dedicated and able to produce high-resolution results in a usable timeframe.

Distributing results from the supercomputer to others will be possible via the public Internet at all times, sometimes with the intermediate support of Cloud computing. Through such a digital infrastructure, it will be possible for clients to access visual simulations via the Internet that are both immersive and interactive, and in this manner the proposed Centre will make use of the many layers of computing at its disposal.

In the case of accessing data files essential to the modelling and simulation efforts of the Centre and that are scattered around the globe in various locations and which are managed by various organizations and institutions, such files can be federated and made accessible to the Centre over the Internet by using a consolidating portal and invoking ontologies to translate between differing file formats and metadata conventions.

9.9
Getting a Complete Picture of the Whole Earth

When looked upon from space, the Earth is seamlessly integrated – a borderless swirl of land mass and oceans – yet our knowledge about the Earth is neither seamless, nor borderless, nor integrated (Figure 9.2). To better understand it, science has sliced the Earth into various time-periods and subsystems. Three centuries of reductionist thinking has advanced by separating the world into manageable pieces – a process of disintegration that has resulted in the academic silos and stovepipes we witness today.

A brief survey of present-day research fields illustrates the fragmented picture: the solid globe beneath our feet is categorized into the elementary layers of inner core, outer core, mantle, crust, asthenosphere and lithosphere. The special

Figure 9.2 NASA's Blue Marble: the Earth, as seen from a height of 45 000 km by the spacecraft Apollo 17, December 7th, 1972 at 10:39 UTC (Coordinated Universal Time).[12]

---

12) A.G. Petsko, *The blue marble*, Genome Biology, Volume 12, Issue 4, p. 1, 2011.

phenomena within these layers then become subfields, such as fault lines, subduction zones, geysers and glaciers, to name but a few. Likewise, in atmospheric science we have the elementary layers of troposphere, stratosphere, mesosphere, thermosphere, exosphere, ionosphere and plasmasphere, followed by the recognition of special phenomena such as radiative transfer, aerosol chemistry, cloud formation, convection, and so on – all of this just a small sampling of the complex research fields and subfields that make the study of the Earth System tractable today.

Were we to also segment the oceans (which occupy more than 70% of the Earth's surface), or to divide up the Earth's biological systems along similar lines, the number of research fields would explode exponentially and would reach well over 100 independent, but somewhat overlapping, domains. Thus, the intrinsic complexity of the Earth is manifest by this vast array of specialized research areas, which has been formed to cope with the intricacies of the Earth's holistic function.

Yet even this is insufficient, for another dimension of knowledge organization is needed to speak of cyclic effects and the time dynamics of most processes. There is a *hydrological cycle* for clouds and rain, rivers and oceans, plant respiration, and erosion of the land, for example. *Carbon and nitrogen cycles* combine geology with life – tracking the profoundly malleable atoms chemically embodied in all living things, while also constituting the various forms of air pollution, atmospheric heat-trapping and rock formation. So many other processes on Earth entail atmospheric circulation and seasonal change. And at an even larger scale, we have the *Milankovitch cycles*, describing the orbital parameters of the Earth–Sun System which dictate the coming and going of ice ages.

In order to understand the multitudinal aspects of our changing Earth – a prerequisite for civilization to become sustainable – many specialized fields and subfields must be brought back together into an integrated whole, and this is indeed the first and key priority of the proposed Centre.

Reintegrating multiple sciences however, is no easy matter, because the multiple specialized sciences have very often evolved in isolation to each other and have frequently deployed differing branches of mathematics, differing time-steps, and differing boundary conditions. The need to build coupling algorithms that successfully bridge such disparities is an essential requirement of the Centre, and this will certainly demand very high-level mathematical talent.

9.10
Influence of the Solar System

It should be pointed out that any effort to model and simulate the Whole Earth System will need to include the physics of large-scale cosmological influences. The Earth is not a closed system, for it is constantly bombarded by radiation and high-energy particles from the Sun. This activity warms the planet, feeds the photosynthetic process, and alters the chemistry of our atmosphere.

Just as studies of the Earth have traditionally been fragmented into multiple domains, so has the study of the Solar System in which the Earth is embedded. Embracing the complexities of our changing planet requires that we include the physics of relevant cosmological forces in all of our simulation studies. State-of-the-art climate models in use today do not include many of the dynamic interactions that actually occur between Earth and Sun, the Solar Wind, for example. The Sun is often simplified in these models as virtually a static and constant energy source, despite the fact that Coronal Mass Ejections (CMEs)[13] from its surface routinely collide and interact with the upper atmosphere of the Earth and interrupt its chemistry as well as the magnetic field that shields all land-based life from deadly radiation and high-energy particles.

On longer time scales, the *many body*[14] gravitational effects of our neighbouring planets produce both tilting of the Earth's axis, as well as variations in Earth–Sun orbital distance, which in turn together produce both ice ages and warmer periods on Earth. Such factors are often included in advanced astrophysics simulations, but remain yet to be integrated into computational modelling of the Earth System itself. In addition to this omission, it should be noted that our Sun is in constant and violent flux, and that the entire 'Space Weather' phenomenon must be taken into account as to its effects on Earth, if we want to arrive at accurate future projections.

A multitude of less-identified space objects constitutes a further subject for inclusion. Indeed, the prevalence of comets, asteroids and meteorites in our Solar System regularly threatens the home planet; and even our own self-created 'space junk' from 50 years of national space programs is a continuous hazard for all of society's communication satellites and future space projects. One theory even has it that the age of dinosaurs was brought to an end by a massive meteorite that struck the tip of the Yucatan peninsula some 65 million years ago.

Clearly, a comprehensive study of Earth Systems requires us to bring all of these considerations together, and to combine a much wider plethora of interacting factors than are currently embraced. The proposed Centre must necessarily incorporate solar physics and astronomical phenomena into its modelling to be sure that the associated risks are fully quantified. Taking an even broader perspective, the Centre would include inter-galactic cosmic rays that emanate from distant supernovae explosions and that also impinge upon Earth, since they too are part of the overall physical forces that determine the ultimate evolution of life on our planet. Figure 9.3 illustrates the step-by-step additional features that have been sequentially integrated into Earth System Models over a recent 30-year period.

---

13) A Coronal Mass Ejection (CME) is a massive burst of solar wind (a stream of charged particles, 'plasma', released from the upper atmosphere of the Sun) and magnetic fields rising above the solar corona (a type of plasma that surrounds the Sun) or being released into space.

14) The many body theory is an area of physics which provides the framework for understanding the collective behaviour of vast assemblies of interacting bodies. In general terms, this theory deals with effects only in systems containing large numbers of constituents.

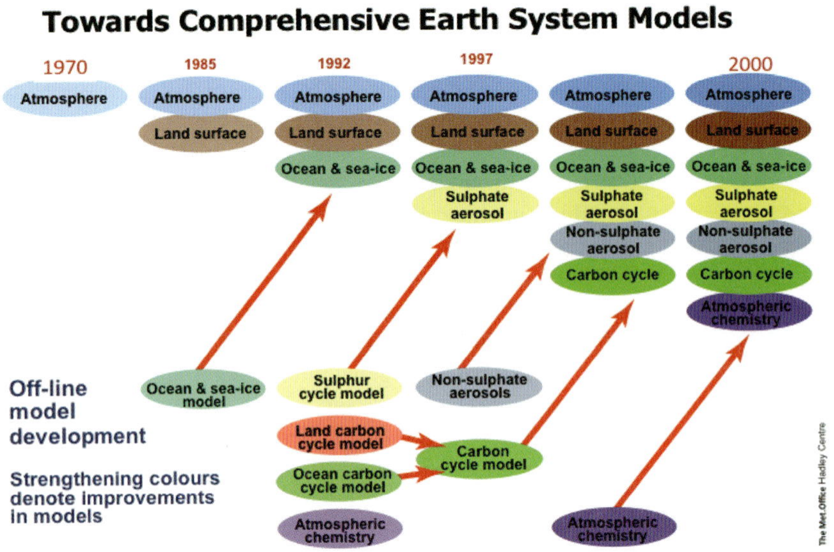

Towards Comprehensive Earth System Models

Figure 9.3 The main additional features that have been sequentially integrated into Earth System Models over a recent 30-year period. (© British Crown copyright, Met Office, 2014).[15]

9.11
Prediction and Uncertainty of Extreme Events

A major contribution to social well-being that the proposed Centre could be expected to make is in the arena of extreme weather prediction and natural disaster risk assessment.

The challenges of predicting the behaviour of complex systems are well known, namely that intricate feedbacks create sensitivity to unknowns that tends to be both subtle and dramatic. The most sensitive of all being those processes that are nonlinear in Nature – whereby small changes of input lead to extremely large changes in output.

It is striking to note that these processes are the very ones that are least understood and most poorly characterized in the best computer models we have today.

As for climate change, uncertainty is the major issue – potential harms of global change have indeed a broad range of possibilities, many of which can be dramatic in their impact. One critical outcome of global warming within climate change is the issue of sea level rise. If major ice sheets in Antarctica and Greenland melt more quickly than current climate models predict, there will be

---

15) ECMWF (European Centre for Medium-Range Weather Forecasts): www.ecmwf.int.

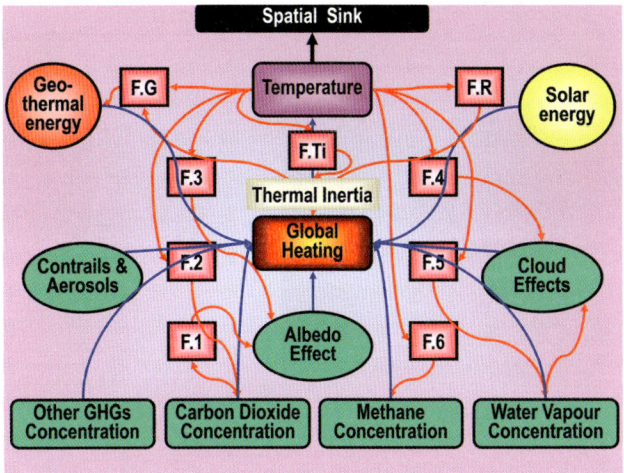

Figure 9.4 Multiple nonlinear feedback loops (some positive, some negative) in the Earth System. (Courtesy of Apollo-Gaia Project, Meridian House).[16]

significant rising of ocean levels worldwide, but little time for social preparation and adjustment. Additionally, the Arctic Ocean will rapidly become ice-free.

Current model outputs give a conservative estimate of 1 or 2 metres sea level rise by the year 2100. The incorporation of nonlinear land-ice feedbacks, however, will likely increase this by several metres, with obvious catastrophic effects on coastal populations worldwide.

Incorporating all relevant feedbacks in our Earth System Models however, will require a substantial enhancement in the sophistication of the numerical models themselves, as well as a strong boost of underlying compute power to bring the results to a timely conclusion.

Current global models have a horizontal resolution accuracy of around 100 km, whereas we will need to improve resolution accuracy down to at least 10 km going forwards. Vertical resolution, on the other hand, will need to be taken to the single-kilometre resolution level. Available compute capacity is a major limiting factor that slows progress on improving resolution accuracy, and indeed model development in general, which factors argue strongly for building the proposed facility as soon as possible, so as to enable fine-grain atmospheric phenomena such as cloud physics, convection, and vorticity in both atmosphere and oceans to be taken into account (Figure 9.4).

Further complicating Earth System prediction efforts today is the deep interconnectedness of natural systems with our socioeconomic systems – we have learned the hard way that the propagation of risk across adjacent domains can trigger Multiple Synchronous Collapse: that is, disruption in one system spreading quickly to disrupt others around it.

16) www.meridian.org.uk/menu.htm.

Figure 9.5 The Great East Asian earthquake and tsunami of March 2011 is a perfect illustration of Multiple Synchronous Collapse. (© Reuters)[17]

A striking example was witnessed in the March 2011 Fukushima nuclear meltdown in Japan, which arose through an unanticipated series of system linkages. In this case, the major slippage of a subduction zone on the ocean floor near Northeast Japan produced a magnitude 9.0 earthquake,[18] creating multiple tsunami waves that were 10 metres high or more when they reached the shoreline of the Tohoku region some 30 minutes later. Water poured into the narrow coastal valleys, increasing the height of water to 40 metres or more at the inland end of those valleys.

On the coastline itself, waves broke over the protective barriers and inundated the various cooling systems of the Fukushima Daiichi nuclear energy plant, causing the meltdown of three reactors and producing several explosions that shook the neighbouring villages (Figure 9.5).

Fukushima is a wake-up call, and we should all realize how serious the phenomenon of risk propagation and Multiple Synchronous Collapse can be. It cannot be approached in a piecemeal manner. Only through synthesis and integration can the systems in close proximity to one another be simulated as a whole. It is in this capacity that the proposed Centre will be uniquely useful. A global effort to bring coupled systems together as a seamless whole is needed to even begin the difficult work of disaster risk reduction in the twenty-first century.

Since the tragic events in Japan of March 2011, the journal *Natural Hazards* in its April 2012 online edition has published a paper entitled *Civil nuclear*

---

17) www.reuters.com.
18) The magnitude is a number characterizing the relative size of an earthquake based on measurement of the maximum motion recorded by a seismograph. The scales most used are: 1 local magnitude (Richter), 2 surface-wave magnitude, 3 body-wave magnitude, and 4 moment magnitude. Scales 1–3 have limited range and do not satisfactorily measure the size of the largest earthquakes. Scale 4 is uniformly applicable to all sizes but more difficult to compute than the others. All scales should yield approximately the same value for any given earthquake.

power at risk of tsunamis in which a total of 23 plants comprising 74 active nuclear reactors are identified as located in areas vulnerable to tsunamis.[19]

9.12
Impact on Cities and Bioregions

Many of the most significant safeguard actions pertaining to sustainability are now underway at the local level – principally at municipal and regional scales. It is almost a truism to say that ecological risks are irrevocably bound to urban issues. Whether dealing with water quality, air pollution, access to a stable food supply or increasingly constrained material inputs for industry, the majority of solutions are being crafted and implemented by city managers, regional planners and their affiliated local institutions.

The recent disaster in Japan reminds us that global threats arise from local disturbances. The same is true in reverse – localized risks are contingent on the larger-scale drivers that influence and constrain them. In the parlance of numerical models, the key issue is the characterization of boundary conditions that represent what flows into a region and how changes in the region flow outward to impact other localities.

Currently there are many efforts to simulate the dynamic linkages within specific bioregions of the planet, one example being the Amazon River watershed where vast forests intersect with numerous waterways to cycle water and nutrients across the South American continent. Efforts such as this depend on highly detailed sensor networks to gather sufficiently precise data to describe all of the key physical processes. And yet the aerosol transfers and large-scale energy fluxes from global atmospheric circulation, ocean mixing and plate tectonics all contribute significantly to what happens locally.

It is here that the proposed Centre can play a critical role for cities and bioregions – downscaling from the global Earth System models so as to reliably inform cross-boundary fluxes into local and bioregional simulations. Downscaling involves finer spatial and temporal resolution than that used by the global model, and the challenge is to form a mathematical bridge across scales. No global-scale climate model in use today is able to characterize localized weather patterns well enough for regional planning purposes.

Contributing to model improvement goes both ways. Global models can increase the effectiveness of high-resolution local and bioregional simulations by strengthening linkages between large-scale patterns and local boundary conditions. Inversely, high-resolution output from local and bioregional simulations can be used as inputs to improve global models. This virtuous cycle allows for the intricacies of local outcomes to be combined with the larger global patterns that arise at the whole-system level, to the benefit of both.

---

19) J. Rodriguez-Vidal *et al.*, *Civil nuclear power at risk of tsunamis*, Natural Hazards, Volume 63, Issue 2, p. 1273, 2012.

Neither can succeed on their own, however. While the appropriate scales for policy implementation are municipal and regional, the flows coming in from global systems are necessary for accurate planning purposes. Continual improvements can be made for the modelling efforts at both scales through ongoing dialogue and collaboration. The proposed Centre would provide an international platform with sufficient neutrality to accelerate this discourse so that model improvements can be developed more quickly.

Additional areas demanding even closer collaboration between local and global efforts include precipitation forecasting, flood forecasting, drought forecasting, as well as closing the gap between current numerical weather forecasting techniques, inter-seasonal, inter-annual, decadal and longer-term climate predictions.

9.13
Towards Urban Resilience

A global movement is now underway that focuses on carbon neutral urban landscapes, regional food safety, green building construction, and the recycling of waste products spearheaded by the world's major cities.

Increasing emphasis is placed on transportation systems, energy distribution, public health, and quality of life for people within these cities.

Every State of the World Report published by the Worldwatch Institute[20] since 2006 has focused on urbanism to draw attention to the remediation of severe ecological deterioration using improved design of cities, old and new.

In a recent Technology, Entertainment, Design (TED) talk, renowned innovator Stewart Brand described how we are in the midst of a great migration of people from rural areas to squatter cities and urban slums in many regions of the world.[21]

Nearly all population growth in this century will take place in these new urban dwellings, with mega-cities emerging rapidly throughout Africa, Asia and Latin America. Already, as of this year 2014, the United Nations has announced that half of the world's population now lives in cities.

Cities are both the progenitors of development pressure contributing to environmental harm as well as the cultural incubators of technological and social innovation, with the latter increasingly paving the way to a sustainable future. Cities are where the action is for policy implementation, as well as being the 'economic engines' of the future.

Ecological design principles are being used to apply biomimicry[22] – inspiration from biological and living systems – to plan urban development through the lens of ecosystem science.

---

20) www.worldwatch.org.
21) www.ted.com/speakers/stewart_brand.html.
22) Biomimicry (from Greek *bios*, 'life' and *mimesis* 'to imitate') is the imitation of the models, systems, and elements of nature for the purpose of solving complex human problems.

The same principles apply here as they do at the bioregional level. Local systems must incorporate global fluxes into their development models. Once again, the proposed Centre can play a contributing role.

As cities grapple with economic and environmental issues, they will increasingly depend on dynamic modelling to identify hazards and reveal opportunities. Urban planners extensively use Geographic Information Systems (GIS) to analyse layers of information that overlap within the same landscape. In this way, they build up detailed databases for utility grids, storm water and sewage systems, transportation networks, and so on – often enabling them to run simulations of their interconnected systems in order to develop *smart eco-city* architectures.

Of particular concern is the balance between urban density and hinterland agriculture, where the rural areas become closely coupled with city planning. Again, this is an area where the whole-system integration of larger-scale processes becomes essential for synergistic results.

The impact of land use changes and water depletion often jeopardize the productivity of agricultural lands. Conflict is a major source of frustration for urban development in such cases, exacerbated in part by the lack of adequate foresight and understanding of these interlocking issues.

City planners and regional planners would benefit equally from an integrated picture of complex interlocking issues in order to do their jobs more effectively, and the proposed Centre would provide a global forum for standards and best practice to emerge and be tested in a cost-free virtual environment. For these reasons and more, the International Centre for Earth Simulation (ICES) works in close collaboration with UK charity 'Ecological Sequestration Trust'.

9.14
Modelling the Whole-Earth System: A Challenge Whose Time has Come!

Looking forward over the next 10 years, one can expect steady continuous improvements in computing capabilities, storage capacity, networking speeds, sensor networks, instrumentation accuracy, satellite coverage, Doppler radar[23] coverage, lidar[24] and a host of new technologies to emerge that will help us monitor the physical state of our planet to a high degree of accuracy. Furthermore, we will see much more attention paid to the state of the planet's biosphere – its flora and fauna – and thereby it is hoped that society will develop a deeper understanding of the interaction between the biosphere and the planet's physicality, to everyone's advantage.

---

23) A Doppler radar is a specialized radar using the Doppler effect (the change in frequency of a wave, or other periodic events, for an observer moving relative to its source) to produce velocity data about objects at a distance. It is used in aviation, sounding satellites, meteorology, police speed guns, radiology, and bistatic radar (surface-to-air missile).
24) Lidar (light and radar) is a remote sensing technology that measures distance by illuminating a target with a laser and analysing the reflected light.

But, where exactly do politico-socioeconomic factors and the study of human dynamics fit into this picture – and to what extent can we couple these factors into the Earth System models as well?

Here, we turn to the current round of the United Nations' Intergovernmental Panel for Climate Change (IPCC). Underway is Assessment Report 5 (AR 5), the first release of which was made in September 2013, and within which report we can find the Coupled Model Inter-comparison Project 5 (CMIP5). This Project brings social scenarios into the modelling picture for the first time, and attempts to show how social and political policy decisions will ultimately affect the world's future climate.

This may seem obvious in a globalized world with an annual US$ 70 trillion interconnected economy, and with over seven billion residents pursuing ever-increasing levels of consumption and living standards. However, until now, such Integrated Assessment Models (IAMs) were not coupled with climate models to any useful degree.

Within this current year of 2014, however, the software as well as the compute capacity to attempt this coupling has come at hand, and first preliminary results are now imminent. These will be watched closely by scientists and policy-makers alike.

What gives the social sciences a stronger hand today more than ever before is the progressive digitization of broadcast and printed material such as TV, radio, film, medical records, court records, family records, and financial records, along with the recent rapid emergence of e-mail and electronic social networking. All such digital media content can now be 'mined' at high speed, allowing social and political theories to be tested in a much quicker and a more concise manner than ever before. Indeed, one could say that we can almost feel the 'pulse of the planet' with the aid of such digital technologies.

Nevertheless, at the current pace there is quite a way to go before we can expect to successfully integrate natural sciences and socioeconomic sciences into a set of models that are cohesively interlinked and which together prove to have good skill at forecasting the key aspects and directions of life on Earth. What could change this situation, however, are breakthrough innovations which provide leaps in insight, base knowledge and systems performance. Such leaps and discontinuous jumps in capability could emerge from fields such as stochastic computing, neuromorphic computing or quantum computing (the latter being hinted by the Nobel Prize in Physics 2012)[25] and where we plumb the secrets of Mother Nature herself to mimic her brilliant ways.

---

25) In 2012, the Nobel Prize in Physics was awarded jointly to Serge Haroche and David J. Wineland ' . . . for ground-breaking experimental methods that enable measuring and manipulation of individual quantum systems': http://www.nobelprize.org/nobel_prizes/physics/laureates/2012.

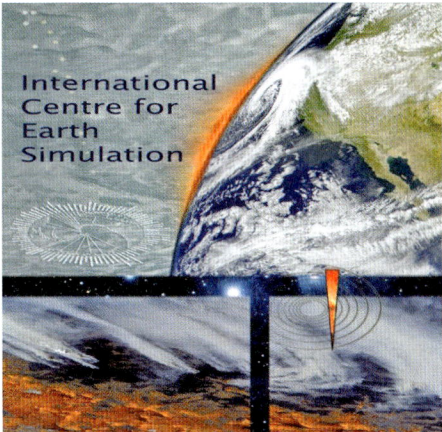

Figure 9.6 Montage created by © Tony and Bonnie DeVarco for the International Centre for Earth Simulation (ICES) Foundation[26] with use of satellite imagery. (Courtesy of NASA's Earth Observatory). The inset diagram is a simplified version of the Tree of Life,[27] showing common names of the major groups.[28] Other insets include the Sun Dagger and the Analemma. The constellation Pleiades peeks through in the background.

In either case, as we plan towards a globally focused, probabilistic, exascale, visually orientated Earth System Simulator, the journey is bound to be exciting and full of learning and valuable surprises (Figure 9.6).

26) www.icesfoundation.org.
27) The concept of a 'Tree of Life' is a common motif in various theologies, mythologies, philosophies, sciences and religions that alludes to the interconnection of all life on our planet and serves as a metaphor for common descent in the evolutionary sense.
28) D. Sadava *et al.*, *Life: The Science of Biology*, 8th Edition, Sinauer Associates and W.H. Freeman, 2008.

10
Radiation Detection in Environment and Classrooms

Michael Campbell

Imagine you are standing in front of a dark-coloured wall about 2 metres high and 2 metres wide. The wall is made of glass, and behind the glass there are long thin metal plates arranged horizontally from floor to ceiling, much like an open venetian blind. The entire wall is sealed and the metal plates are enveloped in an inert gas. Suddenly, there is a loud noise like a fire cracker and small sparks appear between the plates, creating a line of light from the ceiling to the floor. Then, there is a second bang and this time, instead of one line there may be two lines or there may be an upside-down Y-shape. These sparks will continue to go off as long as the 'wall' is primed. This 'spark chamber' detects cosmic particles and provides a very graphic demonstration of the natural background radiation to which we are exposed throughout our lives. In this case, very high voltages have been applied between the plates, and it is the small number of free electrons that have been liberated by the cosmic ray particles passing through the gas which provoke the electric discharge, like tiny sparks of lightning between the plates. An example is shown in Figure 10.1.

This type of detector was used extensively by particle physicists during the 1960s, and a number of important discoveries (such as that of the muon neutrino)[1] were made using such devices. One advantage of this type of detector is

---

1) The Standard Model includes all the known particles present in Nature, differentiating them by statistical properties and physical laws which they obey, into two families: *bosons* (which govern the interactions) and *fermions* (which make up matter). The *bosons*, that is, the carriers of the interaction of the four fundamental forces working in our Universe, are: *W* and *Z*, and the *gluon*, respectively for the *weak force* and the *strong force*, which dominate at the subatomic level, the *photon* for the *electromagnetic force*, which causes the interaction between electrically charged particles, and, although not yet observed, the *graviton* for the *gravitational force*, which regulates the movement of the planets. Only the first three forces are described in the Standard Model. The *fermions* are divided into two groups, *quarks* and *leptons* with their respective antiparticles (identical in every respect except for some physical properties, such as the opposite electric charge). The six types of quark are coupled in three generations: the lighter more stable (*quark up*, *quark down*), the heavier less-stable (*quark charm*, *quark strange*) followed by the *quark top*, *quark bottom*. There are six leptons, subdivided into three generations: the *electron*, the *muon* and the *tau*. To each of these three particles with electric charge and mass, is associated a neutral lepton, the *neutrino*: the *electron neutrino*, the *muon neutrino* and the *tau neutrino*.

From Physics to Daily Life: Applications in Informatics, Energy, and Environment, First Edition.
Edited by Beatrice Bressan.
© 2014 Wiley-VCH Verlag GmbH & Co. KGaA. Published 2014 by Wiley-VCH Verlag GmbH & Co. KGaA.

10 Radiation Detection in Environment and Classrooms

Figure 10.1 An example of a spark chamber. (Courtesy of Fred Hartjes, NIKHEF,[2] the Dutch National Institute for Nuclear Physics and High Energy Physics).

that it brings a real understanding to the lay observer of what cosmic radiation 'looks like' and, with the accompanying sound, it is something which easily stays in the mind's eye. A disadvantage is that it is a rather difficult detector to build, run and maintain. In this chapter, it will be described how, as a spin-off from detector R&D for the Large Hadron Collider (LHC), it was possible to develop a modern cosmic ray detector which would fit into a person's pocket and can be run from a laptop computer or even a smartphone. An explanation will also be provided of how the same technology can be used in many applications outside of High-Energy Physics (HEP).

10.1
The Origins of the Hybrid Pixel Detector

The story of these modern cosmic ray detector detectors begins back in the 1980s with the dream of the CERN physicist Erik Heijne, who dreamed (and wrote) about how the then fledgling integrated circuit technologies could be applied to radiation detection. Heijne imagined that one day, all of the electronics needed to process the signals from a radiation sensor could be packed into a single tiny square of silicon of less than 1 mm square. At that time, each particle detection element would have its own electronics readout card. Heijne

---

2) NIKHEF (Nationaal Instituut voor Kernfysica en Hoge-Energiefysica)

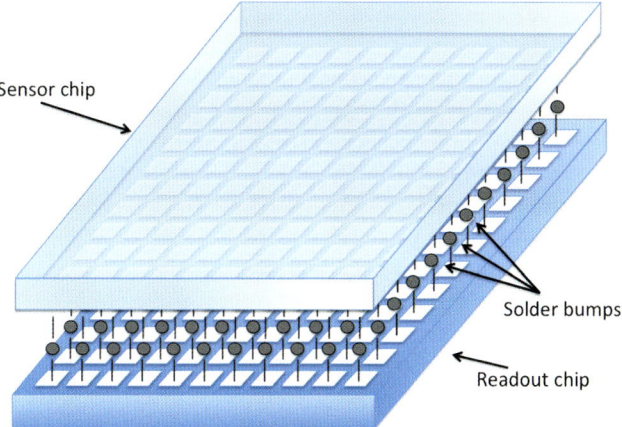

Figure 10.2 Illustration of a hybrid pixel detector, comprising three components: a segmented sensor chip connected using solder balls to an identically segmented readout chip.

also postulated that it would be possible to create a two-dimensional (2D) array of such circuits on a single electronics chip, and each circuit would be connected to its own correspondingly small sensor element. The sensor and readout electronics chips would form a type of sandwich composed of a pixelated sensor slab on one side and a pixelated readout chip on the other side (see Figure 10.2). The sandwich would be filled with an array of microscopic solder balls, with each one connecting a detector pixel to its own unique readout pixel. Consequently, the hybrid pixel detector was born – at least on paper. The challenge back then was first, to understand how to design circuits in this new technology, and second to see if it would be possible to make the circuits small enough to turn the hybrid pixel detector from a dream to reality.

Particle physics stimulates the imaginations of some of the most brilliant minds, and helps humanity to deepen its understanding of the nature of the Universe around us. However, along the way its requirements sometimes provoke technology leaps which themselves spill over to other spheres of scientific endeavour, industry, and to education (see Section 10.4). The development of hybrid pixel detectors for the Large Hadron Collider (LHC) and the subsequent transfer of this technology to other applications in the context of the Medipix project, provides a good example of this process.

10.2
Hybrid Pixel Detectors for High-Energy Physics

At the LHC, protons are squeezed together into bunches and accelerated to close to the speed of light in opposite directions inside a 27 km-long circular vacuum tube. Each bunch of particles at the LHC is actually composed of about

100 billion protons. (Imagine the population of India, multiply it by about 100, and that number of people is the same as the number of protons in a single LHC bunch.) Moreover, at any one time there are a few thousand bunches going around in each direction. When the machine reaches its full design regime, those bunches will be made to 'collide' 40 million times per second right in the middle of the large experiments, although only about 20–100 of the particles in each bunch will actually interact in any given single bunch crossing. Each time the bunches collide, between 1000 and 50 000 particles are produced. Hence, it is easy to understand the importance of not mixing up the bunches, and also that the detectors used should not produce too many false hits as this would confuse the picture.

It was this requirement which motivated the development of hybrid pixel detectors for use at the LHC experiments. Because the sensor and electronic readout are in such close proximity it is possible to have detection with an extremely high signal-to-noise ratio.[3] The hybrid pixel detector readout chips used at the LHC are rectangular in shape and measure 1–2 cm on the side, and there may be up to about 10 000 pixels in a single chip. A scanning electron microscopy image of bump bonds similar to those used is shown in Figure 10.3. Often, the readout chips and sensor chips are almost identical in size; however, in High-Energy Physics (HEP) and for photon science it is common to connect a number of readout chips (4 to 16) to a single large sensor 'ladder'. The ladder dimensions are limited by the diameter of the sensor wafers (the silicon plates from which they are made), and are usually around $1-2\,\text{cm} \times 5-10\,\text{cm}$. The most important feature of a hybrid pixel detector is that it can provide 'noise-free' images of radiation even at very high shutter speeds. In other words, if a hybrid pixel detector system is properly designed and configured, it is possible to open its electronic shutter (just like a sensor for a camera) and the detector will take a clean picture of the ionizing radiation in the environment, without any background blur.

The large pixel detector systems at the LHC are typically composed of hundreds or even thousands of ladders, arranged to form concentric cylinders around the particle interaction point. For every bunch crossing they produce a single clean image of the particle tracks. However, although thousands of particles are produced at each bunch collision, only less than 1000 of the 40 million

---

3) When physicists are looking for new particles they aim for the highest signal-to-noise ratio possible. To make an analogy, imagine you are lying in bed in a house in the countryside surrounded by trees. It is dark outside but it is a warm summer's night and the window is open. The only sound from outside is a gentle rustling of leaves caused by a light breeze. In the distance you hear what might be a twig snapping. You ignore it because it could be just the wind or your imagination. Then, there is a second snap, this time closer. If you hear the snapping several times in close succession you can be pretty sure there is someone or something out there. That is how physicists explore different 'channels' which might point to a new particle, but all the time they must be sure that those 'signals' really represent a new particle and not a coincidence due to background noise. Similarly, when designing circuits made to detect particles, physicists try to achieve the best signal-to-noise ratio possible. In practice, the circuit ignores all signals below a certain level of deposited charge and only detects those which exceed it.

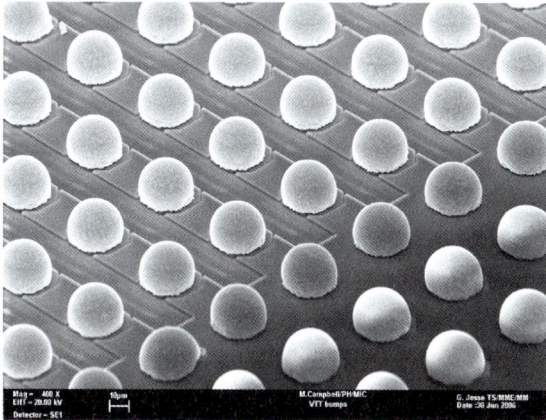

Figure 10.3 An example of bump bonds on the Medipix2 readout chip. The solder balls are about 25 μm in diameter on a pitch of 55 μm. The same bump bonding technology is used by some of the pixel systems at the LHC. (Courtesy of Gudrun Jesse, formerly of CERN).

bunch crossings per second contain 'head-on-head' proton collisions which have enough energy to produce new physics. The detector physicists apply a series of selection criteria, called 'triggers', in order to save only 800 or so interesting bunch crossings per second. For their part in the process, the pixel detectors operate like a type of fast camera taking 40 million frames per second. However, due to space limitations the registered frames are only kept for a few microseconds in on-chip memories before being overwritten. During that time, the 'first-level trigger' is applied to the chip and only the frames of interest are preserved for further analysis.

The ALICE (A Large Ion Collider Experiment), ATLAS (A Toroidal LHC ApparatuS), and CMS (Compact Muon Solenoid) experiments are equipped with inner tracking detectors based on hybrid pixels. The LHCb (Large Hadron Collider beauty) experiment has a photon detector system which also uses hybrid pixels in its readout. All of these systems provide vital data to help physicists in their search for new particles and physics.

As might be imagined, going from Erik Heijne's 'dream' to the large systems operating at the LHC was a long process that involved large teams of dedicated specialists. In fact, there was a gap of about 20 years between the publication of the 'dream' paper and the first data produced by the LHC pixel detector systems.

10.3
Hybrid Pixel Detectors for Imaging: The Medipix Chips

It was during the evaluation and production of the first prototype hybrid pixel detectors in the early 1990s that the potential of the new technology for applications outside of HEP became obvious. In particular, the quality of the bump bonds was tested using radioactive sources effectively taking pictures of the sources themselves. Sometimes, for fun, objects were even inserted between the source and our detector forming X-ray images. However, because the pixels were rectangular in shape the images were not very nice to look at. Also, the readout electronics was designed to select only single frames, and therefore many frames had to be selected and added up to create an X-ray image.

An informal collaboration was launched involving chip designers at CERN and physicist colleagues from Germany, Italy and Scotland in order to make a chip with square pixels and with a form of camera logic. An electronic signal applied to the chip acts like a shutter, defining the time during which the chip is sensitive to incoming particles. A counter in each pixel adds up the total number of hits in that pixel while the shutter is open. The contents of the counters are read out afterwards and used to create an image on a screen. The Medipix1[4] (or Photon Counting Chip; PCC) was produced in 1997. The chip had 4096 square pixels (a matrix of 64×64) for a total sensitive surface of about $11\,\text{mm} \times 11\,\text{mm}$, and was used extensively to test a new sensor material called Gallium Arsenide (GaAs), which was particularly well suited to the detection of X-ray photons in the low-energy part of the diagnostic spectrum, a region used in mammography. Although measurements showed that using such a device could provide a significant reduction in the dose received during a mammographic examination, the complexity and cost of scaling up to a large system were ultimately prohibitive. The chip was also combined with more conventional silicon sensors similar to those used in the physics experiments. Devices such as this were used extensively in applications ranging from 'soft' X-ray imaging to beta-ray imaging, and even as a detector in an electron microscope. The main achievement of this chip was to demonstrate the potential of hybrid pixel detectors in fields beyond HEP. For many purposes, however, the pixel size was slightly too large and the number of pixels in the matrix slightly on the low side. Fortunately, however, microelectronics technology does not stand still and it was possible to take a step further by moving from a chip manufacturing process with a minimum feature size of $1\,\mu\text{m}$ to a process with a minimum feature size of $0.25\,\mu\text{m}$.

In 1965, Gordon Moore – one of the founders of Intel, the company whose processors power millions of PCs and laptops – made a remarkable prediction, that the number of transistors in a given area of silicon would double about every 18 months to 2 years. This prediction became known as *Moore's law*, and it has now been followed by the semiconductor industry for well over 40 years. But, one could ask the question: Why would the industry bother to make the huge effort to continuously upgrade every few years? The answer is very simple: money. Anyone who likes gadgets will know that one can expect with each new generation of a device (such as a smartphone) increased functionality, speed and new features. Yet, the price of the top-of-the-range smartphones does not go up with time, because the cost to produce silicon wafers in large quantities is about

---

4) http://medipix.web.cern.ch/medipix.

the same from one semiconductor generation to the next. Therefore, any company which does not respect *Moore's law* will be overtaken by its competitors whose products will occupy a smaller chip area and be cheaper, or occupy the same chip area and be more powerful. In the Medipix pixel detector studies we have been able to jump on the *Moore's law* bandwagon and create increasingly sophisticated electronics on a fixed pixel area or, alternatively, a similar readout scheme on a smaller area.

Moving from 1 µm Complementary Metal Oxide Semiconductor (CMOS) to 0.25 µm CMOS chip technology was a major step, but we had already accumulated (even in 1999) significant experience with the new technology. The readout chip for the ALICE pixel detector (developed at CERN during the late 1990s) was the first LHC pixel chip to use the new process, and other groups followed our example later. The aim in developing the Medipix2 chip was to reduce the pixel size to 50 µm while increasing the number of pixels to 256 × 256. The success of the Medipix1 chip had stimulated a growing interest in the approach, mostly amongst groups interested in medical imaging applications. A new collaboration, the Medipix2 Collaboration, was formed in 1999 which permitted a large number of groups to pool resources to produce the new chip. Although *Moore's law* says that the cost per unit area of silicon is constant in large volumes, the prototyping costs do increase significantly from one generation to the next. A set of masks (the templates used to create the chips) costs approximately fourfold more in 0.25 µm CMOS than in 1 µm CMOS.

After a period of consultation, the members of the new collaboration settled on a pixel design which was innovative but required a little more area than foreseen. While the circuit which amplifies the signal from the sensor was very similar to that of Medipix1, the decision was to implement not one, but two, thresholds per pixel. This meant that the pixel would be sensitive to charge deposits lying between a lower threshold and an upper threshold, the idea being to take a first step towards spectroscopic or 'colour' X-ray imaging.[5] With the Medipix2 chip it is possible to record hits from one range of X-ray energies in one exposure. Multiple exposures are required to make a true colour image.

The first version of the Medipix2 chip was produced in early 2001 and, after various 'respins' over the following years, the final version was submitted to production in April 2005. During the four years between the submission of the first chip and the final version many lessons were learned. In spite of the long experience accumulated from the ALICE pixel studies, the Medipix2 design was many fold denser. The tools used to verify the designs did not provide full coverage for possible errors and, in particular, many of the effects observed on the chips only became visible on full-sized prototypes. Nonetheless, the Medipix2 chip was a great success and generated a great deal of scientific output.

---

5) In the electromagnetic spectrum our eyes are sensitive to only a very small number of wavelengths, running from red to blue. Like visible light, X-rays also cover a spectrum of energies. As each X-ray is detected one by one, it should be possible to classify them according to energy, and the result would be an X-ray image which retained the energy information of the detected charge. It was possible to consider that as constituting a colour X-ray image.

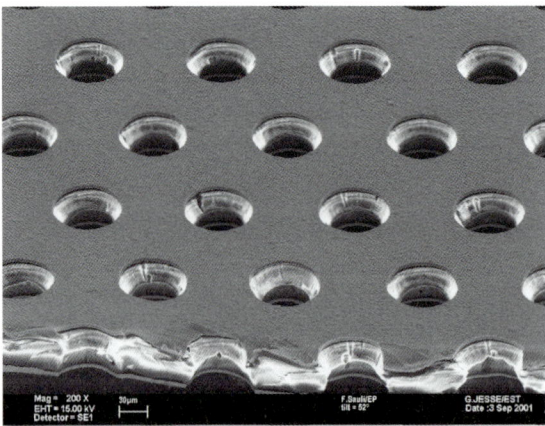

Figure 10.4 Image of a Gas Electron Multiplier (GEM) foil. A high electric field is applied between the conducting upper and lower surfaces. Electrons which arrive in the vicinity of the foil are focussed electrostatically towards the holes. The very high electric field in the holes causes charge amplification to occur. (Courtesy of Gudrun Jesse, formerly of CERN).[6]

It was also in 2005 that the Medipix2 Collaboration was approached by colleagues from the EUDET Collaboration[7] (in particular groups from The Netherlands, Germany and France) who were working with gas detectors. These groups, who asked us to modify the Medipix2 chip for them, were using Gas Electron Multiplier (GEM) foils mounted on top of – but not touching – the Medipix2 chip, but this time with no sensor attached. A scanning electron microscopy image of such a foil is shown in Figure 10.4.

The foils were immersed in a volume of inert gas, and a small electric field was applied between the top of the gas volume and the top of the GEM foil. A much larger electric field is applied across the foil, and there is again a smaller field between the bottom of the foil and the chip. A cosmic ray (or other charged particle) crossing the gas volume would liberate some electrons along its path, and these electrons would drift under the influence of the small electric field to the foil. There, the single charges would be amplified to a larger charge that could be picked up by the pads of the readout chip.

This process is reminiscent of the spark chamber described above, but in this case the charge would be amplified and should not create a discharge. Reading out the gas detector with the Medipix2 chip provided a type of binary image (composed of just 1s and 0s) of a projection of the particle track from three dimensions onto the two dimensions of the pixel matrix. The colleagues from EUDET asked the design team to make the chip sensitive to the particle arrival

6) D. Attié, *TPC review*, Nuclear Instruments and Methods in Physics Research Section A: Accelerators, Spectrometers, Detectors and Associated Equipment, Volume 598, Issue 1, p. 89, 2009.
7) EUDET is a European Union project for Detector R&D towards the International Linear Collider in the FP6 (6th Framework Programme): www.eudet.org.

time, so that they could determine where the charges were created in the depth of the gas volume and thus produce a three-dimensional (3D) image of the track.

Hence, the Timepix chip was born, in which a clock signal is brought to each pixel and the pixel can be programmed to function in one of several different modes. In particle arrival mode, the pixel counts the total number of clock ticks, starting from when the charge from the particle is first detected and stopping when the shutter closes. The timing of the particle arrival tells us where in the gas volume depth it arrived. Alternatively, the counter can be programmed to count the number of clock ticks while a particle hit is being detected. As the time it takes the pixel amplifier to recover from a hit is proportional to the total deposited charge, then in this mode – called 'Time over Threshold' (ToT) – the number of counts in a pixel would indicate how much charge had been deposited in that pixel. This chip was also used in many applications and, in particular, in ToT mode it provided colour images that showed not only the presence of a hit per pixel but also the energy deposited in that pixel. Subsequently, use of the Medipix2 and Timepix chips led to the generation of over 150 peer-reviewed scientific papers and formed a critical technology platform for a large number of PhD theses. As will be seen later, both were also used in a wide range of applications.

From the Timepix images of single X-ray particles it was possible to see that the quantity of charge measured in a pixel is not perfectly uniform, even when the incoming X-ray particles all have the same energy. This is because the pitch of the pixel sensors is usually much smaller than the sensor thickness and, depending on where the incoming hit generates its charge in the pixel volume, more or less of that charge will spill over to the neighbouring pixel. When there are only a few hits in a frame one can look for the individual clusters of hits and try to add up the charge in each cluster to reconstruct the total deposited energy. Whilst this works to a certain level, there may always be charge lost to pixel neighbours which is less than the detectable threshold. Moreover, if there are a large number of hits in the image it becomes impossible to distinguish one cluster from another. On careful consideration of this issue, the idea of using the opportunities offered by an even more advanced CMOS process to solve the problem finally emerged. The trick is to sum the charge deposited in a local region first while, at the same time, deciding which pixel in that region to assign the hit to. As prototyping in the new CMOS processes becomes increasingly expensive, the decision was taken to form a new Collaboration of people and institutes wishing to pursue this new line of investigation: hence, the Medipix3 Collaboration was formed in 2006.

As most HEP groups were fully focussed on building and commissioning the detectors for the LHC, the Medipix design team was the first to study the use of the 130 nm CMOS chip manufacturing technology node for pixel detector readout applications. Following a successful prototyping experience with a small chip, the first full-scale prototype in the new technology was submitted to production in 2008. While the small prototype chip had worked perfectly, and certain features of the full-size chip worked well, there were still issues with yield

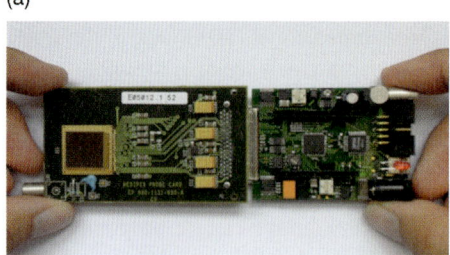 

Figure 10.5 (a) Image of the USB (Universal Serial Bus) readout system which can be used to read out Medipix2 and Timepix chips[8] (Reproduced with permission of Elsevier); (b) Image of the USB Lite version (light version) of the same system[9] (Reproduced with permission of Elsevier).

(the number of chips without defects) and with unexplained variations in behaviour from pixel to pixel. Ultimately, it took several years of study and a full new chip iteration to finally have a chip which would work to its full capacity. However, this learning phase was not time wasted, as the Medipix team were able to provide hints and advice to other groups (e.g. the ATLAS pixel design team) which saved them much time, money and effort. The latest version of the Medipix3 chip is being tested at the time of writing this text and, by all accounts, it is now working extremely well.

Although the development of readout chips is at the core of our work at CERN, every chip requires some means of communicating with a computer. In the days of the Medipix1 chip that task was taken care of by using rather heavy, HEP readout cards which were run from an electronics readout crate. However, over the years the Collaborations have moved gradually towards increasingly miniaturized readouts. Although demanding applications (such as high-rate imaging at synchrotron light sources) require extremely large bandwidths, and therefore sophisticated readout, many applications can be taken care of using the standard USB (Universal Serial Bus) interface. The group of the Institute of Experimental and Applied Physics (IEAP) at the Czech Technical University in Prague has made a particularly strong contribution to this field by developing USB-based readout interfaces. The first version was about the size of a packet of cigarettes (Figure 10.5a) and was used to interface to a small card carrying a single chip. A newer version includes the chip inside the interface box, where the whole device is about the size of a USB pen drive (Figure 10.5b).

---

8) Z. Vykydal et al., *USB interface for Medipix2 pixel device enabling energy and position-sensitive detection of heavy charged particles*, Nuclear Instruments and Methods in Physics Research Section A: Accelerators, Spectrometers, Detectors and Associated Equipment, Volume 563, Issue 1, p. 112, 2006.
9) Z. Vykydal and J. Jakubek, *USB Lite – Miniaturized readout interface for Medipix2 detector*, Nuclear Instruments and Methods in Physics Research Section A: Accelerators, Spectrometers, Detectors and Associated Equipment, Volume 633, Supplement 1, p. S48, 2011.

Another equally important contribution from the IEAP group was the development of some very user-friendly readout software, called 'Pixelman'. Alternative readout systems with high rate and capacity have been developed in various institutions around Europe, such as the European Synchrotron Radiation Facility (ESRF) in Grenoble, France, DESY,[10] the National Research Centre in Hamburg, Germany, the Diamond Light Source near Oxford, UK, and at NIKHEF,[11] the Dutch National Institute for Nuclear Physics and High-Energy Physics in Amsterdam, The Netherlands.

10.4 Applications

As already mentioned, the Medipix family of chips has been used in numerous applications outside of HEP; some examples are provided in the following subsections.

10.4.1 Medical X-Ray Imaging

As the name 'Medipix' would suggest, the first application which was foreseen for the pixel imaging chips was medical imaging. The original idea was that, by detecting every single photon unambiguously, we would reduce to the absolute minimum the number of photons needed to obtain an image with a given contrast. Moreover, the single photons could be attributed a weight according to their importance for the imaging process, because high-energy photons usually carry less information but generate larger charge signals in conventional imagers. Although the first measurements on medical phantoms (models which mimic the body) were very promising, the penetration of the photon-counting technique into the medical X-ray imaging field has been hindered by a lack of availability of uniform highly absorbing sensor material and the challenge of covering a large area seamlessly and at a reasonable cost. That being said, many studies have shown that even with silicon (a relatively poor X-ray absorber) as a sensor material, the technology competes well (at least technically) with commercial solutions based on charge integration. Figure 10.6 shows the X-ray Computed Tomography (CT) images of a living mouse head, acquired using a Medipix3 with silicon sensor and by a conventional flat panel detector.

The improvement of clarity using Medipix3 silicon sensors is obvious even to the untrained observer. Looking further ahead, the combination of high-resolution colour X-ray imaging using chips such as the Medipix3 with modern, metal nanoparticle-based biological tracers may permit the simultaneous imaging of organ shapes and function, provoking a paradigm shift in medical X-ray imaging.

---

10) DESY (Deutsche Elektronen-Synchrotron).
11) NIKHEF (Nationaal Instituut voor Kernfysica en Hoge-Energiefysica).

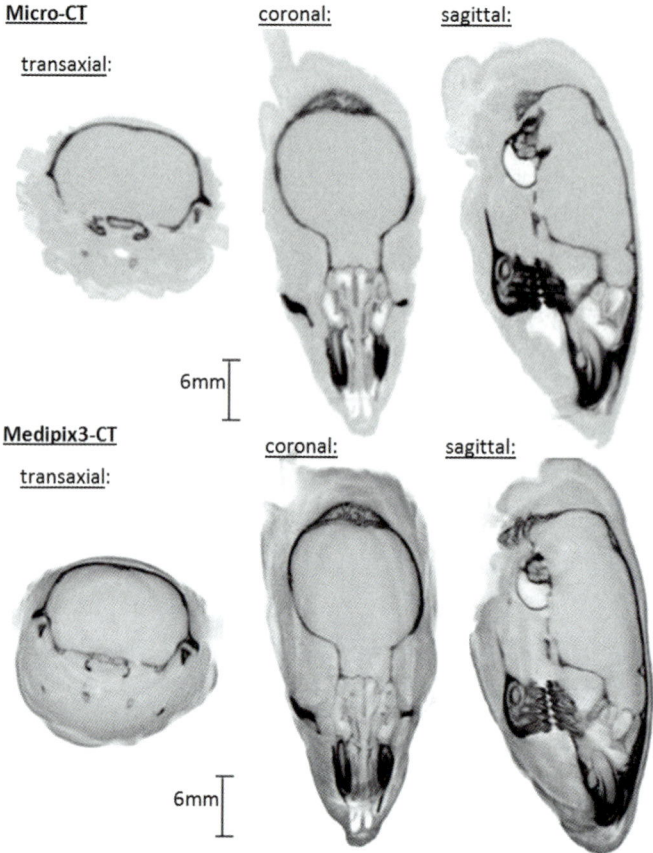

Figure 10.6 Comparison of Computed Tomography (CT) images of a mouse head taken under identical circumstances between a conventional X-ray detector and the Medipix3 chip combined with a Silicon sensor. The dose used for both measurements was about 170 mGy cm^{-1}.[12]

Physicists and clinical colleagues in Christchurch, New Zealand are leading the biomedical research efforts using the Medipix3 chip.

10.4.2
Biology

Another interesting example of the capabilities offered by such a sensitive detector is for the imaging of living, extremely low-contrast organisms. Research groups at the IEAP, together with biologist colleagues, have recorded movies of

12) J. Lübke, *Entwicklung eines iterativen Rekonstruktionsverfahrens für einen Medipix3-Computertomographen*, PhD thesis, University of Freiburg, Germany, 2011.

10.4 Applications

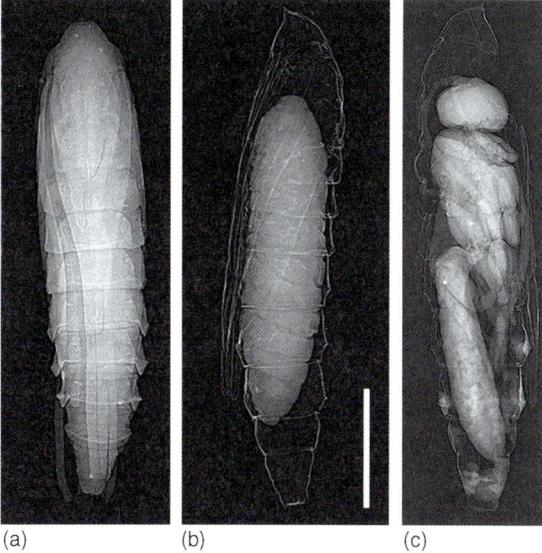

(a) (b) (c)

Figure 10.7 Pupae of the horse chestnut leaf miner examined with microradiography. (a) Living diapausing pupa; (b) Dead diapausing pupa parasitized with a larva of hymenopteran parasitoid of the genus *Pnigalio*; (c) Dead diapausing pupa parasitized with a pupa of the hymenopteran parasitoid of the genus *Pnigalio*. Scale bar = 1 mm. (Courtesy of Vít Sopko, Institute of Experimental and Applied Physics, IEAP).[13] (Reproduced with permission of Formatix Research Centre).

live leaf miner worms chewing their way through horse chestnut leaves. But, the leaf miner also has a natural enemy in the form of a hymenopteran parasitoid, a wasp which lays its egg inside the leaf miner; the egg subsequently hatches, develops and then metamorphoses into a fly, fully consuming its host in the process. Some examples of images taken at various stages of the metamorphosis are shown in Figure 10.7. Note the dimensions!

10.4.3
X-Ray Materials Analysis

An important partner for both the Medipix2 and Medipix3 Collaborations has been the X-ray materials analysis company, PANalytical.[14] Their early commitment to the Medipix2 chip provided the Collaboration with the motivation (and the money needed) to take the chip beyond the 'well working prototype' stage to an industry standard level. Moreover, the rigorous and systematic evaluation of the device in real-life circumstances provided valuable input to the design team.

---

13) A. Méndez-Vilas and J. Díaz (Eds), *Microscopy: Science, Technology, Applications and Education*, Formatex Research Center, 2010.
14) www.panalytical.com.

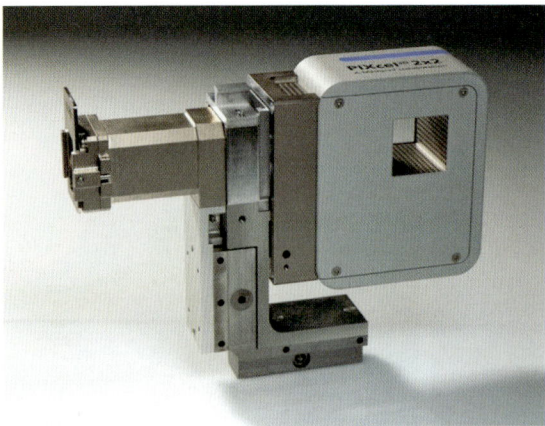

Figure 10.8 Image of the most recent incarnation of the PIXcel detector. (Courtesy of PANalytical B.V.).[15]

An image of the so-called 'PIXcel' detector, as commercialized by PANalytical, is shown in Figure 10.8. Today, PANalytical have well in excess of 500 PIXcel systems running in their machines across the globe.

10.4.4
Gas Detector Readout

Colleagues who were originally in the EUDET Collaboration have made significant progress in designing and constructing new gas detectors compatible with CMOS readout. One important development is the micro pattern gaseous detector called 'GridPix' (see Figure 10.9a). The readout chip is first protected with an amorphous silicon layer to prevent damage from sparks, and a foil is laid on top of the polymer pillars. A high voltage applied between the foil and the chip amplifies the charge, and the amplified charge pulse is detected by the readout chip. One of the most spectacular images produced (of two beta-particles[16] rotating in a magnetic field) is shown in Figure 10.9b. This image shows clearly that, with a 2D detector which records time per pixel, it is possible to create a beautiful 3D reconstruction of the particle tracks. Currently, this concept is being evaluated for use in a future linear collider.[17]

---

15) Ibidem.
16) As unstable atoms decay, they release radiation in the form of electromagnetic waves and subatomic particles. Some forms of this radiation can detach electrons from, or ionize, other atoms as they pass through matter; this is referred to as ionizing radiation. Alpha- and beta-particles, X-rays and gamma-rays are forms of ionizing radiation.
17) H. van der Graaf, *Gaseous detectors*, Nuclear Instruments and Methods in Physics Research Section A: Accelerators, Spectrometers, Detectors and Associated Equipment, Volume 628, Issue 1, p. 27, 2011.

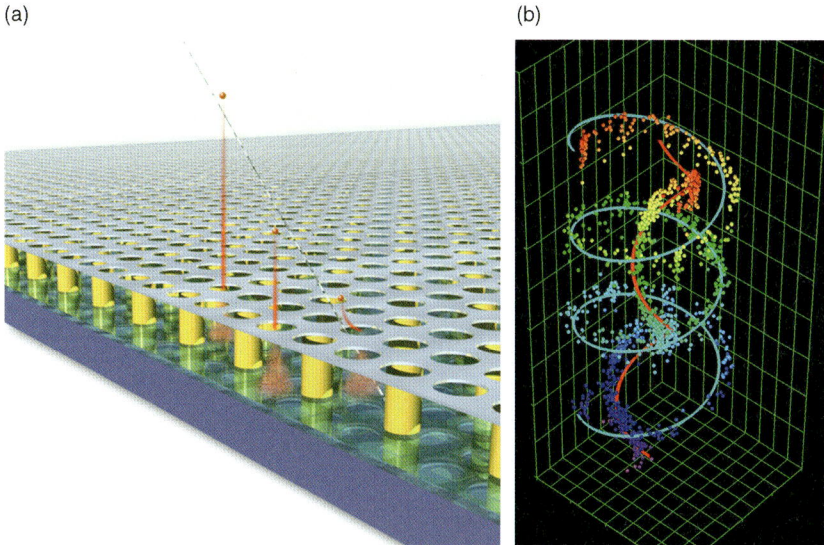

Figure 10.9 (a) Image of the GridPix concept; (b) Two beta-particles from a strontium-90 source curling in a magnetic field, recorded with a Timepix-based GridPix detector. (Courtesy of Harry Van der Graaf, NIKHEF,[18] the Dutch National Institute for Nuclear Physics and High Energy Physics).

10.4.5
Radiation Monitoring

There are many other applications to which the various chips of the Medipix family have been applied, and it is fair to say that many of those were unforeseen at the start of the development. An example of a development which was foreseen from the beginning was led by a group based at the CEA,[19] the French Government-funded Technological Research Organization in Saclay. Their idea was to develop a lightweight and sensitive imaging system to permit the safe decommissioning of nuclear installations. In Figure 10.10, it can be seen that they largely succeeded, and this system is now being produced commercially by the French company, Areva.[20] The Medipix2 Collaboration has no formal link with Areva, but the latter company sources its devices via one of the (at the time of writing) four spin-off companies that provide Medipix2 and Timepix sensors commercially. All four of these spin-off companies originated at academic institutes that participated in the Medipix2 Collaborations, and are based in the Czech Republic, Germany, Spain and The Netherlands. Clearly, such endeavours can lead to new commercial activities.

---

18) NIKHEF (Nationaal Instituut voor Kernfysica en Hoge-Energiefysica).
19) CEA (Commissariat à l'Énergie Atomique et aux Énergies Alternatives).
20) www.areva.com.

Figure 10.10 A 1 s exposure image taken with the Medipix2-based system developed by CEA Saclay, showing the presence of contamination emitting at the level of 300 µSv h^{-1}. (Courtesy of Mhedi Gmar, CEA-LIST, the French Laboratory for System and Technology Integration).[21]

10.4.6
Chemistry

An excellent example of a fully unforeseen application was developed by colleagues at the Chemistry Department in the University of Leiden, The Netherlands, who used a Timepix chip combined with a silicon sensor to replace a commercial imaging camera in a system built to study the growth of graphene flakes on a surface. The technique, known as Low-Energy Electron Microscopy (LEEM), involves scattering very low-energy electrons (ca. 10 eV) from a surface of interest; the scattered electrons are then accelerated towards the detecting device, giving them enough energy to deposit sufficient charge in the silicon sensor that each electron can be detected by the readout chip. As can be seen in Figure 10.11, a spectacular improvement was achieved in image quality as a result of the detection of single electrons using Timepix.

10.4.7
Dosimetry in Space

Probably one point that astronauts worry least about is the amount of ionizing dose they receive during missions. However, the total exposure during long missions is a concern, especially if the moon is to be colonized or Mars visited, when the total accumulated dose will become a major concern. Likewise, astronauts working at the International Space Station (ISS), which is in a relatively low

---

21) LIST (Laboratoire d'Integration des Systèmes et des Technologies).

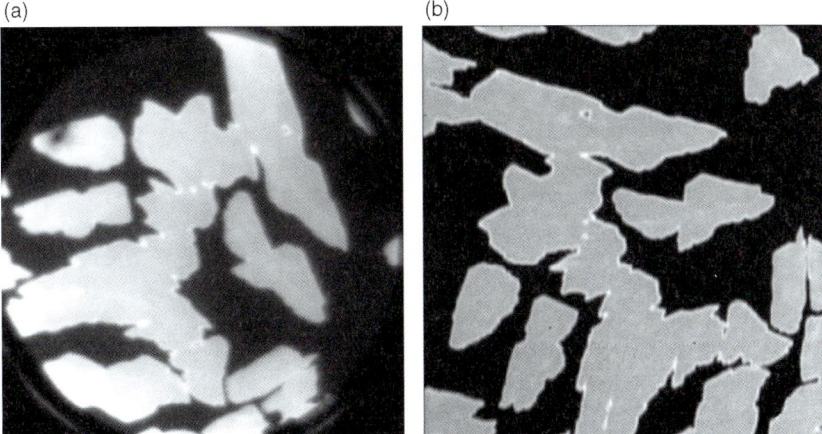

Figure 10.11 Comparison of images taken with (a) a conventional Low-Energy Electron Microscopy (LEEM) detector and (b) a Timepix-based detector.[22] (Reproduced with permission of Elsevier).

Earth orbit, would like to avoid receiving sudden high radiation doses during periods of increased solar activity. The type of radiation present in space is quite different from that to which we are exposed on Earth, because the Earth's magnetic field acts like a shield and causes charged particles to be focussed at the poles (this is the origin of the Northern and Southern lights). Yet, this type of radiation, which is composed mostly of protons and heavily charged particles, produced very characteristic tracks on the Timepix device. Our colleagues from the University of Houston, US, together with the IEAP, have sent six Timepix-USB systems to the ISS for evaluation (see Figure 10.12). These studies are leading the effort towards a new generation of radiation area monitors and personnel dosimeters for space flight crew. Consideration is also being given to the use of Timepix-based chips as part of future Mars Exploration Rover experiments.[23]

10.4.8
Education

In 2006, the CERN Medipix team received a first visit from a group of high-school students from the Simon Langton Grammar School in Canterbury, UK. The students experienced first-hand how to detect radiation in the environment using the USB-based Timepix detector. According to their teacher, the students left full of enthusiasm to get their hands on one of the Medipix systems and to start creating their own experiments with it. The Collaboration agreed to lend them a few systems and, through the efforts and imagination of the students,

---

22) I. Sikharulidze *et al.*, *Low-energy electron microscopy imaging using Medipix2 detector*, Nuclear Instruments and Methods in Physics Research Section A: Accelerators, Spectrometers, Detectors and Associated Equipment, Volume 633, Supplement 1, p. S239, 2011.

23) http://mars.jpl.nasa.gov.

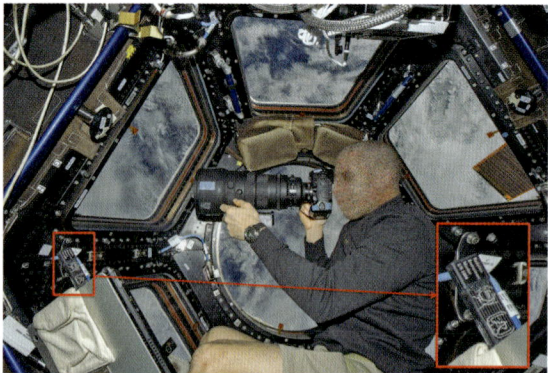

Figure 10.12 Image of the astronaut Christofer Cassidy working near the Timepix USB (Universal Serial Bus) on the International Space Station, ISS. (Courtesy of National Aeronautics and Space Administration, NASA).

they devised the idea of entering a British national science competition with the concept of placing five Timepix chips in the form of a cube to be installed on a satellite to study cosmic radiation. Their entry was rated second in the competition, and was therefore funded. In 2014, the TechDemoSat satellite will be launched from Baikonur, Russia, containing the LUCID (Langton Ultimate Cosmic ray Intensity Detector).[24] The data produced will be beamed back to Earth and then made available to other schools via the LHC computing Grid. Needless to say, the young scientists at Langton are getting a very real taste of how a high-energy physicist feels when his or her experiment starts to take data for the first time. In the meantime, there is an on-going programme to find a means of funding the provision of the USB-based readout system (codenamed: CERN@school) to as many schools as possible throughout the CERN Member States.

10.4.9
Art Meets Science

Sometimes, the beauty of our natural surroundings can be captured by even the most primitive camera, yet in other circumstances it may take a combination of special lighting and shadows and very sophisticated equipment to bring the beauty of a particular object fully into focus. A colleague from the University of Freiburg in Germany has been experimenting with the study of very low-contrast natural objects illuminated using a micro-focus X-ray source. Although these studies began with the aim of providing scientific evidence of the quality of the Medipix3 detector, it soon became apparent that the images in themselves are beautiful, and many have been taken with the aim of highlighting the beauty

---

24) http://www.sstl.co.uk/Blog/February-2013/TechDemoSat-1-s-LUCID–a-novel-cosmic-ray-detector.

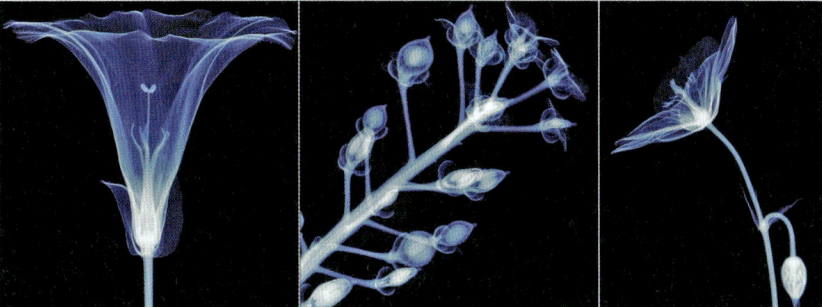

Figure 10.13 X-ray images of various wild flowers. A Medipix3 detector with a 300 μm-thick silicon sensor was used for this measurement. The X-ray flux was produced with a microfocus X-ray tube with a tungsten anode. In order to achieve a larger field of view, the detector was raster scanned. In combination with a 2× geometrical magnification, a total of 6 × 8 tiling positions was necessary to scan each flower. (Courtesy of Simon Procz, University of Freiburg).[25]

of these natural objects. I will leave to the reader the opportunity to appreciate a sample of my colleague's work (see Figure 10.13).

10.5 Back to High-Energy Physics

This chapter has emphasized how the hybrid pixel detector technology, which was initially developed for tracking at the LHC, had been adapted to other applications. At the present time, the Medipix3 Collaboration is developing two new chips which are pointing the way forward to new HEP applications. The Timepix3 chip, as its name suggests, builds on experience with the Timepix and Medipix3 chips, but aims to explore a new data push readout scheme. In other words, instead of waiting for instructions to take and read out data, the chip will start emitting hit information as soon as it starts to detect events. This approach is being evaluated for adaptation to an upgrade of the LHCb vertex detector. The chip will also be used by groups studying new detector concepts for a future linear collider. The Smallpix chip aims to shrink the pixel size, further pushing the limits of the bump bonding interconnect technology towards the dimensions which will be needed at a future linear collider. Looking ahead, the increased cost of prototyping in even more advanced processes will oblige different groups to pool resources, just as Medipix has done. It could be that the Medipix Collaborations form a template on which future HEP chip design efforts are based.

25) S. Procz, *Hochauflösende Computertomographie mit Medipix3-Halbleiterdetektoren*, PhD thesis, University of Freiburg, Germany, 2012.

10.6
Collaboration, Organization and Serendipity

The Medipix Collaborations have grown in an organic way as new applications have been found for the technology. The collaboration meetings provide an open and friendly forum where creativity can be stimulated by the mingling of minds and cultures from different areas of research. The growth of the Collaborations has permitted the pooling of resources and the exploitation of the latest CMOS processes available. It has also allowed their researchers to do more than what was originally foreseen, developing new chips and interconnect techniques which would otherwise have been financially infeasible for individual groups. An important – but often underestimated feature – has been a legal framework for the Collaborations which seeks to permit the creative mingling of ideas while protecting and acknowledging those whose intellectual property is ultimately commercialized. The rigorous legal and commercial support by members of CERN's various Knowledge and Technology Transfer (KTT) departments have played a major role. Many applications were entirely unforeseen in the beginning of the research work. Serendipity has also played an important role in the Collaborations' success, but that statement can be applied to many technical and scientific breakthroughs. The moral of the story is that collaboration between different fields of science can be – and in the case of the Medipix Collaborations has proven to be – a winning formula.

Acknowledgements

Many people have contributed to the success of the Medipix developments, and to mention only a small number of them here would surely lead to injustice to those not named. The author would therefore like to dedicate this text to all past and present members of the Medipix1, Medipix2, and Medipix3 Collaborations. He is personally indebted to all those who have sustained him, even in challenging moments, with their unfailing enthusiasm.

11
Theory for Development

Fernando Quevedo

Theoretical physics and pure mathematics are often ignored in discussions about the importance of science for development and the well-being of society. However, whenever someone talks about the great minds in the history of science, most of the names that are mentioned are the famous theorists, from Isaac Newton to Albert Einstein and, more recently, from Richard Feynman to Stephen Hawking. The lack of appreciation for the relevance of theoretical scientific subjects has a direct impact on the difficulty of making funds available for researchers in these fields. This may be understandable since, usually, the motivation for their research is curiosity rather than a direct will to solve one particular applied problem. Moreover, when the outcome of their research has direct implications for society, recognizing their contribution is not straightforward. Thus, people tend to leave them out in the distribution of credits for the corresponding application.

Experimental physicists – particularly those working in institutions such as CERN – are equally guided by the curiosity to understand our world but, contrary to theorists, they are closer to technology and, although sometimes questioned by policy makers, their impact on society is easier to appreciate.[1] Besides being the biggest experimental facility in the world, CERN also boasts a very active theoretical group that has played a crucial role in the experiments over its 60 years of operation, and has also acted as a catalyst for European theorists to interact and collaborate with each other and with their experimental colleagues. It is clear that without the ideas proposed by theorists, the experimentalists would lack a road map and many of their spectacular successes that have led to direct applications would never have been achieved. As in the case of CERN, one can hardly ignore the relevance of theorists for society.

---

1) For discussions about the importance of fundamental science: J.A. Kay and C.H. Llewellyn Smith, *The Economic Value of Basic Science*, Oxford Magazine, February 1986; *Is 'Big Science' Expensive?*, Science in Parliament, Volume 59, N. 2, p. 2, 2002; *What's the Use of Basic Science*: http://wwwnew.jinr.ru/section.asp?sd_id=94.

From Physics to Daily Life: Applications in Informatics, Energy, and Environment, First Edition.
Edited by Beatrice Bressan.
© 2014 Wiley-VCH Verlag GmbH & Co. KGaA. Published 2014 by Wiley-VCH Verlag GmbH & Co. KGaA.

11.1
The Importance of Theoretical Research Through History

History is full of examples in which theoretical research has been the driving force for many important changes in society. A typical case is that of wireless communication through radio waves, such as in a telephone conversation, which is usually attributed to Heinrich Hertz, Guglielmo Marconi and others. Whilst there is no doubt that their contributions were crucial, the line of credits should extend – at least – to the theoretical ideas of James Clerk Maxwell who, by unifying electricity and magnetism, predicted the existence of electromagnetic waves that can freely propagate through empty space at the speed of light and realized, simultaneously, that light is precisely a manifestation of electromagnetic waves (therefore putting the whole field of optics on a firm ground).

Hertz's experiments were essentially designed to test Maxwell's ideas and, curiously enough, even after succeeding, Hertz believed his results had no practical implications. The ideas of Maxwell, in turn, were influenced by the work of probably the greatest experimental physicist ever, Michael Faraday. Through his experiments on electromagnetism, Faraday had imagined the existence of electric and magnetic fields, but his mathematics skills were not strong enough to be able to synthesize his findings in mathematical terms, and he passed the challenge on to Maxwell. This series of events illustrates the close interdependence of experimental and theoretical ideas, leading after some time to important innovations that have had a major impact on society. On the road towards these innovations, it is impossible to isolate the role of theoretical and experimental developments.

Theorists have proven connections to some of the most important historical social, economic and technical movements and advances. It is impossible to overestimate the impact that the thinking generated by Galileo Galilei and Isaac Newton's mechanical view of the world had for the development of the Age of Enlightenment that had a major impact on the establishment and implementation of democracy and liberalism in political systems around the world.[2] It is also well documented that the close collaboration between theorists, experimentalists and inventors which allowed the understanding of mechanics, electromagnetism and thermodynamics led to the industrial revolution of the eighteenth and nineteenth centuries.

The discovery of quantum mechanics early in the twentieth century by the minds of Niels Bohr, Werner Heisenberg, Erwin Schrödinger, Albert Einstein and others has led to the great technological advancements of the second half of that century, from lasers to transistors and integrated circuits that have innumerable applications in today's society, to the use (for good or bad) of nuclear energy including for defence, energy generation and many medical applications in both diagnosis and treatment such as Magnetic Resonance Imaging (MRI), Computed Tomography (CT) and Radiotherapy.

---

2) T. Ferris, *The Science of Liberty: Democracy, Reason and the Laws of Nature*, Harper Collins, 2010.

It may be less known that the Global Positioning System (GPS), which is employed so ubiquitously today, relies on the level of precision provided by Einstein's theory of gravitation, his acclaimed *General Theory of Relativity*, probably the prime example of a fundamental physical theory. This theory was conceived purely on theoretical reasoning simply requiring mathematical consistency and elegance without any experimental and even less practical motivation. The Positron Emission Tomography (PET) medical scanning technique is also based on the existence of anti-particles called 'positrons', first hypothesized during the 1930s by the great Paul Dirac, who was probably the physicist with the strongest tradition of performing research based on the fundamental principles of beauty and symmetry enacted by mathematics rather than on any immediate or potential application.

The rapid development of current information technology owes its basics to a relatively unknown mathematician working at Bell Laboratories[3] in the middle of the last century, Claude Shannon, who introduced and developed *Information Theory*.[4] Shannon provided, on purely theoretical grounds, a quantitative measure of information, and this turned out to coincide with the concept of entropy, introduced during the previous century by Rudolf Clausius in heat exchange phenomena and interpreted by Ludwig Boltzmann and Josiah Willard Gibbs in the statistical interpretation of thermodynamics as a quantitative measure of the amount of disorder of a system. The enormous impact of Shannon's work includes data compression, storage and communication with use for space exploration, computing science, cryptography, and the invention of compact discs (CDs) and other digital devices.

Another mathematician of the same generation, the tragic figure of Alan Turing, laid the foundations of computer science and artificial intelligence with his vision of an all-purpose computer (the Turing machine),[5] the formal definition of algorithms, and a description of computability. It is also well known that Turing played a leading role in the British project to decipher the German codes in World War II. Turing and other theorists, such as the famous theoretical physicist, John von Neumann, who conducted related ground-breaking studies, deserve greater recognition for the enormous impact that computers are having on our lives.

It was from the mind of probably the greatest theoretical physicist of the second half of the twentieth century, Richard Feynman, that some of the current major developments in applied science originated, namely nanotechnology and some preliminary ideas on quantum computers. In a famous after-dinner speech titled, *There is plenty of room at the bottom*, presented at an American Physical

---

3) Bell Laboratories were formerly known as AT&T (American Telephone & Telegraph Company) Bell Laboratories.
4) J. Gleick, *The Information: A History, A Theory, A Flood*, Pantheon Books, 2011.
5) A Turing machine is a hypothetical device that manipulates symbols on a strip of tape according to a table of rules. This machine can be adapted to simulate the logic of any computer algorithm.

Society (APS) meeting in 1959, Feynman set the ground for the development of nanotechnology with its great potential applications.[6] His idea was that physics at the nanoscale is well understood, and that there is plenty of room to design nanometre-sized devices, manipulating atoms and molecules. Feynman's talk is generally acknowledged as the beginning of nanotechnology, with its many technological and medical applications and, interestingly enough, deriving from the brain of a theoretical rather than an applied physicist.

Feynman was also involved in the origin of quantum computing during the 1980s. When extrapolating nanotechnology ideas to the concept of computation, it is natural to imagine a computer in which the basic elements operate at the quantum scale and are therefore fully subject to the laws of quantum mechanics. Following the remarkable empirical fact known as *Moore's law*, which states that the number of transistors per integrated circuits doubles every 18 months or so, electronic components must at some point become so small to reach the truly quantum domain. If a classical computer is based on two-state 'bits' of digital information, a quantum computer uses a quantum superposition of states or 'qubits'[7] with the potential of performing calculations more efficiently than the classical computers as conceived by Turing.

Once again, quantum information was developed by theoretical minds that even had some philosophical inclination towards theoretical physics, and their research was considered by most of the community to be of no use, not even for theoretical developments. Ideas that are becoming reality in these days, such as teleportation and quantum entanglement, came from the worried brain of Albert Einstein and collaborators in their effort to reject the standard philosophical interpretation of quantum mechanics.

I was fortunate to study for my PhD at the University of Texas at Austin, US, during the mid-1980s where the great theorist, John Archibald Wheeler (who was known for many important contributions to physics, for coining the term 'black hole', and for supervising top scientists such as Feynman, Hugh Everett and others) had a small group of young researchers (including William Wooters, Wojciech Zurek and visitors such as David Deutsch) that were working on the foundational principles of quantum mechanics. At that time, they were considered to be working on such a difficult task that it was closer to philosophy than physics. Very few other groups worldwide, such as Giancarlo Ghirardi and collaborators in Italy and the great John Bell at CERN (he transformed the field by proposing his inequalities later tested experimentally by Alain Aspect in the 1980s) were following similar ideas. It was again, pleasantly surprising for me to realize over the years that this small group of people were pioneers in such an active field with potential applications and in close contact with experiments. As a theorist, this is something that is much appreciated.

---

6) www.zyvex.com/nanotech/feynman.html.
7) In quantum computing, a qubit (quantum bit) is a unit of quantum information which is a two-state quantum-mechanical system.

One of the founders of quantum mechanics, Erwin Schrödinger, who gives his name to the most famous equation in quantum mechanics, decided during the 1940s to make an incursion into the (at the time) completely separate field of biology, and with his ground-breaking book *What is life?*[8] opened the road to the development of genetics, including DNA and its many applications. Several other famous physicists can be added to the list of pioneers of molecular biology, including of course Francis Crick who, together with James Watson, deciphered the structure of the DNA molecule[9] that gave rise to the genetics revolution.

The list of ideas and discoveries can go on and on. The time taken from the new idea or paradigm to the application varies but it is usually the case, especially in the modern era, that an application is not devised or even foreseen by the theorists that originally developed the concept. The case of the laser (light amplification by stimulated emission of radiation) is a good example. Although Einstein introduced the concept of stimulated emission in 1917, it took more than 40 years for the device to be designed, and even more years before its many applications were implemented. That is why a close collaboration and flow of information is needed amongst theorists, experimentalists and applied scientists. This is only achieved if there is a well-established culture of science in a given community or country.

11.2
Knowledge Management and Science for Peace

To create a culture of science, theorists are definitely the key players. First, it is important to appreciate that science stems from a most natural human quality: curiosity and the sense of wonder – the desire to understand the world around us and to wonder about our own existence. It is this curiosity that attracts bright young people to study science. Most of them then tend to choose a specialization in the applied aspects of science, since more jobs are usually available in applied subjects that are not restricted only to academic life. However, to interest and attract these young generations to science, whether pure or applied, theorists play a key role. It is natural to motivate children into science simply by asking them to look at the sky. This fortunately is an 'experiment' available to every human being regardless of their origin. Wondering about the Universe is usually a first step towards a scientific career, both in pure and in applied science.

---

8) E. Schrödinger, *What is Life? With Mind and Matter and Autobiographical Sketches*, Cambridge University Press, 1992 (first published 1944).
9) The Nobel Prize in Physiology or Medicine 1962 was awarded jointly to Francis Harry Compton Crick, James Dewey Watson and Maurice Hugh Frederick Wilkins ' . . . for their discoveries concerning the molecular structure of nucleic acids and its significance for information transfer in living material': http://www.nobelprize.org/nobel_prizes/medicine/laureates/1962.

The great Pakistani theoretical physicist Abdus Salam, Nobel Prize winner in 1979,[10] nurtured a vision that was considered revolutionary in his time. A native of a poor country and educated in a good but isolated school, he wished to create a culture of science in all countries of the developing world by promoting the study of theoretical physics. One of his arguments was that theoretical physics needs only a brain with some appropriate and relatively modest working conditions, without the need to invest in expensive equipment.[11] He followed his vision by creating ICTP (Abdus Salam International Centre for Theoretical Physics), and other institutions, such as The World Academy of Sciences (TWAS) for the advancement of science in developing countries,[12] that have been promoting science in the developing world for almost 50 years.

The mission started by Salam and now continued by other scientists is a very long-term project, often obstructed by complicated socio-political events happening in some of these countries, but so far the overall results have been very positive. Following Salam's vision has helped many of the countries to start developing a scientific culture in which a small and fragile, but usually active and enthusiastic, scientific community promotes scientific studies with its typical characteristics including analytical thinking and problem solving – something that is very much needed in developing countries. The big challenges facing humanity for the future of our Earth, such as climate change, energy and health, have a particularly negative effect on the developing countries, which need the background knowledge to anticipate and attack these problems according to their local conditions. The community and the type of knowledge management promoted by ICTP are crucial for addressing these problems. It starts by having a critical mass of highly qualified individuals who, with the support of the policy makers, should themselves be able to address the problems that affect their own countries or regions. It is encouraging that other institutions promoting higher education in developing countries have also emerged recently, such as AUST (African University of Science and Technology) and AIMS (African Institute of Mathematical Sciences).

For 50 years, ICTP has managed to exploit one key property of science that makes it unique. It is its global aspect. Scientists from more than 185 countries come regularly to exchange and refresh their knowledge, learn from each other not only high scientific standards but also other aspects of culture. In time, our scientists have become science ambassadors by organizing conferences and workshops in developing countries, supporting research networks, supervising students, postdocs, and so on. This mentoring adds a special role for a scientist besides the standard research and teaching duties of academic life and fulfils an

---

10) In 1979, the Nobel Prize in Physics was awarded jointly to Sheldon Lee Glashow, Abdus Salam and Steven Weinberg ' . . . for their contributions to the theory of the unified weak and electromagnetic interaction between elementary particles, including, *inter alia*, the prediction of the weak neutral current': http://www.nobelprize.org/nobel_prizes/physics/laureates/1979.
11) A. Salam, *Ideals and Realities: Selected Essays of Abdus Salam*, World Scientific, 1989.
12) From its creation in 1983 until 2004, TWAS was named Third World Academy of Sciences, and from 2004–2012 as The Academy of Sciences for the Developing World.

aspect of the scientist's life that gives it a more human component. Other busy scientists from famous institutions in rich countries, when given the opportunity, are very generous with their time, knowledge and experience, which they share at the ICTP with their colleagues irrespective of their country, religion and political beliefs. This is a great contribution of science for world peace.

These days, essentially all reasonable policy makers and funding agencies have recognized the urgent need to support science in developing countries. However, they often insist on emphasizing only applied sciences and engineering rather than fundamental science, arguing that the urgent needs of these countries are such that it would be a luxury to invest in fundamental science. This short-term vision fails to appreciate several aspects of science. First, it is impossible to have good engineers if they do not have good lecturers on the basic subjects, and these lecturers have to be active scientists. Second, scientists themselves are the most important resource in establishing a scientific culture in a country. They should not be denied the right to choose the field of study that attracts them the most, since this is the reason they chose to go into science in the first place. Scientists from developing countries have the same right as any other scientist to be able to wonder about the Universe, its origin and composition. This right should not be reserved only for scientists from developed countries. Third, basic science, as argued above, is not as useless as non-experts think. Finally, scientists trained in basic subjects are usually very versatile and, when presented with a challenge, are able to adapt their knowledge to address applied subjects and also to take responsibility for promoting more applied fields. It is interesting to observe that in several emerging countries, even though the founding fathers of physics have often been theoretical physicists, over the years a larger community of applied scientists has developed naturally, approaching with time a natural state of equilibrium in which most of the physicists end up being applied or experimental with a relatively small group of theorists – a healthy and balanced configuration that is naturally achieved without the need to impose strict priorities from above.

It may seem too easy for me as a theoretical physicist to stand up and emphasize the importance of my own field of theoretical physics for development. It is hardly an unbiased situation, but let us heed Mike Lazaridis,[13] founder of Research In Motion (RIM),[14] the company that developed the Blackberry, and clearly an applied voice. According to Lazaridis, in order to make a breakthrough in science people should invest more in the support of fundamental science institutions. (He imagined sending a Blackberry back in time to Maxwell, who would not be able to understand its functioning since he did not know quantum mechanics. However, if instead of a Blackberry he would have sent Maxwell an undergraduate textbook on modern physics, he would be able to learn quantum mechanics and eventually build a smart phone.) Lazaridis actually took action in

---

13) M. Lazaridis, *The power of ideas*, The Science Show, 2012.
14) www.rim.com/index_na.shtml.

this direction by creating the now well-established Perimeter Institute[15] as well as the Institute for Quantum Computation,[16] both in Waterloo, Canada. These are both becoming leading research institutions in fundamental science created by an applied scientist who is convinced of the key role that basic science has in technology and the future of society. This is a very encouraging message to young generations of scientists.

I was fortunate to be present at the CERN auditorium during the recent announcement of the discovery of the Higgs particle, sharing the authentic excitement of basic science discoveries. It is well known that thousands of people around the world worked in unison in order to achieve this great step in the understanding of our Universe. The Nobel Prize was given one year after the discovery, but it was given first to the theorists that developed this idea 50 years before.[17] We do not even know if this discovery will lead to a concrete application in the future, and it may take many decades before we will know. But certainly, since the highest-level technology was needed to make the Large Hadron Collider (LHC) experiment, we know that this discovery has already made a huge impact in technology, not only by the invention of the World Wide Web (WWW) that has changed our world more than any applied subject developed during the past 30 years, but also by just developing the technology required to build and run these complex experiments. This is certainly a great example of collaboration amongst theoretical physicists, experimentalists and engineers, and one that has three (and not only two) sides that should be more appreciated. It is also an inspiring effort from scientists from many different countries and cultures. It is reassuring to know that CERN is making such a great effort to enhance its impact on many non-European countries, especially from the developing world. Collaborations with ICTP and other international institutions are also strengthening their common efforts.

Theoretical physics has also been expanding its domains, especially by the increasing role that computing techniques are playing in science. Theoretical modelling and the simulation of materials at the molecular level, of climate change, telecommunications, astrophysics, particle physics, quantitative biology and applied mathematics, are allowing theorists to have a more direct impact on applied fields with the use of supercomputers, the Grids pioneered by CERN and in general High-Performance Computing (HPC). Theoretical research on climate change is a good example where the modelling of climate in different regions of the planet is crucial to assess the extent of every specific regional challenge. ICTP, being a theoretical institution since its foundation, has been moving in this direction and, since 1998, has expanded its research efforts to include

---

15) www.perimeterinstitute.ca.
16) http://uwaterloo.ca/institute-for-quantum-computing.
17) The Nobel Prize in Physics 2013 was awarded jointly to François Englert and Peter W. Higgs
  '. . . for the theoretical discovery of a mechanism that contributes to our understanding of the origin of mass of subatomic particles, and which recently was confirmed through the discovery of the predicted fundamental particle, by the ATLAS and CMS experiments at CERNs Large Hadron Collider': http://www.nobelprize.org/nobel_prizes/physics/laureates/2013.

climate change, participating actively in the Peace Nobel Prize winning Intergovernmental Panel on Climate Change (IPCC).[18] These initiatives have brought theorists closer to the immediate needs of the population, in particular those from developing countries, and also close to the priorities set by the policy makers, but of course they do not substitute the core of fundamental theoretical research.

An unexpected application of fundamental science is the work of the string theorist Paul Ginsparg and his development of the arXives.[19] This is a repository of by now hundreds of thousands of original scientific articles freely accessible to everybody through the Internet. The world scientific community – especially the poorest part without money to buy very expensive journal subscriptions – can access many of the latest articles much before they are published in peer-reviewed, but expensive, scientific journals. This has not only closed the gap in access to scientific information amongst different institutions and countries but also has changed the way that all scientists work. I happened to be collaborating with Paul on a research article when he developed this idea, and have been a constant user since its beginnings more than 20 years ago, to the extent that, together with many other researchers, I have essentially stopped reading scientific journals since all the relevant information in my field is available in the arXives. Furthermore, this is the way we communicate our results with our colleagues in the most efficient way possible. The day each of us finishes an article on our own research, we upload it to the arXives and it is immediately available on the Web to everybody in the world, including our colleagues who may be working on the same subject. This has helped to accelerate communication and has optimized the way in which we work and perform our own research.

Since its beginning in 1991, the arXives have been expanding to many fields including mathematics, all areas of physics, quantitative biology, computer sciences, and financial mathematics. This is democratizing scientific research in the world. The impact of this system on the way that scientific investigations are conducted is enormous and its implications for society are definitely important, although it is difficult to determine a concrete measure of their magnitude. Some critics of string theory have claimed, sarcastically, that the arXives are the most important outcome of string theory and probably the only one with implications for our world. This may be too strong a claim, but it does show an important spin-off of theoretical physics research with impact on all science, and indirectly on society.

When discussing the direct contribution to society of scientists educated in fundamental theoretical sciences, it is clear that their contributions to educating younger generations should not be overlooked, since preserving their knowledge

---

18) In 2007, the Nobel Peace Prize was awarded jointly to IPCC (Intergovernmental Panel on Climate Change) and Albert Arnold (Al) Gore Jr, ' . . . for their efforts to build up and disseminate greater knowledge about man-made climate change, and to lay the foundations for the measures that are needed to counteract such change': http://www.nobelprize.org/nobel_prizes/peace/laureates/2007.
19) www.arxiv.org.

and experiences for the good of the future generations is probably the most important social activity that a scientist can perform. A country without a community of enthusiastic research workers who are well connected to the international scientific community is a doomed country since, in time, it will lose its ability to train its students to solve challenging and ever-new problems, which is the essence of scientific and technical education. It is important for a country or region to have a continuously renewed critical mass of highly trained scientists in the basic sciences that educate the younger generations, while carrying out their own research.

It is well known that the job market in academia is, unfortunately, very limited and only a selected few theoretical physicists and mathematicians end up with an academic position. Many young PhD and postdoctorate students, as well as most undergraduates, leave academic life for completely different jobs. It is again fortunate that a solid training in problem-solving and analytical thinking is appreciated very highly in the job market. Today, PhDs trained in theoretical physics are in great demand in both the industrial and financial sectors, as ultimately they are also usually the job creators who start companies specializing in software and other applied subjects. Typical examples are the famous mathematician and philanthropist James Simons, who founded Renaissance Technologies,[20] one of the world-leading hedge fund companies, and the theoretical physicist Stephen Wolfram, founder and CEO of the Wolfram Research[21] software company.

Graduates of theoretical physics and mathematics are flexible enough to be able to move to other academic fields such as applied sciences, biology and computing sciences and, within a short period, to start making contributions in a world where multidisciplinary science is becoming increasingly needed. There are also great personalities that have succeeded in the world outside science with impact on politics (Nobel Peace Laureate Andrei Sakharov is a famous example),[22] the arts, literature, administration, and diplomacy.

Theoretical physicists and mathematicians embody the concept that science is not an alternative to humanistic culture, but rather is part of it. Most of these people apply their high standards to all other areas of human endeavour, carrying with them the honesty, generosity and no-nonsense quantitative approaches with which they were imprinted. Yet, this relatively small body of people represents a precious reservoir of high-quality personalities who are generally ready and willing to serve the community, often at their own personal disadvantage.

---

20) www.rentec.com.
21) www.wolfram.com.
22) Andrei Sakharov, the father of the Soviet hydrogen bomb, was awarded the Peace Prize in 1975 ' . . . for his opposition to the abuse of power and his work for human rights': http://www.nobelprize.org/nobel_prizes/peace/laureates/1975. The leaders of the Soviet Union refused him permission to travel to receive the Prize, and his wife, Jelena Bonner, received it on his behalf. Sakharov was subsequently deprived of all his Soviet honorary titles, and the couple was for several years kept under strict surveillance. Only when Gorbachev came to power in 1985 were they allowed to return to Moscow.

Figure 11.1 Basic science educational programmes inspire the curiosity of younger generations, investing in science for the future. (Courtesy of Abdus Salam International Centre for Theoretical Physics, ICTP).[23]

Our society and our countries – not only the rich but especially the poor – need people like that.

Today, investing only in applied science would be unfair to the next generations, since we have been profiting throughout our lives from basic research conducted in the past. Hence, the legacy we can leave to our children's children is a healthy support for basic science that could guarantee better living standards in the future (Figure 11.1).

23) www.ictp.it.

Part III
Economic Aspects of Knowledge Management and Technology Transfer

Knowledge and innovation are the driving forces of today's modern economies. The capability to deal with Big Data is no longer a remote necessity but will play a fundamental role in the near future, as is explained in one of the following chapters. Taking one example from many, translational medicine will grow and computer-based predictive models of disease and multimodal health data will be stored in repositories. Soon, patient records that are created at birth will accompany people throughout their lives.

Knowledge and Technology Transfer (KTT) activities have been set up worldwide to build bridges and foster interactions between firms, universities and institutions, with the final aim of boosting knowledge and technology transfer and thus shortening the time taken for innovative technologies to penetrate into daily life. However, it is important to keep monitoring and analysing the outcomes of the KTT strategies that are adopted, as there are many different motivations for, and diverse obstacles to, the achievement of an efficient realization of technology transfer. The survey made here of Swiss companies and academic institutions contributes to adapting and modifying the strategies for better exploitation.

12
Innovation and Big Data
Edwin Morley-Fletcher

A succession of paradigm shifts has accompanied the varying perceptions of wealth in economics. Starting from those who thought that wealth resided in agriculture, moving then to those who saw wealth as correlative to the division of labour, and further to those who posited it as corresponding to profits, in opposition to rents and wages, as industrialization became, for some, synonymous with exploitation. Marginalism redirected economics away from this connection, positing wealth as the outcome of perfect competition, where profits would wane. Economies of scale were then left as the only factor of increasing returns operating as the engine of economic growth. This pleased at the same time free-traders, empire builders, and socialists and communists, who were both striving to transform the whole of society into a single factory. Schumpeter's 'creative destruction', followed, on the one hand, by Frank Knight's and John Maynard Keynes' distinction between risk and uncertainty, showed a different reading of capitalism, while on the other hand socialism's demise appeared to originate from having pursued accumulation without disruptive innovation. The sudden shock incurred since 2008 by globalization and free flows of capitals proved financial stability to be still an inherently uncertain precondition for securing further growth and a larger diffusion of wealth. Since 1990 knowledge had, however, begun to be theoretically perceived as the endogenous factor that really allows increasing returns. This implied the need to understand which interconnected set of market and nonmarket institutions could best make the innovation process work effectively, and what role should be taken by governments. Being immaterial, knowledge appears to be limitless and cognitive capitalism passage 'from atoms to bits' entails various post-scarcity features. Non-rivalness, non-excludability, cumulativeness, as well as network effects, jointly lay the ground for moving past abundance. Finding the adjacent scarcity allows to discover potential economies of scale which ensue from free (or nearly free) digital goods opening up new markets where to exert monopoly power and reap extraordinary profits. Where the primary remaining scarce resource is human creativity, the relative advantages of peer production emerge, opening to criticism of traditional intellectual property rights. Peer-produced knowledge can be more cost-effective than either markets or hierarchies, provide better information about available human capital, and better allocate creative efforts, by taking into account the highly variable and individuated nature of human intellectual efforts. The ecosystem of technologies is expected to grow in complexity and colonize new areas. Some anticipate an 'intelligence explosion' that will ultimately make humans and their machines merge and become indistinguishable. The HBP (Human Brain Project), IBMs Watson, and MD-Paedigree (Model-Driven European Paediatric Digital Repository) testify to the growing and converging power of Big Data analytics, when applied to healthcare. Amidst signals of growing difficulties in guaranteeing the sustainability of welfare states, rather than a further cost driver, a knowledge-based redesign of healthcare systems needs now to become a political and economic priority for overcoming the residual grip of scarcity which is hindering humankind from attaining its potential of holistic growth.

12.1
The Wealth of Nations: Agriculture, the Division of Labour, or Profits?

A succession of paradigm shifts has accompanied the varying perceptions of wealth in economics.

The Physiocrats[1] believed that agriculture was the only activity generating a 'net product', while industry was deemed to be 'sterile'. Given this belief, that Nature allowed humankind to obtain a surplus only in agriculture, they focused all of their reform efforts in trying to spread what they termed '*la grande culture*'; that is, a reorganization of land plots leading to the establishment of big farms.

A great shift was brought forward by Adam Smith,[2] who subverted the physiocrat approach by persuasively arguing that the 'wealth of nations' resided instead in the division of labour. His forecast was that the latter, by increasing specialization, and having it mediated through the 'invisible hand' of the market, would bring about rising efficiency in production and rising living standards.

However, then came David Ricardo. Expounding on Smith's basis, Ricardo argued that the wealth of nations was founded on profits, in opposition to agricultural rents and working class salaries. While wages were not due to increase in conditions of abundance of population, because of the working of the 'iron law of wages', growth was destined – in Ricardo's forecast – to peter out owing to the scarcity of land and the correlative increase of rents, unless, through free trade, Britain were to take advantage of comparative costs and become the workshop of the world, purchasing most of its food overseas.

---

1) F. Quesnay and the Marquis of Mirabeau, *Philosophie rurale*, 1763; Du Pont de Nemours, *La Physiocratie*, 1768.
2) A. Smith, *An Enquiry into the Nature and Causes of the Wealth of Nations*, 1776.

12.2
Industrialization and/or Exploitation

Saint-Simon saw wealth as depending on the process of 'industrialization' (this was the new term that he coined), which allowed the translation of the scientific and technical revolution of his age into economics. On this basis, he could project a future society in which poverty and war would eventually be eradicated, provided the 'industrial system' were to be applied on sufficiently large scale. In his *Parabole politique*,[3] Saint-Simon posited that if France were to lose its 50 best scientists, artists, military officers, physicians, entrepreneurs and craftsmen, it would take a whole generation to recover, whereas if 30 000 of the nation's nobles and bureaucrats were suddenly to disappear, this would not be such a dramatic loss, given that '. . . current society is truly an inverted world, where in all types of occupations men who are not capable are in charge of directing the capable ones'.[4],[5]

Karl Marx, introduced in his youth to the writings of Saint-Simon by his future father-in-law,[6],[7] stated that the bourgeoisie had 'accomplished wonders' and 'conducted expeditions' that surpassed by far any of those of the past, and was '. . . constantly revolutionizing the instruments of production, and thereby the relations of production, and with them the whole relations of society'. By the '. . . rapid improvement of all instruments of production, by the immensely facilitated means of communication', the bourgeoisie – Marx reckoned – was drawing 'even the most barbarian nations into civilization'. The cheap prices of commodities – he added – were 'the heavy artillery' with which all nations were being compelled, 'on pain of extinction, to adopt the bourgeois mode of production, [. . .] that is to become bourgeois themselves'.[8]

Modern industry, 'by means of machinery, chemical processes and other methods', was '. . . continually transforming not only the technical basis of production but also the functions of the worker and the social combinations of the labour process', revolutionizing at the same time '. . . the division of labour within society, and incessantly throwing masses of capital and of workers from one branch of production to another'.[9]

The flip side, in all this, was for Marx that profits were strictly founded on exploitation – that is, on not recognizing labour's real value, but only its money-mediated market value, where the *labour force* appeared to have a given price, as

---

3) C.H. de Saint-Simon, *L'Organisateur*, 1819.
4) N. Coilly and Ph. Régnier, *La parabole de Saint-Simon*, in *Le siècle des saint-simoniens: du Nouveau christianisme au canal de Suez*, Bibliothèque nationale de France, 2006.
5) R.B. Carlisle, *The Proffered Crown: Saint-Simonianism and the Doctrine of Hope*, Johns Hopkins University Press, 1987.
6) Baron Ludwig von Westfalen, whose daughter Jenny Karl Marx (1818–1883) married in 1843.
7) M. Kowalewsky's personal recollections on Marx (1909), in L. Meldolesi, *L'utopia realmente esistente: Marx e Saint-Simon*, p. 47, Laterza, 1982.
8) K. Marx and F. Engels, *Manifesto of the Communist Party*, 1848.
9) K. Marx, *Das Kapital, Kritik der politischen Ökonomie*, Volume 1, Verlag von Otto Meissner, 1867.

with any other good. This intertwining of exploitation and capital accumulation made capitalism to be an inherently contradictory process, leading inevitably – in Marx's forecast – to under-consumption crises.

12.3
Perfect Competition, the Disappearance of Profits, Economies of Scale

Marginalism[10] redirected the economic discourse away from the dangerous connection between profits and exploitation, positing that ultimately wealth depended on the degree by which society could approach perfect competition, where marginal returns on all forms of resource investment would eventually be equalized.

This way, entrepreneurs' profits would tend to fall in line with capitalists' interests, and vice versa. This link would work to minimize both their rates. Economies of scale[11] could, however, allow for increasing returns operating as an engine of economic growth, but also leading to the monopolization of markets as large firms would develop lower cost structures than smaller firms, driving smaller competitors out of business. Though this was not fully realized then, the economies of scale were therefore contradicting the assumption of perfect competition being the preferred social outcome as the fundamental mechanism for maximizing wealth.

The pursuit of economies of scale could justify free trade as well as, conversely, empire building and 'imperial preference'. As a synonym of modern industry, economies of scale also became a supporting argument for socialism. The greatest economy of scale would in fact be attained by organizing society as 'a single, huge industrial company'[12]: a goal commonly shared by both Kautsky and Lenin,[13] even though they conflicted on all other issues.

---

10) Marginalism started by three forerunners: Stanley Jevons (1835–1882), Karl Menger (1840–1921), and Léon Walras (1834–1910).
11) Alfred Marshall (1842–1924).
12) K. Kautsky, *Das Erfurter Programm*: An earlier draft of this programme, adopted by the Social Democratic Party of Germany during the congress held at Erfurt in 1891, had been commented upon by Friedrich Engels (1820–1895), who, on 18th and 19th June 1891, had famously written that the aim should be 'the transformation of present capitalist production [. . .] into socialist production on behalf of society as a whole and *according to a preconceived plan*' (*A Critique of the Draft Social-Democratic Program of 1891*). Kautsky (1854–1938) had specified that it was 'by no means necessary' that the passage to socialism 'be accompanied with violence and bloodshed. [. . .] Neither [was] it necessary that the social revolution be decided at one blow' since 'revolutions prepare themselves by years or decades of economic and political struggle' (*Das Erfurter Programm*).
13) V.I. Lenin, *The State and Revolution* (1917): 'The whole of society will become a single office and a single factory, with equality of labour and pay'. For Lenin there was, however, a proviso: it should be born in mind that 'this 'factory' discipline, [. . .] is by no means our ideal, or our ultimate goal. It is only a necessary step for thoroughly cleansing society [. . .] and for further progress'.

12.4
Creative Destruction

At this point, however, a young Austrian economist, Joseph Schumpeter, had entered the scene. In a path-breaking book,[14] both the static marginalist vision and Marx's picture of exploitative capitalism were replaced by the process of what Schumpeter termed 'creative destruction', fuelled by a new character: 'The dynamic, innovating entrepreneur as the lynchpin of the capitalist system, responsible not just for technical progress but for the very existence of a positive rate of profit'.[15] As Schumpeter later further clarified,[16] the initial super-profits of the innovator were a sort of 'quasi-rent', derived by a position of temporary monopoly supply of a particular good, which in due course would be eroded by imitators, requiring new innovations from the successful entrepreneur to make him survive.

Lately, Schumpeter's vision has been sharply criticized by the 2006 Nobel laureate for economics, Edmund Phelps.[17] Schumpeter's context was one in which an outstanding German intellectual personality like Max Weber had in his vocabulary 'no room for experimentation, exploration, daring, and unknowability'[18] (so much so – we may add – that moving away from the weberian 'synoptic rationality' assumption with regard to public decision makers has required, some decades later, the fundamental theoretical switch brought forward by the Herbert Simon's[19] 'bounded rationality'[20] concept). It is not surprising, therefore, that Schumpeter still thought that, though 'getting the job done' was a particularly onerous task, the 'likelihood of an 'innovation' [was] as knowable as the prospects faced by established products. There [would not be any] chance of misjudgement, provided there [were] due diligence. An expert entrepreneur's decision to accept a project and a veteran banker's decision to back it

---

14) W.J.A. Schumpeter, *Theorie der wirtschaftliche Entwicklung*, Duncker & Humboldt, 1912.
15) M. Blaug, *Great Economists before Keynes*, Wheatsheaf Books, 1986.
16) W.J.A. Schumpeter, *Business Cycles: A Theoretical, Historical and Statistical Analysis of the Capitalist Process*, McGraw-Hill, 1939; *History of Economic Analysis*, E. Boody Schumpeter (Ed.), Allen & Unwin, 1954.
17) In 2006, Edmund Phelps was awarded the Nobel Prize in Economic Sciences '. . . for his analysis of intertemporal tradeoffs in macroeconomic policy': http://www.nobelprize.org/nobel_prizes/economic-sciences/laureates/2006.
18) E. Phelps, *Mass Flourishing: How Grassroot Innovation Created Jobs, Challenge, and Change*, Princeton University Press, 2013.
19) In 1978, Herbert A. Simon was awarded the Nobel Prize in Economic Sciences: '. . . for his pioneering research into the decision-making process within economic organizations': http://www.nobelprize.org/nobel_prizes/economic-sciences/laureates/1978.
20) H.A. Simon, *Administrative Behaviour: a Study of Decision-Making Processes in Administrative Organization*, Macmillan, 1947; *Models of Bounded Rationality: Behavioural Economics and Business Organization*, The MIT (Massachusetts Institute of Technology) Press, 1982.

[would be] correct *ex ante*, even uncanny, though *ex post* bad luck might bring a loss and good luck an abnormal profit'.[21],[22]

On a more practical note, some years before, an influential American business consultant, Clayton Christensen, had already highlighted the 'dilemmas posed to innovators by the conflicting demands of sustaining and disruptive technologies'.[23],[24] In a book later to be thought of as revolutionary in its own domain, Christensen had showed how 'very capable executives in [. . .] extraordinarily successful companies, using their best managerial techniques',[25] could perfectly face the challenges of developing and adopting sustaining innovations, but had often 'led their firms toward failure',[26] when confronting disruptive technological changes which could allow to reap the real competitive advantage, by 'facilitating the emergence of new markets',[27] and where leadership would eventually be paying off huge dividends. The point was – Christensen stated – that, on one hand, 'innovation is *inherently* unpredictable',[28] because 'not only are the market applications for disruptive technologies *unknown* at the time of their development, they are *unknowable*'; and on the other hand, that 'amidst all the uncertainty surrounding disruptive technologies, managers can always count on one anchor',[29] namely that 'experts' forecasts will always be wrong'.[30]

---

21) E. Phelps, *Mass Flourishing: How Grassroot Innovation Created Jobs, Challenge, and Change*, Princeton University Press, 2013.
22) N. Rosenberg, *Schumpeter and the Endogeneity of Technology: Some American Perspectives*, Routledge, 2000.
23) C.M. Christensen, *The Innovator's Dilemma*, Harper, 2000.
24) A similar distinction, dubbed 'incremental innovation' and 'radical innovation', is at the core of David Sainsbury's *Progressive Capitalism*, though on a much different note. Following the literature on the 'varieties of capitalism' (M. Morishima, *Why Has Japan 'Succeeded?': Western Technology and the Japanese Ethos*, Cambridge University Press, 1982; R. Dore, *Taking Japan Seriously: A Confucian Perspective on Leading Economic Issues*, Athlone Press, 1987; M. Albert, *Capitalisme contre capitalisme*, Seuil, 1991; D. De Giovanni *et al.*, *Le politiche industriali della Cee*, Il Sole 24 Ore, 17, 24, 31 May, 1991; C. Hampton-Turner and F. Trompenaars, *The Seven Cultures of Capitalism*, Doubleday, 1993; K. Seitz, *Die japanisch-amerikanische Herausforderung*, Aktuell Verlag, 1994; S. Berger and R. Dore (Eds), *National Diversity and Global Capitalism*, Cornell University Press, 1996; R. Dore, *Stock Market Capitalism: Welfare Capitalism: Japan and Germany* versus *the Anglo-Saxons*, Oxford University Press, 2000; P.A. Hall and D. Soskice (Eds), *Varieties of Capitalism: Institutional Foundations of Comparative Advantage*, Oxford University Press, 2001; F. Giavazzi and R. Prodi, *Discutono di modelli di capitalismo*, Il Mulino, 2011), Sainsbury posits that Germany and Japan excel in the first type of innovation, while the US in the second, and that 'neither variety of capitalism is consistently better than the other, and the task of economic policy-makers has therefore to be the improvement of their particular variety of capitalism and its constant adaptation to the changing economic and technological opportunities and challenges that it faces' (*Progressive Capitalism: How To Achieve Economic Growth, Liberty and Social Justice*, Biteback Publishing, 2013).
25) C.M. Christensen, *The Innovator's Dilemma*, Harper, 2000.
26) Ibidem.
27) Ibidem.
28) Ibidem.
29) Ibidem.
30) Ibidem.

12.5
Risk and Uncertainty

Whatever the current perception of the limitations of Schumpeter's assumptions regarding the key role of innovation in driving long-term economic growth, it deserves to be noted that the initial focus on the crucial issue of unpredictability had been prompted, already in 1921, by an American economist, Frank Knight. He had in fact posited that it was precisely 'true uncertainty' – which he defined as being 'not susceptible to measurement and hence to elimination' – that was 'preventing the theoretically perfect outworking of the tendencies of competition'.[31] This was, according to Knight, the reason which gave 'the characteristic form of 'enterprise' to economic organization', while also accounting for 'the peculiar income of the entrepreneur';[32] had there not been such uncertainty, the entrepreneur's income would not be "what is left', after the others are 'determined'"[33] – stated Knight – but only the normal return required to pay the manager's wage and the competitive level of interest to creditors.

Though sharply aware of the crucial role of irreducible uncertainty in economic life, John Maynard Keynes objected then that 'to convert the business man into the profiteer' would be a sure way for striking 'a blow to capitalism, because it [would] destroy the psychological equilibrium which permits the perpetuance of unequal rewards. The economic doctrine of normal profits, vaguely apprehended by everyone, is a necessary condition for the justification of capitalism. The business man is only tolerable so long as his gains can be held to bear some relation to what, roughly and in some sense, his activities have contributed to society'.[34]

Still, Keynes' most eminent biographer, Robert Skidelsky, is right in stating that 'uncertainty pervades Keynes's picture of economic life. It explains why people hold savings in liquid form, why investment is volatile, and why the rate of interest doesn't adjust savings and investment. [. . .] Under capitalism, uncertainty is generated by the system itself, because it is an engine for accumulating capital goods whose rewards come not now but later. The engine of wealth creation is at the same time the source of economic and social instability'.[35] Keynes prescription would follow Knight's distinction between risk and uncertainty: in general terms, 'risk could be left to look after itself; the government's job was to reduce the impact of uncertainty'.[36] Controlling demand, while not interfering

---

31) F.H. Knight, *Risk, Uncertainty and Profit*, Houghton Mifflin Co., 1921.
32) Ibidem.
33) Ibidem.
34) J.M. Keynes, *The Collected Writings of John Maynard Keynes*, in R.E. Backhouse and B.W. Bateman, *Capitalist Revolutionary: John Maynard Keynes*, p. 60, Harvard University Press, 2011.
35) R. Skidelsky, *Keynes, The Return of the Master*, Allan lane, 2009.
36) Ibidem.

with the supply of goods,[37] was for Keynes what ought therefore to be the first objective of government policy to maintain economic and political uncertainty within acceptable bounds. A second level of intrinsic uncertainty, related to investment and financial markets, should be countered, in practical terms, by national and international monetary policy, and culturally by spreading the philosophical choice of '... arranging our affairs in such a way as to appeal to the money-motive as little as possible, rather than as much as possible'.[38]

12.6
Accumulation Without Innovation

The disquieting association of profits, innovation and uncertainty, that was coming to maturity in those years, was meanwhile looked at in a dismissive way by those who were pursuing the maximum economies of scale through economic planning. In this sense, it can be noted that when Joseph Stalin promoted a process of accelerated industrialization in the Soviet Union he ended up by providing exactly what was to be the most compelling example of the unintended consequences of trying to use forced accumulation as the key to economic development. As Jeffrey Sachs and John McArthur put it, '... the Soviet economy had very little technological change in the civilian sector for decades and, as a result, came about as close as possible to a case of a high saving rate combined with stagnant technology. It is probably fair to say that it proved a key result [...]: capital accumulation without technological advancement eventually leads to the end of economic growth'.[39] Ironically, the process which had had as grand goal the 'universal development of productive forces',[40] leading to a situation where 'with the all-around development of the individual, [...] all the springs of co-operative wealth [would] flow more abundantly',[41] endogenously turned, through

---

37) A choice which was in line with the traditional and highly influential approach of John Stuart Mill (1806–1873), who famously stated that 'the laws and conditions of the Production of wealth partake of the character of physical truths. [...] We cannot foresee to what extent the modes of production may be altered [...] by future extensions of our knowledge of the laws of nature, suggesting new processes of industry of which we have at present no conception. But howsoever we may succeed in making for ourselves more space within the limits set by the constitution of things, we know that there must be limits. [...] It is not so with the Distribution of wealth. That is the matter of human institution solely. The things once there, mankind, individually or collectively, can do with them as they like' (*Principles of Political Economy*, Book 2, 1852).

38) J.M. Keynes, *The End of Laissez-Faire* (1926), in *Essays in Persuasion*, The Collected Writings of John Maynard Keynes, Volume 9, p. 293, MacMillan-Cambridge University Press, 1989.

39) J.D. Sachs and J.W. McArthur, *Technological Advancement and Long-Term Economic Growth in Asia*, in Chong-En Bai and Chi-Wa Yuen (Eds), Technology and the New Economy, p. 161, MIT Press, 2002.

40) K. Marx, *The German Ideology*, 1845.

41) K. Marx, *Critique of the Gotha Programme*, 1875.

centralized planning, into a system which in 1980 would be bluntly characterized as the *Economics of Shortage* by the great Hungarian economist, Janos Kornai.[42]

12.7
The Real Engine of Economic Growth

When these outcomes became suddenly apparent to all, in the last decade of the twentieth century, the euphoria engendered by the 'victory of capitalism' and the 'demise of socialism' induced many to believe that globalization, in conjunction with the free flow of capitals, would bring about increasing wealth, provided economic growth could be channelled into boundaries of ecological sustainability.

Forgetting about intrinsic economic uncertainty, the theory of efficient market expectations assumed that 'financial markets [were] equivalent to insurance markets',[43] until the financial crisis of 2008 proved once more that financial stability was still an inherently uncertain precondition for securing further growth and a larger diffusion of wealth.

The economic slowdown which followed not only led to increased competition, but also highlighted the growing impossibility for advanced economies to guarantee a continual growth of output just by increasing the inputs to production process, as understood during industrialization. A new paradigm won through: the key driver should be believed to be none of those referred to until then; knowledge, and its continuous translation into innovation and knowledge and technology transfer, were instead to be seen as the real engines of long-term economic growth.

In broad conceptual terms, Hayek[44] had laid the ground for this new way of looking at development by introducing already in the 1930s the idea that, in a complex economy, know-how had necessarily to be highly dispersed and no central planner would ever be able to put it all together;[45] furthermore, in 1968 he had come to dub competition itself as a 'discovery procedure'.[46]

---

42) J. Kornai, *Economics of Shortage*, North-Holland, 1980; *Growth, Shortage and Efficiency*, Blackwell, 1982; *Contradictions and Dilemmas: Studies on the Socialist Economy and Society*, The MIT Press, 1986; *The Socialist System: The Political Economy of Communism*, Clarendon Press, 1992; *Innovation and Dynamism: Interaction between Systems and Technical Progress*, Working Paper N. 33, World Institute for Development and Research, United Nations University, 2010.
43) R. Skidelsky, *Keynes, The Return of the Master*, Allan Lane, 2009.
44) In 1975 Frederick A. Hayek was awarded the Nobel Prize in Economic Sciences jointly to G. Myrdal '. . . for their pioneering work in the theory of money and economic fluctuations and for their penetrating analysis of the interdependence of economic, social and institutional phenomena': http://www.nobelprize.org/nobel_prizes/economic-sciences/laureates/1974.
45) F.A. Hayek, *Socialist calculation*, p. 181 (1935); *Economics and Knowledge*, p. 33 (1937), *The Use of Knowledge in Society*, p. 77 (1945), in F.A. Hayek, *Individualism and Economic Order*, University of Chicago Press, 1948.
46) F. von Hayek, *Competition as a Discovery Procedure* (1968), in F.A. Hayek, *New Studies in Philosophy, Politics, Economics and the History of Ideas*, p. 179, Routledge & Kegan, 1978.

12.8
Endogenous Technological Change

How the reorientation of economic research could technically make its way towards recognizing a fundamental role to knowledge is in its own right a fascinating tale. This tale has been neatly told by, amongst others,[47],[48] David Warsh,[49] showing how it started from the matter of fact assessment by Robert Solow,[50] in 1957, that technological change accounted for seven-eighths of the growth of the US economy while increases in capital stock accounted for only one-eighth of the growth of income per person in the US.[51] The new perception of the problem then slowly moved on, passing first through Lucas[52] and Prescott in 1971,[53],[54] then Martin Weitzman in 1982,[55] followed by Elhanan Helpman and Paul Krugman[56] in 1987,[57],[58] until Paul Romer entered the scene in 1990.[59] Romer clearly posited knowledge as the endogenous factor that economists had hitherto failed to take adequately into account: its accumulation and deepening

---

47) D. Acemoglu, *Introduction to Modern Economic Growth*, Princeton University Press, 2008.

48) P. Aghion and P. Howitt, *The Economics of Growth*, The MIT Press, 2009.

49) D. Warsh, *Knowledge and the Wealth of Nations: A Story of Economic Discovery*, Norton, 2006.

50) In 1987, Robert Solow was awarded the Nobel Prize in Economic Sciences '. . . for his contributions to the theory of economic growth': http://www.nobelprize.org/nobel_prizes/economic-sciences/laureates/1987.

51) R.M. Solow *Technical Change and the Aggregate Production Function*, Review of Economics and Statistics, Volume 39, N. 3, p. 312, 1957; *The last 50 years in growth theory and the next 10*, Oxford Review of Economic Policy, Volume 23, N. 1, p. 3, 2007.

52) In 1995, Robert E. Lucas was awarded the Nobel Prize in Economic Sciences '. . . for having developed and applied the hypothesis of rational expectations, and thereby having transformed macroeconomic analysis and deepened our understanding of economic policy': http://www.nobelprize.org/nobel_prizes/economic-sciences/laureates/1995.

53) R.E. Lucas and E.C. Prescott, *Investment Under Uncertainty*, Econometrica, Volume 39, N. 5, p. 659, 1971.

54) R. Lucas, *On the Mechanics of Economic Development*, Journal of Monetary Economics, Volume 22, Issue 1, p. 3, 1988; *Why Doesn't Capital Flow from Rich to Poor Countries*, American Economic Review, Volume 80, N. 2, p. 92, 1990.

55) M. Weitzman, *Increasing Returns and the Foundations of Unemployment Theory*, Economic Journal, Volume 92, Issue 368, p. 782, 1982.

56) In 2008, Paul Krugman was awarded the Nobel Prize in Economic Sciences '. . . for his analysis of trade patterns and location of economic activity': http://www.nobelprize.org/nobel_prizes/economic-sciences/laureates/2008.

57) E. Helpman & P. Krugman, *Market Structure and Foreign Trade: Increasing Returns, Imperfect Competition, and the International Economy*, MIT Press, 1987.

58) E. Helpman, *The Mystery of Economic Growth*, Harvard University Press, 2004.

59) P.M. Romer, *Endogenous Technological Change*, Journal of Political Economy, Volume 98, N. 5, p. 71, 1990; *Increasing Returns and Long Run Growth*, Journal of Political Economy, Volume 94, N. 5, p. 1002, 1986; *Growth Based on Increasing Returns due to Specialization*, American Economic Review, Volume 77, N. 2, p. 55, 1987; *Crazy Explanations for the Productivity Slowdown*, NBER (National Bureau of Economic Research) Macroeconomics Annual, p. 163, 1987; *New goods, old theory, and the welfare costs of trade restrictions*, Journal of Development Economics, Volume 43, Issue 1, p. 5, 1994.

was the source of increasing returns, which explained why economic growth could be accelerating in rich countries, where the standard of living was diverging rapidly from poorer countries, contradicting the classical economic law of diminishing returns. It was by investing in knowledge that the final output could increase more than proportionately, because of the impetus given this way to the introduction of new technology.

A metaphor that Romer later made use of was plainly taken from everybody's experience in the kitchen: 'To create valuable final products – he stated – we mix inexpensive ingredients together according to a recipe. [. . .] If economic growth could be achieved only by doing more and more of the same kind of cooking, we would eventually run out of raw materials [. . .]. History teaches us, however, that economic growth springs from better recipes, not just from more cooking. [. . .] Every generation has perceived the limits to growth that finite resources and undesirable side effects would pose if no new recipes or ideas were discovered. And every generation has underestimated the potential for finding new recipes and ideas. We consistently fail to grasp how many ideas remain to be discovered. [. . .] Possibilities do not add up. They multiply'.[60]

12.9
The Appropriate Set of Market and Non-Market Institutions

Clearly, this new approach implied the need to understand what interconnected set of market and non-market institutions could best make the innovation process work effectively.[61] This subsequently implied understanding what type of strategies governments should follow in order to foster highly innovative economic systems.

A crucial feature of any innovation strategy would be, for instance, betting on promoting enhanced higher education as a major long-term investment by government. Such a public policy priority would go hand in hand with the need to fostering sufficient scale for a dynamic environment where scientific and technological progress could jointly feed the development of a competitive

---

60) P.M. Romer, *Economic Growth*, in D.R. Henderson (Ed.), *The Concise Encyclopaedia of Economics*, Liberty Fund, 2007.

61) The need to be aware of the 'richness of the institutional alternatives between which we have to choose' had already been most influentially brought forward by Ronald Coase (*The Choice of Institutional Framework*, Journal of Law and Economics, Volume 17, Issue 2, p. 493, 1974; *The Institutional Structure of Production*, The American Economic Review, Volume 82, Issue 4, p. 713, 1992), Oliver Williamson (*The Economic Institutions of Capitalism: Firms, Markets, Relational Contracting*, The Free Press, 1986), and Douglass North (*Institutions, Institutional Changes and Economic Performance*, Cambridge University Press, 1990).

innovation system.[62] As a further feature, the option of an export-orientated, open economy would assume the whole world as a potential market, compensating this way – by the scope of the market – the significant R&D investments which had to be taken into account as necessary 'fixed costs' and 'barriers to entry' of innovation endeavours, and which needed therefore to be recouped through subsequent worldwide sales. Last, but not least, special financing mechanisms beyond the banking sector would prove to be crucial, since banks would normally not fund projects, even if based on excellent ideas, because knowledge constituting their background would be intangible and non-collateralisable. And if banks do not and should not lend for non-collateralized ideas, the innovation process requires then somebody else who will be able to consider that 'creativity and vision are resources':[63] venture capitalists, in the first place, and more generally a capital market making it possible for successful Initial Public Offerings (IPOs) to raise equity. In this way, the openness to innovation processes implies not only specific forms of organization that develop, test, and prove ideas,[64] and 'bureaucratic regulatory environments that [do not] impede capital and labour movements [nor] place unnecessary burdens on firm creation and dissolution',[65] but also a general frame of mind that truly takes into account that the economic death of old sectors is part and parcel of the advance of new sectors.[66]

---

62) Both of the above-reported features are, by the way, not at all new. They can be traced back to *Das Nationale System Der Politischen Ökonomie* (1841), where Friedrich List (1789–1846) had written that '. . . the present state of the nations is the result of the accumulation of all discoveries, inventions, improvements, perfections, and exertions of all generations which have lived before us: they form the mental capital of the present human race, and every separate nation is productive only in the proportion in which it has known how to appropriate these attainments of former generations, and to increase them by its own acquirements. [. . .] There scarcely exists a manufacturing business which has no relation to physics, mechanics, chemistry, mathematics or to the art of design, and so on. No progress, no new discoveries and inventions can be made in these sciences by which a hundred industries and processes could not be improved or altered. In the manufacturing State, therefore, sciences and arts must necessarily become popular' (D. Sainsbury, *Progressive Capitalism: How To Achieve Economic Growth, Liberty and Social Justice*, Biteback Publishing, 2013).

63) E. Phelps, *Mass Flourishing: How Grassroot Innovation Created Jobs, Challenge, and Change*, Princeton University Press, 2013.

64) Ibidem: McKinsey estimated that, from 10 000 business ideas, 1000 firms are founded, 100 receive venture capital, 20 go on to raise capital in an initial public offering of share, and two become market leaders.

65) R.D. Atkinson and S.J. Ezell, *Innovation Economics: The Race for Global Advantage*, Yale University Press, 2012.

66) Ibidem: Atkinson and Ezell remark that 'even though Scumpeter was a European, Europeans are not Schumpeterians. They want the benefit of a knowledge-based economy without the creative destruction that not only accompanies it but also is required to achieve it'.

12.10
Limitless Knowledge

On top of these, other features appear to be further disruptive on a deeper theoretical level. While, with industrial development, economic growth was based on the usage of limited natural resources, knowledge, being immaterial, appears to be limitless. The passage 'from atoms to bits'[67] seems to imply also that the number of bytes we make use of can continue to grow exponentially.

This is not, however, the only post-scarcity element characterizing what has been termed 'cognitive capitalism'.[68],[69],[70] Paradoxically, Marx's insight on machines, being 'organs of the human brain, created by the human hand', which express 'the power of knowledge, objectified',[71] acquires in this context a prophetic echo. So much so, as to announce with an anticipation of 150 years, 'the beginning of a new phase of the division of labour in which the development of fixed capital indicates to what degree general social knowledge has become a *direct* force of production, and to what degree, hence, the conditions of the process of social life itself have come under the control of the *general intellect* and been transformed in accordance with it; to what degree the powers of social production have been produced, not only in the form of knowledge, but also as immediate organs of social practice, of the real life process'.[72]

It is irrelevant to establish, here, whether Marx was positing, in these passages, 'the possibility of a direct transition to communism'.[73] What is significant for us is the fact that the current 'radical transformation of the foundations of wealth'[74] appears to be the consequence of a 'virtualisation of the economy' entering into a condition of growing non-scarcity.

---

67) N. Negroponte, *Being Digital*, Knopf, 1995.
68) C. Vercellone (Ed.), *Capitalismo cognitivo: Conoscenza e finanza nell'epoca post-fordista*, Manifestolibri, 2006.
69) Y. Moulier Boutang, *Le capitalisme cognitif: La nouvelle grande transformation*, Editions Amsterdam, 2007.
70) A. Fumagalli and S. Lucarelli, *A model of Cognitive Capitalism: a preliminary analysis*, European Journal of Economic and Social Systems, Volume 20, N. 1, p. 117, 2007.
71) K. Marx, *The Fragment on Machines*, Grundrisse, 1857–1858.
72) Ibidem, Chapter *The Development of Machinery*: 'In a word – Marx had stated, just a few lines above – it is the development of the social individual which appears as the great foundation-stone of production and of wealth'. In publishing *Das Kapital*, Marx would confine in a footnote the remark that: 'A critical history of technology would show how little any of the inventions of the eighteenth century are the work of a single individual. Hitherto there is no such book'.
73) C. Vercellone, *The Hypothesis of Cognitive Capitalism*, paper presented at the Birkbeck College and at SOAS (School of Oriental and African Studies) Annual Conference on Towards a Cosmopolitan Marxism, 2005.
74) Y. Moulier Boutang, *Le capitalisme cognitif: La nouvelle grande transformation*, Editions Amsterdam, 2007.

The non-rivalness,[75] non-excludability,[76] cumulativeness,[77] and network[78] characteristics of knowledge have 'the potential of creating a 'combinatorial explosion' [allowing it to achieve] 'almost infinitely increasing returns'.[79] Correlatively, a knowledge-based economy would be, however, expected to prevent natural market incentives from achieving allocatively efficient outcomes. But instead of determining a 'tragedy of the commons' – as would run a much cited theory put forward by Garrett Hardin[80] – the final result looks likely to be a 'happy ending', provided it is understood that 'for managing 'knowledge commons' the social regulations which will be needed are to be totally different from those generally used for regulating systems founded on exhaustible resources'.[81],[82]

12.11
Post-Scarcity and Networks

To explain this assumption, Yann Moulier Boutang has recourse to the metaphor of pollination by bees. Google – he writes – employs 19 000 people at its premises in Mountain View, California, while having at the same time 15 000 000 people who work for it for free, producing information, creating a network: 'The value of this network creation activity can be compared to that developed by bees when they pollinate. The value of their honey, and of the wax extracted from their alveoles, when brought to the market, is from 350 to 1000 times lower than the value of the pollination that they perform. Google achieves to draw market profit from human pollination'.[83]

On a different side of the political spectrum, a similar perception of the role of networks – which exist when a product's value to the user increases as the number of users of the product grows and 'each new user of the product derives

---

75) Non-rivalness is the characteristic of indivisible benefits of consumption, such that one person's consumption does not preclude that of another.
76) Non-excludability is the characteristic that makes impossible to prevent others from sharing in the benefits of consumption.
77) Cumulativeness is the characteristic that allows increasing the value of existing knowledge by adding further knowledge to it.
78) Network is the characteristic by which the utility that a user derives from consumption of a good increases with the number of other agents consuming the good.
79) D. Foray, *L'économie de la connaissance*, La Découverte, 2000.
80) G. Hardin's thesis was that, when sharing a resource, individuals acting independently and having in mind only each one's self-interest, would act contrary to the group's long-term best interests and deplete the common resource (*The Tragedy of the Commons*, Science, Volume 168, N. 3859, p. 1242, 1968).
81) D. Foray, *L'économie de la connaissance*, La Découverte, 2000.
82) S.J. Liebowitz and S.E. Margolis, *Network Externality: An Uncommon Tragedy*, Journal of Economic Perspectives, Volume 8, N. 2, p. 133, 1994.
83) Y. Moulier Boutang, *L'abeille et l'économiste*, Carnets Nord, 2010.

Table 12.1 Examples of scarcity and abundance thinking.[84]

	Scarcity	Abundance
Rules	Everything is forbidden unless it is permitted	Everything is permitted unless it is forbidden
Social model	Paternalism ('We know what's best')	Egalitarianism ('You know what's best')
Profit plan	Business model	We'll figure it out
Decision process	Top-down	Bottom-up
Management style	Command and control	Out of control

private benefits, but also confers external benefits (network externalities) on existing users'[85] – has led to focus on the concept that 'today the most interesting business models are in finding ways to make money around Free',[86] that is to make use of network effects and of the correlated economies of scale in providing for free (or almost free) a digital good to create a new market where they can exert a strong monopoly power and reap extraordinary profits.[87],[88]

Abundance thinking – as Chris Anderson has written – is not only discovering '. . . what will become cheaper, but also looking for what will become more valuable as a result of that shift, and moving to that. It's the engine of growth, something we've been riding since even before David Ricardo defined the 'comparative advantage' of one country over another in the eighteenth century. Yesterday's abundance consisted of products from another country with more plentiful resources or cheaper labour. Today's also consists of products from the land of silicon and glass threads'.[89]

Chris Anderson has tried to describe some of the conceptual passages implied by the shift from scarcity to abundance as shown in Table 12.1. His basic assumption is that the new way to compete within the Free economic environment will be 'to move past the abundance to find the adjacent scarcity. If software is free, sell support. If phone calls are free, sell distant labour and talent that can be reached by those free calls (the Indian outsourcing model in a nutshell). If your skills are being turned into a commodity that can be done by software (hello, travel agents, stockbrokers, and realtors), then move upstream to more complicated problems that still require the human touch. Not only can you compete with Free in that instance, but the people who need these custom solutions are often the ones most willing to pay highly for them'.[90] A blunter forecast has recently been posited by Jeremy Rifkin,[91] who sees the new emerging

84) C. Anderson, *Free: The Future of a Radical Idea*, Random House, 2009.
85) W.H. Page and J.E. Lopatka, *Network Externalities*, Encyclopaedia of Law and Economics, 1999.
86) C. Anderson, *Free: The Future of a Radical Idea*, Random House, 2009.
87) O. Bomsel, *Gratuit! Du déploiment de l'économie numérique*, Gallimard, 2007.
88) O. Shy, *The Economics of Network Industries*, Cambridge University Press, 2001.
89) C. Anderson, *Free: The Future of a Radical Idea*, Random House, 2009.
90) Ibidem.
91) J. Rifkin, *The Zero Marginal Cost Society: The Internet of Things, the Collaborative Commons, and the Eclipse of Capitalism*, Palgrave MacMillan, 2014.

technology infrastructure – the Internet of Things (IoT) – as having the potential in the years ahead of pushing increasing segments of economic life to near zero marginal cost, making goods and services nearly priceless, abundant, and less and less subject to market forces. This plummeting of marginal costs – he argues - is going to spawn a hybrid economy – residually capitalist market and growingly Collaborative Commons – in which capitalism will have an increasingly streamlined role, primarily as an aggregator of network services and solutions, and as a niche player. As more and more people – says Rifkin – will learn to live 'beyond markets', this will trigger far reaching implications for society: social capital and crowdsourcing becoming as important as financial capital and banking, access trumping ownership, sustainability superseding consumerism, cooperation ousting competition, 'sharable value' increasingly replacing 'exchange value', traditional educational institutions being substituted by free Massive Open Online Courses (MOOCs), etc.

12.12
Intellectual Property Rights

Will Free destroy the financial incentives needed to push individuals and firms to undertake the costs, risks, and efforts of developing new knowledge? Particularly in research – it had been stated – 'there are huge incentives to free-ride the system: let someone else finance the big breakthroughs and let us focus our money on development, where big payoffs are not far away. Even amongst governments, there is now an incentive to free-ride and let some other government somewhere else in the world pay for the basic research'.[92]

According to Christensen, the argument that Free represents an attack against intellectual property rights 'straddles the line between *libre* and *gratis*. [. . .] The history of intellectual property [is] based on the long traditions of the scientific world, where researchers freely build on the published work of those who came before. In the same vein, the creators of the patent system (led by Thomas Jefferson) wanted to encourage sharing of information, but they realized that the only way people thought they could get paid for their inventions was-to hold them secret. So the founding fathers found another way to protect inventors – the seventeen-year patent period. In exchange for open publication of an invention (*libre*), the inventor can charge a license fee (not *gratis*) to anyone who uses it for the term of the patent. But after that term expires, the intellectual property will be free (*gratis*). [. . .] However, a growing community of creators doesn't want to wait that long. They're choosing to reject these rights and release their ideas [. . .] under licenses such as Creative Commons [. . .] believe that real

---

92) L. Thurow had squarely advocated that, '. . . just as the industrial revolution began with an enclosure movement in England that abolished common land and created private land, the world now needs an organized enclosure movement for intellectual property rights' (*Building Wealth: The New Rules For Individuals, Companies, and Nations in a Knowledge-Based Economy*, HarperCollins, 1999).

Free – both *gratis* and *libre* – encourages innovation by making it easier for other people to remix, mash up, and otherwise build on the work of others'.[93]

On the whole, innovative societies have adopted pragmatic compromises, such as that basic scientific discoveries, in general, are not patentable, and that patents are limited to specific new technologies, and – as we have just seen – are given for a limited period of time. Still, the indivisibility, uncertainty and externalities characteristics of knowledge-generating activities may determine a situation of 'market failure': on the one hand, markets provide too few incentives to introduce new innovations, and the production of information may well, therefore, be insufficient from a social point of view; on the other hand, innovators and creators face a consistent risk of incomplete returns for their efforts.

12.13
Governments' Support of Scientific Research

This problem of appropriability has widely led governments to support basic scientific discovery through direct subsidization of primary research in universities, government research laboratories, and even private companies that qualify for government grants. But in fact the state does more than just fix market failures: 'Large-scale and long-term government investment has been the engine behind almost every general purpose technology in the last century'.[94],[95],[96],[97],[98] This is true for Information and Communications Technology (ICT), but also 'three-quarters of the new molecular biopharmaceutical entities owe their creation to publicly funded laboratories', and 'in the past ten years, the top ten companies in this industry have made more in profits than the rest of the Fortune 500[99]

---

93) C. Anderson states: 'Generation Free – he further posits – doesn't assume that as go bits, so should go atoms. They don't expect to get their clothes or apartments for free; indeed they're paying more than ever for those. Give the kids credit: They can differentiate between the physical and the virtual, and they tailor their behaviour differently in each domain' (*Free: The Future of a Radical Idea*, Random House, 2009).

94) V.W. Ruttan, *Is War Necessary for Economic Growth?: Military Procurement and Technology Development*, in M. Mazzucato, The Entrepreneurial State: Debunking Public vs. Private Sector Myths, p. 62, Anthem Press, 2013.

95) F.L. Block and M.R. Keller (Eds), *State of Innovation: The U.S. Government's Role in Technology Development*, Paradigm, 2011.

96) E. Helpman (Ed.), *General Purpose Technologies and Economic Growth*, The MIT Press, 1998.

97) B. Jovanovic and P.L. Rousseau, *General purpose technologies*, in P. Aghion and S. Durlauf (Eds), Handbook of Economic Growth, p. 1184, Elsevier, 2005.

98) R.G. Lipsey et al., *Economic Transformations: General Purpose Technologies and Long-Term Economic Growth*, Oxford University Press, 2005.

99) The Fortune 500 is an annual list compiled and published by *Fortune* magazine that ranks the top 500 US closely held and public corporations as ranked by their gross revenue after adjustments made by *Fortune* to exclude the impact of excise taxes companies incurred. The list includes publicly and privately held companies for which revenues are publicly available. The first Fortune 500 list was published in 1955.

companies combined'.[100] Yet, in general, 'nothing at all is paid in royalties. It is simply assumed that the public investment is meant to help create profits for the firms in question, with little to no thinking about the obvious distorted distribution of risk and reward this presents'.[101],[102]

This has led Mariana Mazzucato to devote a chapter of her 2013 book, *The Entrepreneurial State*, to 'socialization of risk and privatization of rewards', and to raise the question 'can the Enterpreneurial State eat its cake too?'[103],[104]

Whatever will be the prevailing answer, and whether research funding may come to look more like venture capital, maintaining a stake in the future exploitation of results, the key theoretical issue which is at stake implies reaching a much deeper understanding of the way markets, organizations, and 'commons-based peer production' reciprocally position themselves with regard to their specific comparative advantages.

12.14
The Remaining Scarce Resource is Human Creativity

Yochai Benkler is an author who has sharply analysed how a 'networked information economy' can prove more economically efficient than one in which innovation is encumbered by the strict enactment of exclusive intellectual property rights, when the marginal cost of reproducing most information is effectively nearing zero. In a context in which physical capital costs for fixation and communication of knowledge are low and widely distributed, and where existing information is itself a public good, the primary remaining scarce resource – he states – is human creativity, and it is in this context that the relative advantages of peer production emerge, highlighting how, under certain circumstances, peering can in fact be a more cost-effective institutional form than either markets or hierarchical organizations.[105]

Benkler has provided a simple table showing how ideal organizational forms can be deducted as a function of relative social cost (Table 12.2).

---

100) M. Mazzucato, *The Entrepreneurial State: Debunking Public vs. Private Sector Myths*, Anthem Press, 2013.
101) M. Angell, *The Truth about the Drug Companies*, Random House, 2004.
102) M. Mazzucato and G. Dosi (Eds), *Knowledge Accumulation and Industry Evolution: The Case of Pharma-Biotech*, Cambridge University Press, 2006.
103) M. Mazzucato, *The Entrepreneurial State: Debunking Public vs. Private Sector Myths*, Anthem Press, 2013.
104) L. Burlamaqui *et al.*, *Knowledge Governance: Reasserting the Public Interest*, Anthem, 2012.
105) Y. Benkler, *The Wealth of Networks: How Social Production Transforms Markets and Freedom*, Yale University Press, 2006.

Table 12.2 Ideal organizational forms as a function of relative social cost.[106]

	Property system more valuable than implementation costs	Property system implementation costs higher than opportunity cost
Market exchange of x more efficient than organizing or peering of x	Markets (Farmers' markets)	Commons (Ideas and facts; roads)
Organizing x more efficient than market exchange or peering of x	Firms (Automobiles; shoes)	Common property regimes (Swiss pastures)
Peering of x more efficient than organizing or market exchange of x	Proprietary 'open source' efforts (Xerox's Eureka) Peer production processes[a] (NASA[107] Click workers)	

a) 'Cost' would include the negative effects of intellectual property on dissemination and downstream productive use.
(Reproduced with permission of The Yale Law Journal).

12.15
Different Organizational Modes for Overcoming Uncertainty

Behind this taxonomy, there is the taking into account of recent fundamental shifts in economic reasoning, which Yochai Benkler summarizes into four conceptual moves:

1) Ronald Coase[108] asked why clusters of individuals operate under the direction of an entrepreneur, a giver of commands, rather than interacting purely under the guidance of prices, and answered that using the price system is costly (in terms of 'transaction costs').[109]
2) Assuming that the cost of organization increases with size, Coase consequently posited that 'any given firm will cease to grow when the increased complexity of its organization makes its internal decision costs higher than the costs that a smaller firm would incur to achieve the same marginal result'.[110] Coase could assume, this way, to have a 'natural' – that is internal to the theory – limit on the size and number of organizations.

---

106) Y. Benkler, *Coase's Penguin, or, Linux and The Nature of the Firm*, The Yale Law Journal, Volume 112, Issue 3, p. 400, 2002.
107) NASA (National Aeronautics and Space Administration).
108) In 1991, Ronald H. Coase was awarded the Nobel Prize in Economic Sciences '. . . for his discovery and clarification of the significance of transaction costs and property rights for the institutional structure and functioning of the economy': http://www.nobelprize.org/nobel_prizes/economic-sciences/laureates/1991.
109) Coase, *The Nature of the Firm* (1937), now in *The Firm, the Market, and the Law*, University of Chicago Press, 1988.
110) Ibidem.

3) Harold Demsetz's treatment of property rights[111] added a specific element of explanation: property in a resource emerges if the social cost of having no property in that resource exceeds the social cost of implementing a property system in it.
4) Benkler posits that, when 'the social cost of using existing information as an input into new information production is zero', 'the digitization of all forms of information and culture has made the necessary physical capital cheaper by orders of magnitude than in the past', and 'the dramatic decline in communications costs radically reduces the cost of peering', the centrality of human capital to information production and its variability becomes 'the primary source of efficiency gains from moving from markets or hierarchical organization to peering'.[112] Hence, the conclusion that 'commons-based peer production creates better information about available human capital and can better allocate creative effort to resources and projects'.[113]

Different organizational modes have different strategies for overcoming persistent uncertainty: markets reduce uncertainty regarding allocation decisions through prices that act as clear and comparable signals for choosing which use of the relevant factors would be most efficient; firms or hierarchical organizations resolve uncertainty by instituting a process by which information about which actions to follow is ordered. Both are, however, 'lossy' mediums, in the sense – specifies Benkler – that 'much of the information that was not introduced in a form or at a location that entitled it to 'count' toward an agent's decision is lost'.[114]

12.16
Information and Allocation Gains of Peer Production

It is with reference to this intrinsic 'lossiness' that peering may end up having lower information opportunity costs than markets or hierarchies. In particular, Benkler suggests that the primary source of gains – which he calls 'information gains' – peer production offers is its capacity to collect and process information about human capital.

'Human intellectual effort – he posits – is highly variable and individuated. People have different innate capabilities, personal, social, and educational histories, emotional frameworks, and ongoing lived experiences. These characteristics

---

111) H. Demsetz, *Toward a theory of property rights*, American Economic Review, Volume 57, N. 2, p. 347, 1967; *Ownership, Control, and the Firm: The Organization of Economic Activity*, Blackwell, 1988.
112) Y. Benkler, *Coase's Penguin, or, Linux and The Nature of the Firm*, The Yale Law Journal, Volume 112, Issue 3, p. 404, 2002.
113) Y. Benkler, *The Penguin and the Leviathan: How Cooperation Triumphs over Self-Interest*, Crown Business, 2011.
114) Y. Benkler, *Coase's Penguin, or, Linux and The Nature of the Firm*, The Yale Law Journal, Volume 112, Issue 3, p. 409, 2002.

make for immensely diverse associations with, idiosyncratic insights into, and divergent utilization of, existing information and cultural inputs at different times and in different contexts. Human creativity is therefore very difficult to standardize and specify in the contracts necessary for either market-cleared or hierarchically organized production'.[115]

In addition to these information gains, peering has also potential allocation gains, enabled by the large sets of resources, agents, and projects available to this form of organization. 'As peer production relies on opening up access to resources for a relatively unbounded set of agents, [. . .] this is likely to be more productive than the same set could have been if divided into bounded sets in firms'. Furthermore, while the use of a rival resource would have excluded the use by others, this is not true for a purely non-rival good like information, where the allocation gain can be attained by 'allocating the scarce resource – human attention, talent, and effort – given the presence of non-rival resources to which the scarce resource is applied with only a probability of productivity'.[116]

Benkler's conclusion entails, therefore, a criticism of any attempt to apply strong intellectual property rights with regard to peer-produced information: 'Since the core of commons-based peer production entails provisioning without direct appropriation and since indirect appropriation – intrinsic or extrinsic – does not rely on control of the information but on its widest possible availability, intellectual property offers no gain, only loss, to peer production'.[117]

The idea that openness and connectivity may, in the end, be more valuable to innovation than purely competitive mechanisms is shared also by Steven Johnson, who states that 'we are often better served by *connecting* ideas rather than we are by protecting them. [. . .] good ideas may not want to be free, but they do want to connect, fuse, and recombine. They want to reinvent themselves by crossing conceptual borders. They want to complete each other as much as they want to compete'.[118]

Ideas – he writes – 'rise in liquid networks where connection is valued more than protection',[119] and on this ground 'perhaps 'commons'" is the wrong word [. . .], though it has a long and sanctified history in intellectual property law. The commons' metaphor doesn't suggest the patterns of recycling and exaptation[120] and recombination that defer so many innovation spaces. I prefer another metaphor from nature: the [coral] reef',[121] which is the most extraordinary engine of biological innovation: 'Coral reefs make up about one-tenth of one percent of

---

115) Y. Benkler, *Coase's Penguin, or, Linux and The Nature of the Firm*, The Yale Law Journal, Volume 112, Issue 3, p. 412, 2002.
116) Y. Benkler, *Coase's Penguin, or, Linux and The Nature of the Firm*, The Yale Law Journal, Volume 112, Issue 3, p. 421, 2002.
117) Ibidem.
118) S. Johnson, *Where Good Ideas Come From: The Seven Patterns of Innovation*, Penguin, 2011.
119) Ibidem.
120) The term is taken from S.J. Gould and E. Vrba, *Exaptation: A Missing Term in the Science of Form*, Paleobiology, Volume 8, N. 1, p. 4, 1982, where they suggested that 'characters, evolved other usages (or for no function at all), and later, p. 6, 'coopted' for their current role, be called 'exaptations'.
121) S. Johnson, *Where Good Ideas Come From: The Seven Patterns of Innovation*, Penguin, 2011.

the earth's surface, and yet roughly a quarter of the known species of marine life make their homes there',[122] and what makes them so inventive is not 'the struggle between the organisms but the way they have learned to collaborate'.[123]

12.17
An Ecosystem of Technologies Leading to the Singularity?

By analogy – says Johnson – Kuhn's paradigms of research are like coral reefs, 'raised by myriads of tiny architects',[124] which provide fertile environments for new developments and represent 'the scientific world's equivalent of a software platform: a set of rules and conventions that govern the definition of terms, the collection of data, and the boundaries of enquiry for a particular field'.[125] Since modern scientific paradigms are rarely overthrown, but rather tend to be built upon, the outcome of such a process is a stacked platform that supports new paradigms above the old ones, and, 'in a funny way, the real benefit [of it] lies in the knowledge you no longer need to have'.[126]

Such an ecosystem of technologies is expected to grow in complexity and colonize new areas, just like life itself. Kevin Kelly calls it the 'technium', and describes it as a living, evolving organism that has its own unconscious needs and tendencies: just as water 'wants' to flow downhill and life tends to fill available ecological niches, so technology is similarly due to expand and evolve.[127]

In a visionary book published in 2005,[128] the Director of Engineering at Google, Ray Kurzweil, predicted that 2045 will be the year of the singularity,[129] when computers meet or exceed human computational ability and when their ability to recursively improve themselves can lead to an 'intelligence explosion' that will affect all aspects of human culture and technology, ultimately making humans and their machines merge and become indistinguishable from each other. 'Once a planet yields a technology-creating species – Kurzweil stated – and that species creates computation (as has happened here), it is only a matter of a few centuries before its intelligence saturates the matter and energy in its vicinity, and it begins to expand outward at least the speed of light (with some suggestions of circumventing this limit). Such a civilization will then overcome gravity (through exquisite and vast technology) and other cosmological forces – or, to be fully accurate,

---

122) Ibidem.
123) Ibidem.
124) This was the expression used by Charles Darwin admiring the Keeling Islands in April 1836 (*The Voyage of the Beagle*, Penguin, 1989).
125) S. Johnson, *Where Good Ideas Come From: The Seven Patterns of Innovation*, Penguin, 2011.
126) Ibidem.
127) K. Kelly, *What Technology Wants*, Viking, 2010.
128) R. Kurzweil, *The Singularity Is Near: When Humans Transcend Biology*, Duckworth, 2005.
129) Mathematics defines a 'singularity' as a point at which an object changes its nature so as to attain properties that are no longer the expected norms for that class of objects.

it will manoeuvre and control these forces – and engineer the universe it wants. This is the goal of the singularity'.[130]

12.18
Big Data Analytics and Data-Intensive Healthcare

While the forecast of such technological singularity remains highly controversial, in 2009 Microsoft Research published a book of essays, entitled *The Fourth Paradigm: data-intensive scientific discovery*.[131] One of the essays was dedicated to *The Healthcare Singularity and the Age of Semantic Medicine*,[132] on the assumption that medicine would be likely to be the first Big Data domain where a threshold moment, dubbed the 'Healthcare Singularity', will be reached, making medical knowledge become 'liquid' and its flow from research to practice ('bench to bedside') become frictionless and immediate.[133]

In the few years which have passed, a number of projects and realizations have made similar goals look now much nearer. The Human Brain Project (HBP)[134],[135] has been awarded 1 billion euros Future and Emerging Technologies (FET) flagship funding for the next 10 years by the European Commission, and significant larger funding is accruing on its US counterpart, the Brain Research through Advancing Innovative Neurotechnologies (BRAIN).[136] IBM's Watson,[137] the artificially intelligent supercomputer system capable of answering questions posed in natural language, which won the American television quiz show *Jeopardy!*, can process 500 gigabytes (the equivalent of a million books) per second, and uses natural language capabilities, hypothesis generation, and evidence-based learning to support medical professionals as they make decisions. MD-Paedigree (Model-Driven European Paediatric Digital Repository),[138] the clinically-led FP7 IP (7th Framework Programme Integrated Project) that won the Best Exhibit Award at ICT 2013 European Conference in Vilnius, Lithuania, in 2013, is

---

130) R. Kurzweil, *The Singularity Is Near: When Humans Transcend Biology*, Duckworth, 2005.
131) T. Hey et al. (Eds), *The Fourth Paradigm: Data-Intensive Scientific Discovery*, Microsoft Research, 2009.
132) M. Gillam et al., *The Healthcare Singularity and the Age of Semantic Medicine*, Microsoft Research, 2009.
133) Ibidem.
134) H. Markram et al., *The Human Brain Project: A Report to the European Commission*, HBP-PS (Human Brain Project- Preparatory Study) Consortium, EPFL (École Polytechnique Fédérale, Lausanne), 2012.
135) R. Kurzweil, *How To Create A Mind: The Secrets Of Human Thought Revealed*, Viking, 2012.
136) *BRAIN 2025: A Scientific Vision, Brain Research through Advancing Innovative Neurotechnologies (BRAIN)*, Working Group Report to the Advisory Committee to the Director, National Institutes of Health, 5 June 2014. See the New Scientist comment on the reported $4.5 billion forecasted cost: US brain-map project could dwarf its European rival: www.newscientist.com/article/dn25708-us-brainmap-project-could-dwarf-its-european-rival.html#.U6WP9vmSzpV.
137) J. E. Kelly III and S. Hamm, *Smart Machines: IBM's Watson and the Era of Cognitive Computing*, Columbia University Press, 2013.
138) MD-Paedigree (Model-Driven European Paediatric Digital Repository) Newsletter, N. 1, 2013: www.md-paedigree.eu.

applying Big Data analytics to build a scans-based, genomic and meta-genomic repository of routine patients in four disease areas. The 'model-driven' repository is meant to provide, as key functionalities: (1) similarity search for clinicians, allowing them to access 'patients 'like' mine' (and finding decision support for optimal treatment also based on comparative outcome analysis); (2) similarity search for patients, allowing them to make use of social media for getting in touch with 'patients 'like' me' (while providing them also with comparative outcome analysis for patient decision support); (3) model-based patient-specific simulation and prediction; and (4) patient-specific clinical workflows.

Big Data analytics, and specifically Big Data healthcare, are becoming more and more a generalized concern.[139],[140],[141],[142],[143],[144]

Due to the staggering, and ever-growing, size and complexity of biomedical Big Data, computational modelling in medicine is likely to provide the most intriguing insights into the emerging complementary and synergistic relationship between computational and living systems. Biomedical research institutions and related industries, as well as the whole healthcare and pharmaceutical sectors, must therefore be considered as key stakeholders in the process leading to the next generation of data-centric systems. Whether these are systems capable of learning from data, or data-analysis products and applications capable of translating medical knowledge discovery into widespread medical practice, they will put predictive power in the hands of clinicians and healthcare policy makers.

At the same time, Big Data applications in medicine radically change the capacity for going beyond the average patient population, searching for specific cohorts of patients fitting into very peculiar niches of their own. Such an approach can show the way to truly personalized medicine. Applying the *Long Tail*[145] insight to biomedical research and to drug discovery, will lead to people receiving treatments and drugs specifically targeted to their own genomic, proteomic, and metagenomic characterization. Tailoring treatments, drugs and

---

139) N. Sliver, *The Signal and the Noise: The Art and Science of Prediction*, Penguin, 2012.
140) B. Kayyali et al., *The big-data revolution in US health care: Accelerating value and innovation*, McKinsey Company, 2013.
141) Intel IT Centre, *Big Data Analytics*, Peer Research Report, Intel Corporation, 2013.
142) E. Morley-Fletcher, *Big Data Healthcare: An overview of the challenges in data intensive healthcare*, discussion paper, Networking Session, ICT 2013 Conference, 2013.
143) V. Mayer-Schönberger and K. Cukier, *Big Data: A Revolution That Will Transform How We Live, Work and Think*, Murray, 2013.
144) T. Davenport and J. Kim, *Keeping Up with the Quants: Your Guide To Understanding and Using Analytics*, Harvard Business Review Press, 2013.
145) C. Anderson developed the concept that, when transaction costs are greatly lowered, 'the biggest money is in the smallest sales', whereby a series of small niches cumulatively achieve a much larger amount than the traditional focus on selling the preferred selection of block-buster items. The 'long tail' can be extremely lengthened as a result. Consumers really can find and choose whatever they want, no matter how popular or sought after the item is, and retail niches which were not economically viable in the past, can now better fulfil the market. An analogy to epidemiological studies does apply here (*The Long Tail: How Endless Choice is Creating Unlimited Demand*, Random House, 2006).

research to everyone's individual needs is precisely the point of the long tail approach. Thus, focus on the long tail in healthcare should allow medicine to better address all those diseases and ailments suffered by a relatively small number of people or by a large number of people whose common conditions have numerous underlying causes. Furthermore, Big Data will allow for 'long-tail medicine' drugs with enhanced personalized information content, based on customized algorithms tackling the individual disease conditions that can be best addressed only by personalized treatment.

Amidst signals of growing difficulties in guaranteeing the long-term sustainability of all European welfare states, the acceleration of technological innovation in healthcare has long appeared as a further cost driver. Leaving aside whether such acceleration could clear the way to the hoped-for 'singularity', a knowledge-based redesign of healthcare systems (including how to assess value in people's lives), needs now to become a political and economic priority for overcoming the residual grip of scarcity which is hindering mankind from attaining its potential of holistic growth. Eventually, wealth is health.

13
Universities and Corporations: The Case of Switzerland

Spyros Arvanitis and Martin Woerter

In this study, Knowledge and Technology Transfer (KTT) between academic institutions and the business sector is understood to be any activities aimed at transferring knowledge or technology that may help either the company or the academic institute – depending on the direction of transfer – to further pursue its activities. The information is derived from comprehensive surveys amongst representative samples of: (i) Swiss enterprises; and (ii) departments and institutes of Swiss institutions of research and high education in the years 2005 and 2011, respectively. The descriptive analysis covers detailed information about who is undertaking KTT with whom; which channels/forms and services of mediating institutions are frequently used; what is the main motivation for KTT activities; and what are the obstacles for KTT from both sides of transfer. In the more analytical section, which is based on econometric analysis, important driving factors for KTT activities are identified, both from the points of view of the firms and of the science institutions, to: (i) investigate the relevance of KTT activities for firm innovation performance and labour productivity, respectively; (ii) study the relationship between KTT activities and technological proximity of firms and universities; (iii) identify factors that may influence a firm's decision in favour of either exploration or exploitation of knowledge; (iv) analyse the impact of exploration and exploitation of knowledge on the innovation performance of a firm; and (v) explore different strategies of knowledge and technology transfer.

13.1
Background

We are confronted with a 'paradox' picture in the field of science, technology and innovation. On the one hand, Switzerland shows an outstanding performance in university research whilst, on the other hand, the innovation performance of Swiss firms is stagnating or in some areas has even decreased since the mid-1990s, even if the Swiss economy is still ranked top amongst European countries. One possible reason for this discrepancy could be traced back to deficits with regard to KTT from university to industry. Some observers

From Physics to Daily Life: Applications in Informatics, Energy, and Environment, First Edition.
Edited by Beatrice Bressan.
© 2014 Wiley-VCH Verlag GmbH & Co. KGaA. Published 2014 by Wiley-VCH Verlag GmbH & Co. KGaA.

expressed the opinion that the interface between universities and business firms has to be improved and, as a consequence, KTT activities should be intensified.[1] This is probably a step in the right direction, even if the comparison with respect to university collaborations in innovation with Germany and France (see Table 13.1), as well as the comparison with several European countries (see Table 13.2), shows that Swiss firms hold a good position as to this type of cooperation. Until 2005, however, Switzerland lacked comprehensive data and information on the extent, motivations, challenges, and channels of KTT in Switzerland, with research having been conducted only on selected topics.[2] Since 2005, more information has been collected and more research has been conducted, and this chapter reports on these new investigations. In accordance with the investigation of Dosi[3] on technological trajectories, KTT is defined as follows: KTT between academic institutions and the business sector is understood as being any of the activities aimed at transferring knowledge or technology that may help either the company or the academic institute to further pursue them.

---

1) W. Zinkl and H. Huber, *Strategie für den Wissens- und Technologietransfer an den Hochschulen in der Schweiz*, Mandat im Auftrag der SUK (Schweizerischen Universitätskonferenz), Hauptbericht: Strategie und Politik im WTT (Wissens- und Technologietransfer), 2003.
2) Swiss studies focusing on KTT aspects: A. Balthasar (*Vom Technologietransfer zum Netzwerkmanagement: Grundlagen zur politischen Gestaltung der Schnittstelle zwischen Wissenschaft und Industrie*, Verlag Rüegger, 1998) analysed the occupational impact of information networks of developers in the fields of machine tool building and plastics processing in Switzerland. In the same field, B. Wilhelm (*Mythos 'Wissenshalden Hochschulen'-Zur Neuorganisation des Wissens- und Technologietransfers*, Volkswirtschaft Magazin für Wirschaftspolitik, Volume 75, Issue 1, p. 48, 2001) performed international comparisons between Switzerland, Austria and Baden-Württemberg. A. Thierstein et al. (*Gründerzeit, Unternehmensgründungen von Absolventen der Ostschweizer Hochschulen*, Schriftenreihe des Instituts für Öffentliche Dienstleistungen und Tourismus der Universität St. Gallen, Beiträge zur Regionalwirtschaft, Haupt Verlag, 2002) investigated spin-offs/start-ups from universities in the eastern part of Switzerland. A. Berwert et al. (*THISS-Technische Hochschulen und Innovationen: Start-ups und Spinn-offs unter besonderer Berücksichtigung von Aus-und Weiterbildung und Supportstrukturen*, in F. Horvath-Hrsg, Forum Bildung und Beschäftigung, Workshop-Dokumentation: Nationales Forschungsprogramm Bildung und Beschäftigung, NFP 43, Arbeitsbericht, N. 29, p. 22, 2002) investigated the establishment and development of start-ups and spin-offs from technical universities in Switzerland. W. Zinkl and H. Huber (see footnote 1) pointed at challenges for technology transfer offices in Swiss universities. P. Vock et al. (*Technologietransferaktivitäten 2002-Umfrage bei Hochschulen und öffentlich finanzierten Forschungsorganisationen*, Zentrum für Wissenschafts-und Technologiestudien, CEST, Centre d'Études de la Science et de la Technologie, 2004) carried out a survey on codified forms of KTT (patents, licences) based on information from technology transfer offices of the universities. Finally, F. Barak (*Wissens- und Technologietransfer als Interaktion: Theoretische Überlegungen und Fallbeispiele aus der Schweiz*, Peter Lang Verlag, 2011) investigated, based on several case studies, the role of modern communication technologies (Internet) for technology transfer in Switzerland.
3) G. Dosi, *Technological Paradigms and Technological Trajectories. A suggested interpretation of the determinants and directions of technical change*, Research Policy, Volume 11, Issue 3, p. 147, 1982.

Table 13.1 Firms with university collaborations: France, Germany, and Switzerland.[a]

Sectors	France		Germany		Switzerland	
	2004	2008	2004	2008	2005	2008
Manufacturing						
High-tech	10	32	12	34	20	20
High-/medium-tech	10	16	10	25		
Medium-/low-tech	4	9	6	14		
Low-tech	2	8	2	6	8	8
Services						
Knowledge-intensive	3	8	6	14	9	7
Traditional	1	2	2	3	1	1
Firm size (employees)						
≤49	1	3	3	10	4	3
50–99	4	5	4	11	7	8
100–249	6	10	4	13	13	11
250–499	11	17	6	23	17	20
≥500	21	31	15	35	26	19

a) Percentage of all firms by sector and firm size class.[4]

Table 13.2 Innovation collaborations with science institutions 2008.[a]

Country	Percentage
Switzerland	9
Austria, Belgium, Denmark	8, 9, 8
Germany, Finland, France	7, 13, 4
United Kingdom, Italy, The Netherlands	6, 2, 5
Norway, Sweden, Spain	5, 6, 2

a) Percentage of all firms with 10 or more employees.[5]

Considering KTT in a broad sense means also considering the interaction of public research institutions and private enterprises, but actually this would make it necessary to investigate the behaviour of both firms and scientific institutes at universities with respect to KTT. In the study at hand, attention is focused on the firms' side. The 'stylized' model from Bozeman,[6] modified and adapted for the specific needs of this research, guides the analysis (see Figure 13.1). The model includes five broadly defined entities: (i) the transfer agent (characteristics

---

4) France, Germany: S. Robin and T. Schubert, *Cooperation with Public Institutions and Success in Innovation: Evidence from France and Germany*, Research Policy, Volume 42, Issue 1, p. 149, 2013.
5) Eurostat (Community Innovation Survey); Swiss Innovation Survey, 2008.
6) B. Bozeman, *Technology Transfer and Public Policy: A Review of Research and Theory*, Research Policy, Volume 29, Issue 4–5, p. 627, 2000.

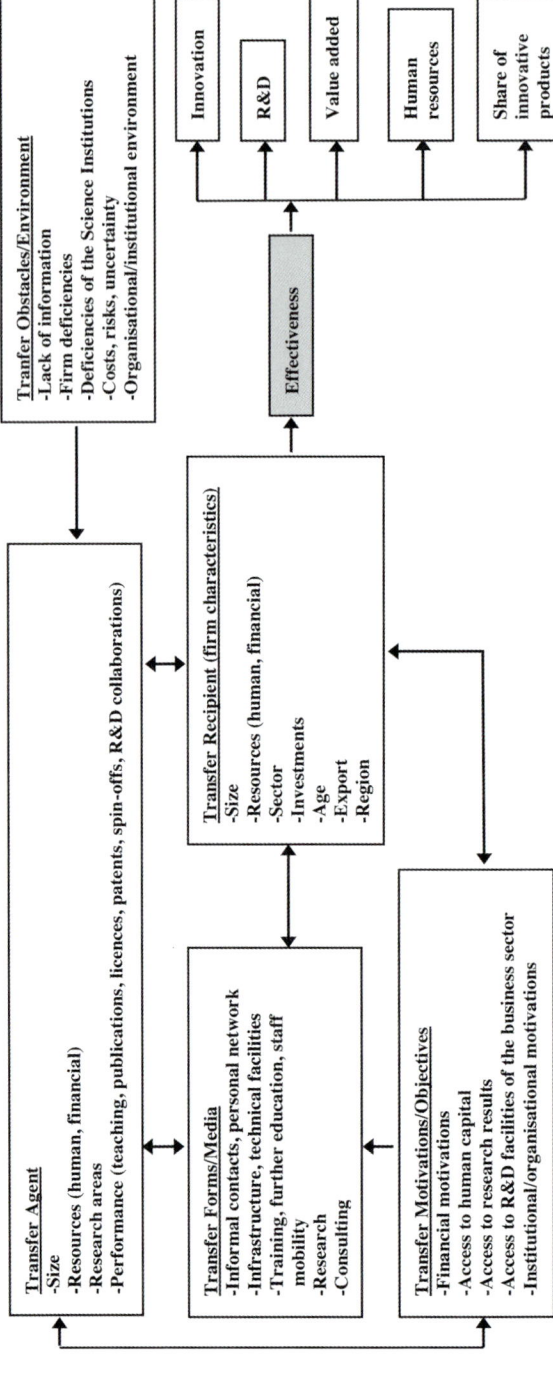

Figure 13.1 'Stylized' model for Knowledge and Technology Transfer.[7] Reproduced with permission of Elsevier.

7) B. Bozeman, *Technology Transfer and Public Policy: A Review of Research and Theory*, Research Policy, Volume 29, Issue 4–5, p. 627, 2000.

of scientific institutes); (ii) the transfer recipient (the characteristics of firms); (iii) the transfer forms or media (e.g. informal contacts, personal exchange, and research cooperations); (iv) transfer motives or objectives (e.g. access to human capital or research results); and (v) transfer obstacles (e.g. firm deficiencies, organizational/institutional obstacles). Their interaction determines whether – and to what degree – KTT takes place, and how effective it is with respect to several criteria (e.g. R&D abilities of firms, value added, share of new products, and skill level).

On behalf of the ETH Board,[8] two comprehensive surveys have been carried out amongst Swiss firms (in 2005 and 2011) in order to fill the stylized model from Bozeman with firm-level data. In addition, two surveys amongst the institutes of all institutions of higher education in Switzerland (Universities and Universities of Applied Sciences) and the Federal Research Organizations (e.g. Paul Scherrer Institute, PSI) have been conducted.[9]

The descriptive analysis enabled us to quantify the different entities in the 'stylized' model. In this way, detailed information has been obtained about: who is undertaking KTT; which channels/forms and services of mediating institutions are frequently used; what is the main motivation for KTT activities; and what are the obstacles for KTT. Furthermore, there is information about the characteristics of a firm, for example size, human resources, sector affiliation, age, and export activities (see Section 13.2).

The descriptive analysis for the institute data was undertaken in Section 13.3. In addition to the descriptive analysis, the data enabled more detailed multivariate econometric investigations to be conducted (see Section 13.4). First, important driving factors for KTT activities, both from the point of view of the firms and the point of view of the science institutions, have been identified. Second, the relevance of KTT activities for a firm's innovation performance and labour productivity, respectively have been investigated. The KTT effectiveness in Figure 13.1 refers primarily to the economic impact of KTT on the business sector. Third, we also explored different strategies to conduct transfer activities, since it is clear that not every transfer strategy exerts the same performance effect; transfer strategies are based on the firms' capabilities, and not every firm can pursue the most effective strategy. Fourth, the relationship between KTT activities and technological proximity of firms and universities has been studied. Finally, the factors that may influence a firm's decision to use transfer contacts in favour of either exploration or exploitation of knowledge have been isolated, and the impact of exploration and exploitation of knowledge on the innovation performance of a firm has been analysed.

---

8) The ETH Board is the authority that supervises the two Swiss federal technical universities and the Swiss federal research organizations.
9) Information on the data collection and the composition of the samples of enterprises and scientific institutes that were used in this study are found in S. Arvanitis et al., *Knowledge and Technology Transfer between Universities and Private Enterprises in Switzerland 2011*, Study on Behalf of the ETH Board, KOF (Konjunkturforschungsstelle) Studies, N. 37, 2012.

The US experience,[10],[11] also in the field of technology transfer, has been pathbreaking for other technologically advanced countries due to its leading position in technology and science after the Second World War. A key turning point in the university to business technology transfer was the *Bayh–Dole Act*, which created a uniform public patent policy, allowing publicly funded research to be patented by universities.[12] As a consequence of this institutional change, the patenting activities of US universities were increased significantly after 1980. In fact, patenting by universities in the US rose from fewer than 250 patents before 1980 to more than 2000 at the end of the 1990s. From other sources, it is known that since passage of the *Bayh–Dole Act*, research collaborations between universities and the business sector played an important role in the development of new technologies in many sectors.[13],[14] However, Mowery attributes a smaller contribution of the *Bayh–Dole Act* to the expansion of university patent activities than do other researchers, and concludes that the main effect of the *Bayh–Dole Act* was to increase the marketing efforts of universities already active in patenting.[15] Independently of this issue, the fact remains that the commercialization of scientific results acquired by university research has become an important driver of innovation in the US, at any rate at a significantly larger extent than in European countries. It has been estimated that about US$ 30 billion of economic activity and 250 000 jobs each year can be attributed to commercializing academic innovation.[16],[17] Further, more than 10% of the new products and processes introduced in several industries could not have been developed without the contributions of academic research.[18] Although no such figures are available for European countries, the general impression is that technology transfer activities in these countries is less developed than in the US.[19]

10) Ibidem.
11) B. Bozeman and M. Decter et al., *University to business technology transfer – UK and USA comparisons*, Technovation, Volume 27, N. 3, p. 145, 2007.
12) The *Bayh–Dole* Act or *Patent and Trademark Law Amendments Act* is US legislation dealing with intellectual property arising from federal government-funded research, *Public Law*: 96–517, 12th December 1980.
13) B. Bozeman, *Technology Transfer and Public Policy: A Review of Research and Theory*, Research Policy, Volume 29, Issue 4–5, p. 627, 2000.
14) See footnote 8.
15) D.C. Mowery et al., *The growth of patenting and licensing by U.S. universities: an assessment of the effects of the Bayh–Dole act of 1980*, Research Policy, Volume 30, Issue 1, p. 99, 2001.
16) B. Bozeman, *Technology Transfer and Public Policy: A Review of Research and Theory*, Research Policy, Volume 29, Issue 4–5, p. 627, 2000.
17) B. Bozeman and M. Decter, *University to business technology transfer – UK and USA comparisons*, Technovation, Volume 27, N. 3, p. 145, 2007.
18) E. Mansfield, *Academic Research and Industrial Innovation: An Update of Empirical Findings*, Research Policy, Volume 26, Issue 7–8, p. 773, 1998.
19) *Benchmarking Industry-Science Relationships*, OECD (Organization for Economic Co-operation and Development), France, 2002.

13.2
KTT Activities in the Swiss Economy: The Main Facts from the *Firm's* Point of View

The descriptive analysis in this section aims at quantifying the different entities in the 'stylized' model shown in Figure 13.1. Detailed information is presented about: who is undertaking KTT (firms by industry and firms size class) with whom (Cantonal Universities, Universities of Applied Sciences, institutions of the ETH Domain); which forms of KTT (general information, education, research, technical infrastructure, consulting); which channels and which services of mediating institutions are frequently used; which motives drive firms; and with which obstacles firms are confronted when pursuing KTT activities.

According to the survey results, about 21% of all firms in the sample were involved in KTT activities with universities and other research institutions between 2008 and 2010 (Table 13.3). During the period of 2002–2004, the corresponding figure was 22%; hence, the share of firms with KTT activities has remained approximately constant over the past six years. The share of firms with KTT activities is almost the same in the manufacturing and service sectors (28% and 25%, respectively). Only 4% of firms in the construction sector are involved in KTT activities. Firms in high-tech manufacturing (i.e. the most innovative part of manufacturing industries) and in knowledge-based services (banking, insurance, and business services such as engineering and computer software) show the highest incidence of KTT activities. Especially, firms in the chemical industry (including pharmaceuticals), in vehicles, in electronics/instruments, and in business services are most often involved in KTT. There is a

Table 13.3 Incidence of KTT activities; percentage share of firms according to sector, sub-sector and firm size class.

	2002–2004	Abroad 2002–2004	2008–2010	Abroad 2008–2010
Sector				
Manufacturing	25.1	13.2	28.0	8.7
Construction	10.1	4.1	4.3	1.1
Services	26.7	8.3	24.6	4.7
Subsector				
High-tech	28.3	18.9	44.6	15.9
Low-tech	23.4	10.1	16.7	3.8
Modern services	27.2	9.2	35.2	6.1
Traditional services	26.2	7.4	10.6	2.9
Firm size (employees)				
Small (5–49)	19.4	7.7	16.2	2.5
Medium (50–249)	33.7	11.9	34.7	11.6
Large (> 250)	44.9	18.3	57.3	27.8
Total	22.2	8.6	21.1	5.0

Table 13.4 Incidence of KTT activities; percentage share of firms according to region.

Swiss region	2002–2004	2008–2010
Lake Geneva	12.9	13.0
Swiss Midlands	22.4	18.5
Northwestern Switzerland	21.7	23.7
Zurich	35.0	25.6
Eastern Switzerland	19.9	23.9
Central Switzerland	21.4	28.9
Ticino	7.4	7.4
Total	22.2	21.1

significant increase in the share of KTT-active firms in high-tech manufacturing and knowledge-based services, but there is also a discernible decrease in the respective share in low-tech manufacturing and traditional services.[20]

There is a considerable size-dependency with respect to KTT activities. The percentage of small firms with KTT activities is 16%, while that of medium-sized firms is considerably higher (35%), and the respective percentage for large firms is 57%. This means that half of all large firms in Switzerland (i.e. with 250 and more employees) are involved in KTT activities.

The incidence of KTT activities according to region in Table 13.4 shows that, since 2002–2004, the share of firms from Zurich has significantly decreased, while that of central Switzerland has increased. The share of all other regions has remained almost constant.

13.2.1
Forms and Partners of KTT Activities

The KTT-active firms were asked to assess the importance of 19 different single forms of KTT activities on a five-point Likert scale,[21] ranging from 1 (not important) to 5 (very important). These 19 single forms were classified in five categories: (i) informal contacts with a primarily general informational character; (ii) educational activities including joint master thesis and doctoral projects; (iii) activities related to the use of technical infrastructure including firm assignments, for example, for specific measurements that could be conducted only

---

20) S. Arvanitis et al., Knowledge and Technology Transfer (KTT) Activities between Universities and Firms in Switzerland – The Main Facts: An Empirical Analysis Based on Firm-Level Data, ICFAI (Institute of Chartered Financial Analysts of India) Journal of Knowledge Management, Volume 5, N. 6, p. 17, 2007.

21) A Likert scale, named after its inventor psychologist R. Likert, is a psychometric scale (study field concerned with theory and technique of psychological measurements: i.e. of knowledge, abilities, attitudes, personality traits, and education) used in research employing questionnaires to scaling responses in surveys. The term is often interchanged with rating scale, that is a set of categories to elicit information on a quantitative or a qualitative attribute.

Table 13.5 Main categories of forms of KTT activities; firms' percentage with KTT activities.[a]

	Information	Infrastructure	Education	Research	Consulting
Total 2011	62.9	13.9	59.3	17.1	14.8
Total 2005	56.6	11.9	52.3	17.8	15.3

a) Firms' percentage reporting 4 or 5 on a five-point Likert scale for any of the KTT activities single forms in a certain main category of KTT activities forms.

with the specialized equipment of research institutions; (iv) research activities; and (v) consulting.

About 60% of KTT-active firms in Switzerland found: (i) informal, personal contacts that aim at gaining some general information on technological opportunities; and/or (ii) a wide spectrum of educational activities as the most important forms of KTT activities (Table 13.5); 17% had a focus on research activities, 15% on consulting, and 14% on the utilization of university infrastructure facilities. There was a slight increase in informational and educational activities as compared to the period of 2002–2004. The proportion of firms engaged in research and consulting activities has remained approximately constant during the past six years.

The percentage of firms reporting high importance of a single form of KTT activities is shown in Table 13.6. In the main group information, the firms' preferences are distributed almost equally amongst informal contacts, attending conferences and publications. The most frequently reported educational activities leading to KTT are attending university training courses (41%), employing graduates in R&D (25%), and joint degree (mostly Master's) thesis (24%). Only 6% of KTT-active firms find joint PhDs important.

Firms also reported the institution(s) [institutions of the ETHD (ETH Domain), CUs (Cantonal Universities) and UAS (Universities of Applied Sciences)] with which they interacted. Many firms reported more than one institution; 70% of all KTT-active firms reported an interaction with institutions of the ETHD, 56% with UAS, and significantly less with CUs (43%). There were significant differences as to the former period (Table 13.7). The percentage of all three types of science institution has increased since 2002–2004, but the increase for the ETHD and the UAS was discernibly larger than that for the CUs.

The percentages of firms with KTT activities with a certain type of science institution as KTT partners by sector, subsector and firm size class are shown in Table 13.8. The business partners of the institutions of the ETHD and the UAS are evenly distributed amongst the three sectors of manufacturing, construction, and services. The CUs have a considerably higher proportion of partners in the service sector than in manufacturing and construction. The institutions of the ETHD cooperate to a greater extent with high-tech firms and firms from the more knowledge-intensive service industries than the other two groups of science institutions. Large firms seem to engage more frequently in KTT activities than smaller ones, independent of the type of science institution.

Table 13.6 Main categories of single forms of KTT; firms' percentage with KTT activities 2011.

KTT main forms[a]/Single forms[b]	Percentage
Information	**62.9**
Informal contacts	29.3
Attending conferences	37.0
Reading of, reference to publications	34.6
Infrastructure	**13.9**
Joint laboratories	5.2
Use of university technical infrastructure	12.7
Education	**59.3**
Employing graduates in R&D	25.3
Contact of graduates with university	12.1
Students' participation in firm R&D	16.0
Diploma thesis	24.4
PhD	6.0
University researchers' participation in firm R&D	6.3
Joint courses	7.6
Teaching of firm researchers at university	17.0
Attending university training courses	41.0
Research	**17.1**
Joint R&D projects	15.8
Long-term research contracts	4.5
Research consortium	4.3
Consulting	**14.8**
Expertise	9.3
Consulting	13.0

a) Firms reporting 4 or 5 on a five-point Likert scale for any of the KTT activities single forms in a certain main category of KTT activities forms;
b) Firms reporting 4 or 5 on a five-point Likert scale for a KTT activity single form.

Table 13.7 Percentage of KTT firms that have a certain science institution as partner.

Science Institutions	2002–2004	2008–2010
ETHD (ETH Domain)	**57.0**	**70.0**
ETH (Eidgenössische Technische Hochschule)	31.5	58.4
EPFL (École Polytechnique Fédérale, Lausanne)	19.1	19.1
PSI (Paul Scherrer Institute)	7.9	10.8
EAWAG (Eidgenössische Anstalt für Wasserversorgung, Abwasserreinigung und Gewässerschutz)	3.2	13.6
EMPA (Eidgenössische Materialprüfungs- und Forschungsanstalt)	25.4	30.9
WSL (Forschungsanstalt für Wald, Schnee und Landschaft)	7.5	8.2
CUs (Cantonal Universities)	**38.0**	**42.8**
UAS (Universities of Applied Sciences)	**56.0**	**68.6**

Table 13.8 Percentage of KTT firms that have a certain science institution as partner; according to sector, subsector and firm size class.

	ETHD	CUs	UAS
Sector			
Manufacturing	71.2	36.8	70.4
Construction	74.2	34.1	60.6
Services	69.0	47.2	68.1
Subsector			
High-tech	76.9	36.2	75.1
Low-tech	60.8	38.0	61.8
Modern services	74.8	49.9	68.1
Traditional services	43.0	35.6	68.1
Firm size (employees)			
Small (5–49)	65.1	37.4	67.0
Medium (50–249)	76.5	49.3	69.6
Large (>250)	79.9	55.4	76.2
Total	**70.0**	**42.8**	**68.6**

The percentages of firms that pursue KTT activities with partners from a given group of science institutions, according to the main forms of KTT activities, are shown in Table 13.9. The sum of the percentages along a row in Table 13.9 is 100%; thus, these percentages reflect the 'KTT portfolio' or the 'degree of KTT specialization' of the institutions or groups of institutions. Many firms reported not only more than one institution, but also more than one group of KTT activities. As a consequence, the sum of the contacts as reported in Table 13.9 is in general larger than the number of firms contacting a certain institution or group of institutions.

It is worth noting that the research institutions of the ETHD, with the exception of EMPA (the Swiss Federal Laboratories for Materials Science and Technology), have a greater frequency of informational contacts with firms compared to ETH, the Swiss Federal Institute of Technology, Zurich or EPFL, the Federal Polytechnic School, Lausanne. In contrast, it can be observed that EPFL and EMPA have considerably more transfer activities dedicated to collaborative research projects. The relative share of research contacts of EPFL (13.2%) is considerably higher than that of ETH (6.9%). Further, it can be seen that EMPA shows not only the highest share of research contacts but also the highest share of educational contacts.

In the case of the CUs, there are relatively many informational contacts with the University of Lausanne and frequent infrastructure-related contacts with the University of St Gallen. Education-related transfer activities are also frequently found at the University of St Gallen, while research collaborations have a relatively greater frequency at the University of Fribourg, while the University of Italian Switzerland (USI)[22] stands out through its consulting activities.

---

22) USI (Università della Svizzera Italiana).

Table 13.9 Percentage of KTT firms that have a certain science institution as partner; according to the main form of KTT activities.

Science Institutions	Information	Infrastructure	Education	Research	Consulting
ETHD					
ETH	46.0	15.5	15.9	6.9	15.7
EPFL	44.2	9.9	13.2	13.2	19.4
PSI	52.6	7.8	18.1	9.5	12.1
EAWAG	56.5	13.0	17.4	7.2	5.8
EMPA	36.5	7.2	26.1	17.9	12.3
WSL	71.4	7.1	17.9	0.0	3.6
CUs					
Berne	57.8	19.6	7.8	6.9	7.8
Basle	48.5	18.2	12.1	9.1	12.1
Fribourg	51.7	15.5	8.6	12.1	12.1
Geneva	52.4	23.8	11.9	4.8	7.1
Lausanne	64.1	15.4	10.3	5.1	5.1
Neuchâtel	56.1	12.2	12.2	7.3	12.2
St. Gallen	40.9	30.6	19.9	1.6	7.0
Italian Switzerland	54.5	4.5	13.6	9.1	18.2
Zurich	55.5	16.8	12.4	6.6	8.8
UAS					
Berne	44.8	23.0	13.9	9.1	9.1
Northwestern Switzerland	43.7	22.1	12.7	7.5	14.1
Eastern Switzerland	44.7	21.3	20.0	5.3	8.7
Central Switzerland	39.3	25.0	16.4	10.0	9.3
Western Switzerland	38.4	20.9	15.1	11.6	14.0
Italian Switzerland	44.9	14.3	14.3	12.2	14.3
Zurich	44.7	18.0	14.9	7.5	14.9

The figures in every row add up to 100%.

The UAS show a rather homogeneous picture as far as the informational transfer contacts are concerned, with the reported frequency lying between 38% and 45%. Further differences can also be observed in terms of infrastructure and education. The firms reported relatively frequent contacts with Central Switzerland and Berne with regards to infrastructure-related contacts, and Eastern Switzerland has frequent education-related contacts. Research collaborations are of similar frequency at the UAS of Southern Switzerland, Western Switzerland, and Central Switzerland. Consulting, like informational contacts, is quite equally distributed.

In the ETH Domain the three largest institutions – ETH, EPFL and EMPA – also show the largest shares in all five categories of KTT activities. It is worth pointing out here that ETH Zurich and EPFL have equal shares of research-related contacts, though ETH has almost twice the number of personnel and resources. In the case of the CUs, the University of St Gallen can be seen to have an above-average proportion of contacts with respect to informational, educational, infrastructure-related and consulting activities, while the University of Zurich has an above-average proportion of research-orientated contacts. Amongst the UAS, the UAS of Northwestern Switzerland has the most frequent informational, infrastructure-related and research-related transfer contacts amongst all UAS. Eastern Switzerland also has an above-average record as regards education-related contacts.

13.2.2
Technological Fields of KTT-Active and R&D-Active Firms

The technological profile of KTT-active firms and firms with R&D activities are shown in Table 13.10.

KTT-active firms are most frequently found in software, simulation and artificial intelligence, followed by environmental technologies and new materials, and energy technology. This clearly mirrors the technological profile of R&D-active firms. However, there are some technological fields that have a comparably greater percentage of KTT-active firms than R&D-active firms. Such a greater affinity to technology transfer can be found in the fields of nanotechnology, biotechnology, medical technology, transport technology, and energy technology.

Table 13.10 Technological fields of activities.

	Firms' percentage with KTT	Firms' percentage with R&D
Nanotechnology	2.9	2.8
New materials	22.7	27.4
Microelectronics/semiconductor technology	5.3	10.5
Laser technology/optoelectronics/displays	7.8	11.9
Software/simulation/artificial intelligence	38.4	41.9
Telecommunication/information technology	15.9	19.8
Biotechnology/gene technology	3.6	2.1
Medical technology/health engineering	13.9	10.8
Flexible computer-integrated manufacturing technology	11.7	16.3
Transport technology/traffic engineering/logistics	19.4	16.5
Energy technology	22.6	21.7
Environmental technologies	33.1	31.4
Geological technologies	4.7	4.8
Mathematical models of finance	1.7	4.6

13.2.3
Mediating Institutions and Motives for KTT Activities

Although, from the point of view of firms, the relevance of all five types of KTT-mediating institutions has increased since 2002–2004, only a small number of KTT-active firms still seem to be aware of the mediating services of these institutions (see Table 13.11). However, the Innovation Promotion Agency, CTI (Commission for Technology and Innovation) is the most important institution, especially for medium-sized high-tech manufacturing firms. Transfer Offices (TOs) are next in importance, and particularly small firms and/or firms in modern services emphasized their usefulness. The Swiss National Science Foundation (SNSF), European Framework Programmes (EFPs), and other European Research Programmes (ERPs) are less important, especially amongst small (except SNSF) and medium-sized firms. For large firms, EFPs and other ERPs are of similar importance to the TOs.

Information on firms' motives for knowledge transfer activities is useful for focussing state innovation policy on the relevant issues that may not be the same for every category of firms with respect to industry affiliation and firm size class. Table 13.12 shows the main motives for transfer activities from the firms' perspective in comparison with the 2005 survey. There are no discernible differences between the two periods.

Access to human capital is by far the most important motive for technology transfer with universities, followed by financial motives, access to research results, and institutional motives (Table 13.13). Not surprisingly, the most

Table 13.11 Single mediating institutions' importance; firms' percentage by sector, sub-sector and firm size class 2011.[a]

	TOs	CTI	SNSF	EFPs	ERPs
Sector					
Manufacturing	12.2	20.6	4.6	5.9	3.4
Construction	1.0	12.9	0.0	1.0	1.0
Services	14.3	18.5	7.0	6.6	6.7
Subsector					
High-tech	12.3	24.7	5.9	8.2	5.0
Low-tech	12.0	13.0	2.3	1.9	0.6
Modern services	16.2	20.9	7.6	8.1	8.3
Traditional services	6.0	7.7	4.4	0.0	0.0
Firm size (employees)					
Small (5–49)	15.1	15.9	8.0	7.9	6.5
Medium (50–249)	9.6	24.0	2.6	1.8	1.9
Large (>250)	10.5	21.5	3.2	10.2	9.5
Total: 2011 (2005)	**12.9 (9.5)**	**19.0 (11.0)**	**5.8 (3.6)**	**6.1 (3.2)**	**5.3 (1.5)**

a) Firms reporting 4 or 5 on a five-point Likert scale, according to sector, subsector and firm size class.

Table 13.12 Main categories of motives for KTT activities; KTT-active firms' percentage pursuing a given category of motives.[a]

	Access human capital (tacit knowledge)	Access research results (codified knowledge)	Financial motives	Institutional/organizational motives
Total 2011	65.1	28.9	33.0	28.1
Total 2005	65.9	29.3	41.1	25.0

a) Firms reporting 4 or 5 on a five-point Likert scale for any single motive in a certain main group of motives.

important single motives are further education, training possibilities, access to human capital, and the recruitment of graduates. All other human capital related motives are amongst the top categories. Project characteristics require cooperation, access to research results for developing new processes, and access to

Table 13.13 Main categories of motives and single motives for KTT activities; firms' percentage with KTT activities.

Single motives[a]/main groups of motives[b]	
Access to human capital (tacit knowledge)	**65.1**
Access to specific skills in addition to internal know-how	40.3
New research ideas	16.9
Further education, training possibilities	44.8
Recruitment of graduates	36.0
Access to basic research	16.4
Access to research results (codified knowledge)	**28.9**
Access to patents/licences	2.3
Access to research results for subsequent internal use	11.2
Access to research results for developing new products	16.1
Access to research results for developing new processes	17.7
Access to R&D infrastructure	8.3
Financial motives	**33.0**
Cost-saving in R&D	9.9
Reduction of technical R&D risks	10.8
Time-saving in R&D	15.7
Insufficient firm R&D resources	11.3
Project characteristics require cooperation with scientific institutions	21.7
Institutional/Organizational motives	**28.1**
Building up a new research field	2.2
R&D outsourcing as a strategic measure	6.1
R&D cooperation as a condition for public funding	15.1
Improvement of firm image through cooperation with scientific institutions	12.6
Indirect access to competitors' knowledge	5.1

a) Firms reporting 4 or 5 on a five-point Likert scale for a certain single motive.
b) Firms reporting 4 or 5 on a five-point Likert scale for any single motive in a certain main group of motives.

Table 13.14 Motivations for firms to acquire technologies from universities: UK, US and CH.[a]

	2000/ TOs UK (n = 32)	2000/ TOs US (n = 57)	2005/ Firms CH (n = 669)	2011/ Firms CH (n = 469)
Access to new ideas/technologies	81.0	91.5	29.3	28.9
Preventing competition from acquiring technology	50.0	23.5	–	–
Reduction in R&D costs	72.0	72.0	10.3	9.9
Greater speed to market with new technology	59.5	43.5	13.4	15.7
Building links with universities	37.5	70.0	–	
Recruitment and retention of staff	22.0	34.5	15.5	36.0

a) Percentage of TOs and firms reporting 4 or 5 on a five-point Likert scale.[23]

research results for developing new products and are motivating factors of medium importance. However, access to human capital dominates the motive profile of transfer-active firms.

Information on the KTT motives for firms in UK and US, compared with Switzerland when possible, is shown in Table 13.14. The data for the two countries are older than those for Switzerland, but it is assumed that such information is of structural nature and consequently is rather stable over time, provided that no significant institutional or policy changes took place during the reference period. This was the case for all three countries. The data with respect to firm motivations for KTT are not quite comparable, due to the fact that the US and UK data were based on information from TOs while the Swiss data came directly from the involved firms. For the US and the UK, access to new ideas/technologies seems to be the most important KTT motive, whilst for Switzerland it is the second most important. The motives' pattern appears to be rather heterogeneous amongst the three countries. For Switzerland, the second most important motive is the recruitment and retention of staff, while for the US the next most important is greater speed to market, and for the UK building links with universities.

13.2.4
Impact of KTT Activities as Assessed by the Firms

Although access to human capital provides the greatest motivation for transfer activities, the greatest impact refers to the development of new processes and the development of new products; that is knowledge exploitation outcomes, followed by recruitment and education (Table 13.15).

New processes are predominantly developed by large firms in the modern service sector, while new products are most frequently the result of transfer

23) M. Decter et al., University to business technology transfer – UK and USA comparisons, Technovation, Volume 27, N. 3, p. 145, 2007.

Table 13.15 Impact of KTT activities; firms' percentage with KTT activities according to sector, subsector and firm size class.

	New project initiation	New product development	New process development	Scientific publications	Patents	Licenses	Human capital: recruitment	Human capital: education
Sector								
Manufacturing	31.6	48.3	38.7	18.1	13.5	3.8	37.1	24.2
Construction	15.0	26.8	41.8	8.0	1.0	1.0	20.9	9.0
Services	18.0	35.5	51.4	22.4	4.8	4.1	39.5	37.7
Subsector								
High-tech	34.8	49.7	34.4	19.8	18.1	3.1	43.2	24.7
Low-tech	25.8	45.6	46.4	15.0	5.3	5.1	26.2	23.3
Modern services	15.2	28.8	52.6	25.9	5.5	4.7	39.5	41.8
Traditional services	30.3	65.3	45.7	6.9	1.7	1.7	39.7	19.3
Firm size (employees)								
Small (5–49)	16.0	38.6	50.0	23.1	4.9	4.3	27.2	33.0
Medium (50–249)	30.0	37.8	38.1	10.9	9.0	1.5	51.6	25.8
Large (>250)	43.7	55.9	51.2	34.7	24.0	9.9	58.7	41.8
Total	**22.9**	**39.8**	**46.3**	**20.1**	**7.8**	**3.9**	**37.8**	**31.4**

activities for large high-tech manufacturing firms. Interestingly, most frequently large firms detect a positive impact of the transfer activities. New projects initiation (knowledge exploration outcomes) is seen to be considerably less frequent than knowledge exploitation outcomes.

13.2.5
Obstacles to KTT Activities

Impediments to KTT activities are a further topic that is especially relevant not only for the directly involved parties (i.e. universities and enterprises) but also for public policy that aims at the enhancement of the interactions between science institutions and the business sector. Are there factors that impede KTT activities between firms and universities?

All the firms were asked to assess the importance of 26 different possible single obstacles to KTT activities, pooled into five main groups of obstacles: lack of information; firm deficiencies; science institution deficiencies; costs/risks and uncertainty; and organizational/institutional obstacles. Table 13.16 compares the firms' obstacle profile from the 2005 survey with the 2011 survey results. Only small differences can be detected: firm deficiencies, costs/risks and uncertainty, and science institutions deficiencies remain the most important obstacles. There are minor differences amongst sectors, subsectors and firm size classes. Large

Table 13.16 Main categories of obstacles to KTT activities; percentage of all firms perceiving a certain category of obstacles as important according to sector, subsector and firm size class.[a]

	Lack of information	Firm deficiencies	Science institution deficiencies	Costs/risks uncertainty	Organizational/institutional obstacles
Sector					
Manufacturing	27.8	50.4	39.6	43.4	30.3
Construction	22.3	50.2	39.9	37.7	24.4
Services	25.0	55.1	43.2	44.2	33.3
Subsector					
High-tech	31.5	47.5	42.4	49.6	35.4
Low-tech	25.3	52.5	37.6	39.1	26.9
Modern services	20.7	53.5	38.2	43.5	29.6
Traditional services	30.7	57.2	49.7	45.3	38.3
Firm size (employees)					
Small (5–49)	25.6	53.4	40.9	42.6	30.2
Medium (50–249)	24.1	52.0	44.2	43.0	31.7
Large (>250)	20.9	39.4	37.8	38.3	31.6
Total: 2011 (2005)	25.2 (24.1)	52.7 (49.2)	41.4 (42.0)	42.6 (42.4)	30.5 (24.5)

a) Firms' percentage reporting 4 or 5 on a five-point Likert scale for any of the single obstacles in a certain main group of obstacles.

firms seem to perceive transfer obstacles (except the institutional) less frequently than the other firm size classes.

Table 13.17 highlights single obstacles and compares the profile of KTT-active firms, KTT-inactive firms, and all firms. Firm deficiencies are most frequently

Table 13.17 Main categories of obstacles and single obstacles; percentage of firms with/without KTT activities.

Obstacles to KTT activities	KTT	No KTT	All firms
Lack of information	**21.6**	**26.1**	**25.2**
Difficulty in getting information about R&D from science institutions	11.3	20.0	18.2
Difficulty in finding contact persons	15.3	20.6	19.5
Lack of resources for 'interface' (e.g. transfer office)	7.3	18.7	16.3
Firm deficiencies	**43.3**	**55.2**	**52.7**
Lack of qualified staff	21.8	21.8	21.8
Lack of technical equipment	10.2	20.3	18.2
Lack of interest in scientific projects	10.2	34.5	29.4
Firms' R&D questions are not interesting for science institutions	25.4	43.6	39.8
Science institution deficiencies	**36.7**	**42.7**	**41.4**
Lack of scientific staff for transfer activities	4.6	19.1	16.1
Lack of entrepreneurial spirit	13.9	17.7	16.9
R&D orientation of science institutions is uninteresting for firms	18.4	33.9	30.7
Possible R&D results cannot be commercialized	19.9	30.4	28.2
Costs/risks uncertainty	**44.3**	**42.1**	**42.6**
Secrecy with respect to firms' know-how is not guaranteed	14.7	17.8	17.1
Need for comprehensive additional follow-up work in order to implement public R&D results	19.3	19.9	19.8
Lack of firm financial resources for transfer activities	25.9	33.9	32.2
Science institutions' lack of financial resources for co-operation on an equal basis with firms	13.0	21.8	19.9
Insufficient efficiency of university staff compared to firms' staff	10.5	17.9	16.3
Technological dependency on external institutions	8.2	15.9	14.3
Uncertainty about outcomes of co-operation	13.1	20.7	19.1
Institutional/organizational obstacles	**31.1**	**30.4**	**30.5**
Costly administrative and approval procedure	18.3	24.4	23.1
Lack of administrative support for joint R&D projects on the university's part	10.2	17.2	15.7
Lack of administrative support for the commercialization of R&D outcomes by the university	7.6	17.2	15.1
Problems with property rights	9.6	17.4	15.8
Problems with project management at universities (e.g. communication problems)	7.0	18.0	15.7
Different understanding of priorities	14.4	19.3	18.3
Lack of trust on the firm's part	4.6	15.9	13.5
Risk of losing reputation on the firm's part	1.1	15.3	12.3

perceived as a category of severe impediments to KTT activities with science institutions (53% of all firms). Firm's research questions not being interesting for science institutions and a lack of interest for scientific projects are the most frequently reported single obstacles in this category, with about the same percentages as in the 2002–2004 period. The obstacle categories cost/risks and uncertainty and science institutions deficiencies are somewhat less important than firm deficiencies. Finally, the lack of information and organizational or institutional or obstacles are a severe problem only for 25% and 31% of all firms, respectively.

The obstacle profile described above reflects to a large extent also the obstacle profile of KTT-inactive firms. However, there are differences compared to KTT-active firms. KTT-active firms are predominantly prevented from intensifying their KTT activities through cost/risks, uncertainty-related obstacle categories, followed by firm deficiencies, and science institutions deficiencies (see Section 13.3 for a detailed analysis of the obstacles hampering KTT activities).

With regard to single-obstacle categories, it can be seen that for all firms and no KTT firms, the top three obstacles are: the firms' R&D questions are not interesting for science institutions; a lack of financial resources to transfer activities; and R&D orientation of science institutions is uninteresting for firms. For firms with KTT activities there is a similar profile; however, the order is slightly different (first: lack of financial resources; second: firms' R&D questions are not interesting for science institutions) and a lack of qualified staff are also amongst the top three obstacles. The obstacle R&D orientation of science institutions is uninteresting for firms, and is not a very important obstacle for transfer-active firms. It is worth noting that KTT-active firms have a less intense perception of all obstacles compared to KTT-inactive firms and the all-firms category.

Although, in this case the data for the US and the UK in Table 13.18 are older than those for Switzerland, it is assumed that such information is of a structural

Table 13.18 Obstacles in university to business technology transfer: UK, US and CH.[a]

	2000/TO	2000/TO	2005/Firms	2011/Firms
	UK (n = 32)	US (n = 57)	CH (n = 669)	CH (n = 469)
Differing financial expectations	50.0	36.0	19.8	19.9
Communication problems	56.0	44.0	—	—
Need for more technical support	53.5	38.0	9.4	10.2
Cultural differences between university and company	62.5	64.5	—	—
Funding for further development	81.0	77.5	26.0	25.9
Lack of entrepreneurs in universities	37.5	42.0	17.2	13.9

a) Percentage of TOs and firms, respectively, reporting 4 or 5 on a five-point Likert scale.[24]

24) M. Decter et al., University to business technology transfer – UK and USA comparisons, Technovation, Volume 27, N. 3, p. 145, 2007.

Table 13.19 Barriers to university to business technology transfer: UK and CH.

	2007		2005	2011
	UK (n = 503)		CH (n = 669)	CH (n = 469)
	SMEs	Large firms	All firms	All firms
University research is extremely orientated towards pure science	31	36	20.7	18.4
Concerns over lower sense of urgency of university researchers versus industry researchers	69	59	16.3	14.4
Mutual lack of understanding on expectations and working practices	34	34	—	—
Industrial liaison offices tend to oversell research	50	49	—	—
Potential conflicts with university regarding intellectual property rights and concerns about confidentiality	57	54	9.2	9.2
Rules and regulations imposed by universities	58	53	9.4	18.3
Absence or low profile of industrial liaison officers in the university	27	24	9.2	4.6

Percentage of firms reporting 4 or 5 on a five-point Likert scale.[25]

nature and, as a consequence, is rather stable over time, provided that no significant institutional or policy changes took place during the reference period. Nevertheless, a certain pattern of barriers emerges from this information.

For all three countries a lack of funding for further development of new products and new processes is the most relevant barrier to technology transfer, even if not to the same extent. For the US and the UK TOs, the cultural differences between university and company are the second most important barrier. Swiss firms, for which no information on cultural differences is available, consider differing financial expectations the second most relevant obstacle of KTT activities.

Table 13.19 contains some comparable data on barriers from the point of view of UK and Swiss firms. As shown, there are considerable differences both with respect to the ranking of problems and with respect to the extent of these problems.

On the whole, Swiss firms appear to be confronted with many fewer obstacles to KTT activities than UK firms. About 18–20% of Swiss firms found that universities are too science-orientated for their needs, but this was a secondary problem for UK firms (31% of Small Medium Enterprise; 36% of large firms). For UK firms the most important barrier is concerns over a lower sense of urgency of university researchers versus industry researchers. This is also a problem for Swiss firms, but at a much lower level. However, for the firms of both

[25] J. Bruneel et al., Investigating the factors that diminish the barriers to university-industry collaboration, Research Policy, Volume 39, Issue 7, p. 858, 2010.

countries rules and regulations imposed by universities are the second most important barrier, even if at quite different levels.

13.3
KTT Activities in the Swiss Economy: The Main Facts from the *Science Institution* Point of View

The views of the business sector and the science institutions with respect to KTT activities are not expected to be identical. It is reasonable to think that there are conflicts between them referring to differing goals and differing ways of doing things.

13.3.1
Incidence and Forms of KTT Activities

A slightly larger proportion of institutes reported (reference period: 2008–2010) KTT activities in 2011 than in 2005 (reference period: 2002–2004): 88% versus 84% (Table 13.20). The positive change was largest in the ETHD, while in the other two groups of science institutions this proportion remained almost constant. UAS show a remarkable greater incidence of KTT activities with foreign partners in the period 2008–2010.

Concerning the forms of KTT activities, the pattern of the main groups of KTT activities also remained relatively stable over time (Table 13.21). However, there was a distinct decrease in the proportion of institutes reporting informational activities (2011: 69%; 2005: 79%). This effect can be traced back to the ETHD (minus 14%) as well as the CUs (minus 10%). A slight increase was reported for consulting activities.

13.3.2
Mediating Institutions and Obstacles of KTT Activities

Some interesting changes can be noted since the earlier period that reflect the increase in relevance, and presumably also in the effectiveness of the mediating

Table 13.20 Incidence of KTT activities according to group of science institutions.[a]

	2002–2004	2002–2004 Foreign partners	2008–2010	2008–2010 Foreign partners
ETHD	80.9	72.1	92.3	78.9
CUs	79.9	72.8	78.6	64.3
UAS	96.6	64.4	100.0	90.5
Total	**83.8**	**70.5**	**88.4**	**75.6**

a) Percentage of all firms.

13.3 KTT Activities in the Swiss Economy: The Main Facts from the Science Institution Point of View

Table 13.21 Main categories of forms of KTT activities according to groups of science institutions.[a]

2011 (2005)	Information	Infrastructure	Education	Research	Consulting
ETHD	60.4 (74.5)	18.8 (12.7)	75.0 (80.0)	77.1 (78.2)	43.8 (43.6)
CUs	69.1 (78.9)	20.0 (22.5)	78.2 (71.1)	63.6 (66.7)	43.6 (47.8)
UAS	78.6 (82.5)	7.1 (14.0)	97.6 (94.7)	92.9 (86.0)	78.6 (56.1)
Total	**69.0 (78.7)**	**15.9 (17.4)**	**82.8 (80.2)**	**76.6 (75.2)**	**53.8 (49.0)**

a) Firms' percentage reporting 4 or 5 on a five-point Likert scale for any single form of a given main form of KTT activities (2011 compared to 2005).

Table 13.22 Importance of single mediating institutions according to group of science institutions.[a]

2011 (2005)	TOs	CTI	SNSF	EFPs	ERPs
ETHD	29.2 (22.4)	52.1 (39.7)	27.1 (13.8)	47.9 (13.8)	20.8 (16.0)
CUs	33.3 (16.0)	31.6 (17.0)	22.8 (31.9)	26.3 (14.9)	17.5 (36.2)
UAS	19.1 (12.3)	57.1 (33.3)	9.5 (5.3)	28.6 (8.8)	11.9 (12.3)
Total	**27.9 (16.8)**	**45.6 (27.8)**	**20.4 (19.6)**	**34.0 (12.9)**	**17.0 (20.6)**

a) Institutes' percentage reporting 4 or 5 on a five-point Likert scale (2011 compared to 2005).

services of the CTI and the university TOs in Table 13.22, much more clearly than the firm view (see Table 13.11). The better position of the TOs can be traced back primarily to the positive changes in the CUs. The CTI increase reflected a stronger involvement in CTI projects in all three groups of science institutions. The same can also be said of the EFPs. The importance of such programmes nearly tripled. As to the SNSF, a stronger involvement was found for the ETHD and the UAS.

The relevance of two groups of obstacles has decreased significantly since 2002–2004: lack of information and costs/risks and uncertainty (Table 13.23).

Table 13.23 Main categories of obstacles according to group of science institutions.[a]

2011 (2005)	Lack of information	Problems in teaching/research	Firm deficiencies	Science institution deficiencies	Costs/risks uncertainty	Organizational/institutional obstacles
ETHD	13.5 (24.6)	23.1 (21.7)	19.2 (27.0)	17.3 (17.5)	19.2 (41.3)	3.9 (12.7)
CUs	20.6 (28.8)	29.9 (24.3)	18.6 (15.5)	20.0 (21.6)	16.2 (20.9)	13.2 (10.0)
UAS	21.4 (31.0)	28.6 (13.0)	19.1 (13.8)	14.3 (15.5)	42.9 (43.1)	14.3 (10.3)
Total	**18.5 (28.3)**	**27.0 (20.9)**	**18.9 (18.2)**	**17.7 (19.0)**	**24.1 (32.0)**	**10.5 (10.8)**

a) Institutes' percentage reporting 4 or 5 on a five-point Likert scale for any single obstacle in a certain category of obstacles (2011 compared to 2005).

Table 13.24 Ranking of importance of obstacles to university to business technology transfer: The Netherlands.[26]

University researchers	2006
	rank (n = 575)
Joint R&D hindered by conflicts between academic researchers and commercial researchers	1.0
Difficult to find appropriate industrial partners for joint R&D projects	1.5
Companies do not want to cooperate with universities but to absorb knowledge	2.0
Cooperation is hindered by cultural differences	3.0
Transfer of knowledge to industry is too costly for universities	4.0
Universities are not willing to spend time and money in knowledge transfer to industry	5.0
Contract research results only in more revenues but not to new knowledge for the university	6.0
The industry is not interested in the knowledge developed at the university.	7.0

Reproduced with permission of Elsevier.

During the 2008–2010 period, three further categories of problems were perceived as serious obstacles to KTT activities to about the same extent as in the 2002–2004 period: firm deficiencies, science institution deficiencies, and organizational/institutional obstacles. An increase was found only for the obstacle problems in teaching and research (caused by involvement in KTT). On the whole, the obstacles seem to be less severe than in the former period. The problems in teaching and research are relevant primarily in the UAS. The decrease in the category costs/risks and uncertainty comes primarily from the ETHD.

Finally, Table 13.24 presents information on a series of barriers of KTT activities from the point of view of Dutch university researchers. Possible conflicts between academic and industry researchers are the most important barrier, while finding appropriate industrial partners is the second most important obstacle. The latter is a relevant barrier also for Swiss institutes. In contrast to the Dutch findings, the lack of interest of industry for the scientific knowledge developed at the universities is the second most important obstacle for Swiss institutes.

13.4
Analytical Part: Exploration of KTT Activities in Switzerland

In the second part of this chapter, a number of specific topics that are related with KTT activities, and which would contribute to a better understanding of the economics of KTT activities, are explored. A common trait of the following studies is that all of them are based on econometric analysis.

26) V. Gilsing et al., *Differences in technology transfer between science-based and development-based industries: Transfer mechanisms and barriers*, Technovation, Volume 31, Issue 12, p. 638, 2011.

In what follows, we want to improve their understanding of the main driving forces for transfer activities and the consequences of transfer activities within the firms, as well as the adequacy of transfer strategies in order to enhance the innovation performance of a firm. Such knowledge would not only help to identify measures to increase or intensify transfer activities, it will also help to identify the potential of transfer activities.

13.4.1
Drivers of KTT Activities from the Point of View of the *Firm*[27]

In a first part-study, the factors that determine a firm's propensity to KTT activities (as defined in this chapter) are investigated. Not only is the overall inclination to such activities interesting, but so is the propensity for each of the five specific fields of KTT activities taken into consideration in this study: general information; education; research; technical infrastructure; and consulting.

As already mentioned, the analysis is guided by the 'stylized' model from Bozeman.[28] In this model, the KTT between transfer agent and transfer recipient covers a wide spectrum of joint activities, including research and education, and is influenced by environmental conditions and perceived obstacles. For an analysis at firm level, KTT activities are envisaged as a result of firms' decision (i) to become involved in KTT activities with science institutions and (ii), given this basic decision, to choose a specific form of KTT activities.

What do we know about the factors that determine this kind of firm decisions? Most of the existing Industrial Organization (IO) literature on this subject focuses on the determinants of R&D cooperation between firms that are direct competitors. Further transfer activities – for example, those related to human capital – are not considered. Theoretical literature focuses primarily on the effect of imperfect appropriability of results of innovation activities on the incentives to innovate, when firms cooperate in R&D.[29],[30] There is a twofold incentive problem. On the one hand, the existence of imperfect appropriability (above a critical level of the underlying knowledge spillovers) increases the incentives to cooperate, because of the profits resulting from internalizing the external losses caused by imperfect appropriability.[31] On the other hand, imperfect appropriability also increases the incentives to utilize spillovers resulting from the R&D

---

27) S. Arvanitis et al., *Knowledge and Technology Transfer Activities between Firms and Universities in Switzerland: An Analysis Based on Firm Data*, Industry and Innovation, Volume 18, Issue 4, p. 369, 2011; *Knowledge and Technology Transfer between Universities and Private Enterprises in Switzerland 2011*, Study on Behalf of the ETH Board, KOF Studies, N. 37, 2012.
28) B. Bozeman, *Technology Transfer and Public Policy: A Review of Research and Theory*, Research Policy, Volume 29, Issue 4–5, p. 627, 2000.
29) M. Spence, *Cost Reduction, Competition and Industry Performance*, Econometrica, Volume 52, N. 1, p. 101, 1984.
30) C. D'Aspremont and A. Jacquemin, *Cooperative and Non-cooperative R&D in Duopoly with Spillovers*, American Economic Review, Volume 78, Issue 5, p. 1133, 1988.
31) R. De Bondt, *Spillovers and Innovative Activities*, International Journal of Industrial Organization, Volume 15, Issue 1, p. 1, 1997.

investment of a cooperating partner and encourages free-riding on the R&D efforts of the cooperating firms by outsiders.[32),33)] However, when firms are not direct competitors (e.g. suppliers of complementary goods), or when one partner is a science institution, imperfect appropriability of the benefits of generated knowledge is not an important issue for cooperation.

A further important concept is the notion of 'absorptive capacity'. In 1989, Cohen and Levinthal emphasized the importance of a firm's own R&D efforts for developing the ability to absorb and utilize external knowledge.[34)] A high absorptive capacity is thus a precondition for cooperation between firms and science institutions.

In 2005, Veugelers and Cassiman mentioned a series of further factors which might possibly influence industry–science cooperation and are related with the specific nature of the know-how being transferred in such collaborations.[35)] First, due to the specific characteristics of scientific knowledge, a specific profile of firms can be expected to pursue such collaborations. Science institutions offer new technological knowledge which is mainly needed in innovation activities characterized by high technological uncertainty and, at the beginning, a low demand for the innovation outcomes. As a consequence, only firms within specific industries using specific technologies (e.g. biotechnology or nanotechnology) will have a strong interest in collaborations with science institutions. This is a strong, but rather too narrow-focussed hypothesis which cannot cover the wide spectrum of effective KTT activities. Second, R&D cooperation between universities and the industry is characterized by high uncertainty, high information asymmetries between partners, high transaction costs for knowledge exchanges requiring the presence of absorptive capacity, and high spillovers to other market actors.[36)]

Based on the above notions from IO literature, and the experiences of previous empirical studies, a series of possible determinants of KTT activities are postulated in the next paragraphs.

13.4.1.1 Determinants of KTT Activities of Firms

A first group of determinants is related to the resource endowment of the enterprises with human capital, physical capital and knowledge capital. It is particularly expected that firms with high human capital and knowledge capital intensity, leading to a high knowledge absorptive capacity, would possess the profile needed for KTT activities with science institutions. Such firms would be

---

32) C. Shapiro and R. Willig, *On the Antitrust Treatment of Production Joint Ventures*, Journal of Economic Perspectives, Volume 4, Issue 3, p. 113, 1990.

33) P. Greenlee and B. Cassiman, *Product Market Objectives and the Formation of Research Joint ventures*, Managerial and Decision Economics, Volume 20, Issue 3, p. 115, 1999.

34) W. Cohen and D. Levinthal, *Innovation and Learning: The Two Faces of R&D*, The Economic Journal, Volume 99, N. 397, p. 569, 1989.

35) R. Veugelers and B. Cassiman, *R&D Cooperation Between Firms and Universities. Some Empirical Evidence from Belgian Manufacturing*, International Journal of Industrial Organization, Volume 23, Issue 5–6, p. 355, 2005.

36) Ibidem.

most frequently found in high-tech innovative manufacturing (e.g. pharmaceutical industry, electronics) and in knowledge-based service industries (e.g. software industry). Physical capital intensity would be a complementary measure for absorptive capacity especially for manufacturing firms.

There are several firm characteristics which are expected to be significantly related to KTT activities. The degree of exposition to international competition should be positively related to KTT, since international firms should have greater know-how requirements for international-orientated firms. The firm size should be also positively related with KTT, since it is likely that there are positive scale effects with respect to the utilization of scientific knowledge. Also, firm age should be positively related, since older firms are likely to possess a longer experience in cooperative projects. The firm status as a subsidiary of a foreign mother-company does not, a priori, have any clear effect.

The propensity to engage in KTT activities is also influenced by the field(s) of technology in which a firm is active. Given its technological profile, a firm intending to become involved in KTT activities would have to consider the costs of this involvement. Possible costs would include: high transaction costs due to deficiencies on the interface between firm and science institution, either on the side of the firm or the science institution; great information asymmetries; great financial risks due to the uncertainty of research outcomes; property rights problems; and potential costs arising from technological dependence on a science partner.

13.4.1.2 Empirical Evidence

Overall Propensity to KTT Activities of Firms

The factors that determine the propensity of Swiss firms to interact with public science institutions in Switzerland (universities and other research institutions) were explored in two studies that were based on the data collected in the two surveys of 2005 and 2011. A summary of the results of the econometric estimations of the propensity equation is given in Table 13.25.[37]

The variables for human capital intensity and the propensity to R&D activities have highly significant positive coefficients. Both variables are closely related to a firm's ability to absorb new knowledge from its environment. Also, capital intensity – the third variable in the model referring to a firms' resource endowment – is also relevant for distinguishing between firms with KTT activities and those without.

Export intensity, taken as a measure of a firm's degree of exposure to international competition, shows also a significantly positive effect, which is much smaller than that for human capital, R&D intensity, and capital intensity. The variable for firm age has a significantly positive coefficient, thus indicating that older firms are more strongly inclined to become involved in KTT activities

---

37) The effects of capital intensity, export intensity and domestic ownership showed the same signs but were statistically insignificant in the estimation based on the 2005 data.

Table 13.25 Drivers of KTT activities from the point of view of Swiss firms 2005/2011.

Determinants	Effect on KTT
Gross investment per employee	Positive/no effect[a]
Share of employees with tertiary-level education	Positive
R&D activities yes/no	Positive
Export intensity	Positive/no effect[a]
Firm size	Positive (non-linear)
Firm age	Positive
Domestic ownership	Positive/no effect[a]
Technological field	
Nanotechnology	Positive
Biotechnology	Positive
Environmental technologies	Positive
New materials	Positive
Software/simulation/artificial intelligence	Positive
Obstacles to KTT activities	
Lack of information	Positive
Firm deficiencies	Negative
Deficiencies of universities	Negative
Cost, risks	Positive
Organizational/institutional problems	No effect

a) No effect for the 2005 data. 'Positive' or 'negative' effects refer to the respective coefficients of the probit-estimates of the propensity equation under the condition that they are statistically significant at the 10%-test level.[38]

than younger firms, presumably because they have a greater experience in cooperating with science institutions than the younger firms. There is a significant difference between domestic and foreign firms with respect to KTT activities: foreign-owned firms seem to be less inclined to cooperation with domestic science institutions than domestic firms. Finally, there is a positive nonlinear relationship between firm size and the propensity to KTT activities, with larger firms appearing to be more strongly inclined to KTT activities than smaller firms.

Firms with KTT activities seem to have a focus on biotechnology/genetic engineering, nanotechnology, new materials, software/simulation/artificial intelligence and environmental technologies (in decreasing ranking order as to marginal effects). For other technological fields,[39] no differences could be found between KTT-active firms and firms without KTT activities. For telecommunication/information technology, a negative effect is found, indicating that R&D in these fields is conducted primarily in the firms' own R&D departments.

---

38) S. Arvanitis et al., *Knowledge and Technology Transfer Activities between Firms and Universities in Switzerland: An Analysis Based on Firm Data*, Industry and Innovation, Volume 18, Issue 4, p. 369, 2011; *Knowledge and Technology Transfer between Universities and Private Enterprises in Switzerland 2011*, Study on Behalf of the ETH Board, KOF Studies, N. 37, 2012.

39) Microelectronics/semiconductor technology, laser technology/optoelectronics/displays, medical technology/sanitary engineering, flexible manufacturing technology, transport technology/traffic engineering/logistics, energy technologies, geological technologies.

Relevant from the policy point of view are the results for the obstacle variables. According to the econometric results, deficiencies of firms with respect to the utilization of the potential of KTT activities seem to be an important category of obstacles. Such deficiencies are according to the firms' assessment of the lack of interest in scientific projects, the lack of qualified staff, and the lack of technical equipment. Furthermore, many firms feel that their research questions are not interesting for research institutions. A further relevant group of obstacles refers to the deficiencies of universities (primarily a lack of entrepreneurial spirit, a fear that R&D results could not be commercialized, and research orientation being not interesting for firms). These two groups of obstacles appear to be the most important source of mismatching between enterprises and universities (negative effects). Organizational and institutional problems seem to be of minor importance. Finally, for cost and risks it is found that firms already engaged in KTT activities seem to be more aware of such problems than firms without these activities (positive effects that are contrary to our expectation of negative effects). This is because knowledge and technology transfer involves a learning process for the firm.

Propensity to Specific Forms of KTT Activities of Firms

Based on the data for 2011, a more detailed investigation of each of the five main forms of KTT activities (informal informational contacts, educational and research activities, utilization of university technical facilities, and consulting) showed that there are some significant differences amongst the five types of KTT activities with respect to the various determinants. However, it must be taken into consideration that, in the case of the five specific KTT forms, firms with a strong engagement in one type of activities are compared with firms showing a strong engagement in any of the other four types of activities.[40]

Particularly, the variables for resource endowment are not equally important for all five specific forms of KTT activities. The econometric results based on the 2011 data showed that the human capital intensity is relevant for educational and for informal informational contacts, but not for consulting or infrastructure-orientated activities and – rather astonishingly – not for research activities. The existence of R&D activities is particularly relevant for research, educational and infrastructure-orientated activities. Firms with a high export intensity show a specific interest for research, educational, and infrastructure-related transfer activities. The capital intensity effects can be traced back to firms with a stronger focus on infrastructure and educational activities, while the effect of foreign-owned firms can be traced back primarily to firms focusing on educational activities (foreign firms being less acquainted with the domestic system of higher education than domestic firms). The propensity to conduct research activities

---

40) While in the estimates for KTT propensity firms with KTT activities are compared with firms that are not involved in such activities.

appears to be higher for younger firms as compared to the other forms of KTT activities. Finally, the positive effect of firm size is more relevant for educational and research activities.

An interesting pattern with respect to the relative importance of the five main forms of KTT activities, for the different technological fields taken into consideration in this study, has been found. For each of the five main forms of KTT activities only the technologies listed below have positive and statistical significant coefficient:

1) Informal informational activities: nanotechnology; energy technologies.
2) Activities related to technical infrastructure: laser technology/optoelectronics/displays; energy technologies.
3) Educational activities: new materials; software/simulation/artificial intelligence; biotechnology/genetic engineering; geological technologies.
4) Research activities: biotechnology/genetic engineering; energy technologies; geological technologies.
5) Consulting activities: laser technology/optoelectronics/displays; flexible manufacturing technology; energy technologies.

Firms with a focus on educational and/or research activities are engaged in similar fields. Energy technologies seem to be a relevant technological field for all five categories of KTT activities. For all other technologies not mentioned above (see Table 13.10), no differences amongst the firms pursuing different forms of KTT activities are discernible.

Based on the data for 2005, it is found that not all activities are affected in the same manner by the various categories of obstacles. Research activities are hampered by a lack of information as to possible contacts and projects in cooperation with the universities by organizational and institutional obstacles. Informational and educational activities are impeded primarily by firm deficiencies. Finally, firms with a focus on consulting found that costs and risks, as well as organizational and institutional problems, were severe impediments of KTT activities.

13.4.2
KTT Activities Determinants from the *University* Point of View[41]

The factors that determine the scientific institutes' propensity to KTT activities are analysed in a further part-study.

There is little theoretical research on the financial incentives facing faculty and the allocation of effort across types of research.[42] In 2003 and 2004, respectively,

---

41) S. Arvanitis *et al.*, *University-Industry Knowledge and Technology Transfer in Switzerland: What University Scientists Think about Co-operation with Private Enterprises*, Research Policy, Volume 37, Issue 10, p. 1865, 2008.
42) M.C. Thursby *et al.*, *Are There Effects of Licensing on Academic Research? A Life Cycle View*, Journal of Economic Behaviour and Organization, Volume 63, Issue 4, p. 577, 2007.

Beath *et al.*, and Jensen and Thursby studied faculty research incentives in the framework of a principal agent model where the university is the principal and the faculty member is the agent.[43],[44]

Beath's analysis is static and investigates the potential for the university to ease its budget constraints by allowing academic scientists to conduct applied research on a consulting basis. They argue that, by allowing academics to supplement their income, universities may be able to hold down academic salaries. Furthermore, universities can effectively 'tax' the income that academics raise through applied research or consultancy, for example through the imposition of 'overhead charges'. This model offers some insights with respect to the financial incentives for conducting applied research in cooperation with the industry.

By contrast, the model of Jensen and Thursby is dynamic and analyses the effect of patent licensing on research and the quality of education. The latter effect is a function itself of research outcomes and hence future stocks of knowledge, as well as the share of patentable knowledge that can be used in education. In this model, an academic scientist derives utility from just doing research as well as the prestige associated with successful research. Jensen and Thursby show that, with these two effects in a scientist's utility function, the opportunity to earn license income may not change an academic scientist's agenda. This result provides, according to their assessment, one explanation for the fact that little change can be observed in the ratio of basic and applied research publications of academic scientists.

Later, Thursby *et al.* discussed, in the framework of a life cycle model of an academic scientist's career, the implications of licensing on research. In this context, the utility function of academic scientists contains on the one hand a motive for generating new knowledge, and on the other hand a financial motive for additional income. An important issue in the debates over university licensing is whether the associated financial incentives compromise the research mission of the university by diverting academic scientists from basic research. In the various versions of the model, it has been considered that the academic scientist faces a fixed teaching load and chooses the amount of time to devote to research (basic or applied) and the amount of time to take as leisure.

In 2007, Hellman developed an interesting formal theory of the search and matching process between scientists and firms.[45] At the core of the model is the problem that scientists rarely know what industrial applications may exist for their scientific discoveries. At the same time, firms are often unaware what scientific discoveries might help them with their needs. The author calls this the 'science to market gap'. The model allows addressing the role of patents in bridging the science to market gap. The gap can be bridged when scientists and firms

---

43) Beath *et al.*, *Optimal Incentives for Income-generation in Universities: The Rule of Thumb for the Compton Tax*, International Journal of Industrial Organization, Volume 21, Issue 9, p. 1301, 2003.

44) R. Jensen and M. Thursby, *Patent Licensing and the Research University*, NBER (National Bureau of Economic Research), Working Paper, N. 10758, 2004.

45) T. Hellman, *The Role of Patents for Bridging the Science to Market Gap*, Journal of Economic Behaviour and Organization, Volume 63, Issue 4, p. 624, 2007.

engage in a process of search and communication. Since patenting affects the distribution of rents, it has an effect on the relative search intensities of firms and scientists. Patenting scientific discoveries helps scientists to 'push' their discoveries out to the business sector. However, it may also dampen the firms' incentives to 'pull' discoveries out of scientists. Thus, the net effect of patenting depends on the relative ease of bridging the science to market gap through 'push' or 'pull'.

The model also examines the importance of universities' technology transfer offices. In principle, such offices allow for task specialization. Scientists benefit from delegating search activities, which may free them up to pursue further research. However, the model explains that such delegation typically requires patenting. In introducing the role of transfer offices, it is assumed that they are more efficient at searching for industrial partners than scientists, and this may be reasonable in many cases, but not in all. Indeed, if this is not the case the formation of a spin-off may be an alternative way that guarantees efficiency, because in a spin-off the scientist always internalizes all benefits from research. A last discussion point refers to the lack of an analysis of the dynamic implications of the commercialization of research output.

On the whole, the existing theoretical literature delivers a number of factors, mainly of motivational character ('push' and 'pull' factors in Hellman), which determine the propensity of academic scientists to engage themselves in commercialization activities that provide additional income. There exists some form of trade-off between financial motives in favour of commercialization, and hence the perspective of additional income and the inherent motives of a scientist who primarily pursues research goals and the reputation associated with research achievements. In addition, some basic characteristics of institutes or departments, such as the field (e.g. engineering versus social sciences), the size, or the existence of a strategic orientation towards research seem to exert a significant influence on the KTT propensity of academics. As a consequence, an empirical investigation would at least contain measures for anticipated costs and benefits of various channels of KTT between universities and the business sector, and measures of the allocation of working time in basic and applied research as well as teaching.

13.4.2.1 Determinants of the Propensity to KTT Activities of Universities

Putting together the information found in the reviewed theoretical and empirical literature, it is possible to identify the following factors explaining the propensity to KTT of university faculties:

- Allocation of university resources (research, teaching); most theoretical studies come to the conclusion that there is no close relationship between the propensity to KTT activities and the orientation (basic versus applied) of institute research activities.
- Extent of external funds, exerting also a positive influence.
- Size of faculty or university, having mostly a positive influence.
- Type of scientific field, engineering and natural sciences showing a stronger inclination to KTT activities than other disciplines.

- Existence of TOs having mostly a positive influence.
- A series of motives that could be grouped in four main categories (access to: industrial knowledge, additional resources, institutional or organizational motives and specialized technology; pursuing higher research efficiency – cost and time savings). Motives influence positively the propensity to KTT activities.
- A series of obstacles that could be grouped into six categories (deficiencies of the firms; different interests and attitudes to research; lack of confidence to business world and risk of damaging scientific reputation; endangering scientific independence and neglect of basic research; lack of human resources for KTT). Obstacles influence negatively the propensity to KTT activities.

13.4.2.2 Empirical Evidence

Overall Propensity to KTT Activities of University Institutes

The most important findings of this econometric study refer to the overall propensity to KTT activities with private enterprises. A summary of the results of the econometric estimations of the propensity equation is given in Table 13.26.

Table 13.26 Drivers of KTT activities from the point view of university institutes 2005.[a]

Determinants	Effect on KTT
Applied research (share of working time as compared to basic research)	Positive
Teaching (share of working time)	Negative
Budget share of third-party funds from business sector	Positive
Size of the institute: 100 institute employees and more	Positive
Obstacle: main activities (research, publication) hindrance/neglect; scientific independence impairment	Negative
Institute discipline (reference: mathematics/physics)	
Engineering	Positive
Natural sciences	Positive
Economics, management	Positive
Medicine	Positive

a) 'Positive' or 'negative' effects refer to the respective coefficients of the probit-estimates of the propensity equation under the condition that they are statistically significant at the 10%-test level.[46]

46) S. Arvanitis et al., Knowledge and Technology Transfer (KTT) Activities between Universities and Firms in Switzerland-The Main Facts: An Empirical Analysis Based on Firm-Level Data, The ICFAI Journal of Knowledge Management, Volume 5, N. 6, p. 17; University-Industry Knowledge and Technology Transfer in Switzerland: What University Scientists Think about Co-operation with Private Enterprises, Research Policy, Volume 37, Issue 10, p. 1865; Is There Any Impact of University-Industry Knowledge Transfer on Innovation and Productivity?–An Empirical Analysis Based on Swiss Firm Data, Review of Industrial Organization, Volume 32, Issue 2, p. 77; Do Specific Forms of University-Industry Knowledge Transfer Have Different Impacts on the Performance of Private Enterprises?–An Empirical Analysis Based on Swiss Firm Data, The Journal of Technology Transfer, Volume 33, Issue 5, p. 504, 2008.

Institutes with a stronger orientation to applied research and/or lower teaching obligations are also more strongly inclined to become involved in KTT activities. The same is valid for institutes which have already experience with industry cooperations, as reflected by a high share of external funds in an institute's budget. Further, there was a significantly positive correlation of institute's research output measured as number of PhDs with the propensity to KTT activities. This correlation becomes statistically insignificant if the number of scientific publications is used as a measure of institute research output. Thus, being involved in KTT activities is at best positively correlated with a high research performance, and at worst not at all related with the level of scientific output.

Determinants of Patenting, Licensing and Formation of Spin-Offs of University Institutes

A further part-study explored econometrically the factors determining the propensity of institutes and/or departments of Swiss universities and other research institutions to patent new inventions, license new technologies, and launch new firms, respectively. These are – according to the literature – the most important ways of commercializing of university research output. A summary of the results of the econometric estimations of the equations for patenting, licensing and founding of spin-offs is given in Table 13.27. The degree of orientation to applied research and the extent of teaching obligations are not relevant determinants for university patenting. Applied research intensity shows no effects, and teaching is negatively related to spin-offs. The firm size is positively correlated with the propensity to launch new firms. Applied research intensity and teaching are negatively correlated with licensing. Particularly for patenting activities, the access to industrial knowledge for gaining practical experiences and possible applications of research findings is an important motive for KTT activities. A further important finding is that all three types of activities are hampered by the same category of reported obstacles, reflecting the perception of academics of an industry research profile that does not correspond well to their own needs and interests. There is a positive firm size effect with respect to the formation of spin-offs and partly with respect to patenting, but not for licensing. TOs appear to be important for university knowledge. Scientific disciplines matter commercialization only for patenting.

13.4.3
Impact of KTT Activities on Innovation and Labour Productivity[47]

A further interesting topic concerns the possible impact of KTT activities on a firm's economic performance, particularly on firm innovative activities. For an

---

47) S. Arvanitis et al., *Is There Any Impact of University-Industry Knowledge Transfer on Innovation and Productivity? – An Empirical Analysis Based on Swiss Firm Data*, Review of Industrial Organization, Volume 32, Issue 2, p. 77; *Do Specific Forms of University–Industry Knowledge Transfer Have Different Impacts on the Performance of Private Enterprises? – An Empirical Analysis Based on Swiss Firm Data*, The Journal of Technology Transfer, Volume 33, Issue 5, p. 504, 2008.

Table 13.27 Drivers of patenting; licensing; founding of spin-offs of Swiss universities 2005.[a]

Determinants	Effect on patenting	Effect on licensing	Effect on founding of spin-offs
Applied research (share of working time as compared to basic research)	No effect	Negative	No effect
Teaching (share of working time)	No effect	Negative	Negative
Firm size: 40–99 institute employees	No effect	No effect	Positive
Firm size: 100 institute employees and more	Positive	—	Positive
Obstacle: Differing interests, attitudes to research	No effect	Negative	Negative
Motive: Access to industrial knowledge	Positive	No effect	No effect
Mediation of TOs	Positive	Positive	Positive
Affiliation to a Federal Research Institution	No effect	No effect	Positive
Institute discipline (reference: mathematics/physics)			
Engineering	Positive	No effect	No effect
Natural sciences	Positive	No effect	No effect
Economics, management	—	No effect	No effect
Medicine	Positive	No effect	No effect

a) 'Positive' or 'negative' effects refer to the respective coefficients of the probit-estimates of the propensity equation under the condition that they are statistically significant at the 10%-test level.[48]

enterprise, KTT is not an aim in itself but a means for achieving a higher performance.

The main hypothesis of this study is that KTT activities would improve the innovation performance of firms and also – either directly or indirectly via innovation output – their economic performance in the narrow sense, for example average labour productivity. This KTT effect could be traced back to an increase of technological opportunities anticipated by firms due to university-industry knowledge transfer. This would include effects from a wide palette of KTT activities such as exchanging scientific and technical information, various educational activities (e.g. recruitment of R&D personnel from the universities, joint PhDs, specialized training courses), consulting, use of technical infrastructure, and, of course, co-operation in research. The prominent role of technological opportunities as a major supply-side determinant of innovation is often emphasized in the literature.[49],[50] A further hypothesis is that R&D activities that are closely

---

48) S. Arvanitis et al., *Knowledge and Technology Transfer (KTT) Activities between Universities and Firms in Switzerland-The Main Facts: An Empirical Analysis Based on Firm-Level Data*, The ICFAI Journal of Knowledge Management, Volume 5, N. 6, p. 17, 2008.
49) A.K. Klevorick et al., *On the Sources and Significance of Interindustry Differences in Technological Opportunities*, Research Policy, Volume 24, Issue, p. 185, 1995.
50) S. Arvanitis and H. Hollenstein, *Industrial Innovation in Switzerland: A Model-based Analysis with Survey Data*, in A. Kleinknecht (Ed.), *Determinants of innovation. The message from new indicators*, Macmillan Press, 1996.

related to knowledge generation would be more strongly enhanced by the interaction with universities than would activities that are near to the market launching of a new product (e.g. construction of prototypes, test production, market tests for new products, etc.).

13.4.3.1 Empirical Evidence

In order to investigate the role of KTT for performance, it is possible to identify the impacts of a wide spectrum of KTT activities (i) on several innovation indicators in the framework of an innovation equation with variables of endogenized KTT activities (overall activities, specific forms of activities such as educational and research activities, consulting and informal contacts) as additional determinants of innovation, and (ii) on labour productivity in the framework of a production function with KTT activities as an additional production factor. A summary of the results of the econometric estimations of the innovation and productivity equations is found in Table 13.28.

It has been found that KTT activities improve the innovation performance of firms, both in terms of R&D intensity and sales of innovative products. The positive effect of overall KTT activities can be traced back mainly to research and educational activities. This could be shown by several methods: the innovation equation approach with endogenized KTT variable as well as three matching methods, which enable us to compare 'structurally similar' firms with KTT activities and without such activities. Furthermore, KTT activities seem to

Table 13.28 Impact of KTT activities on firm innovation and firm economic performance 2005.[a]

	R&D expenditures/sales	Sales share of modified products	Sales share of new products	Labour productivity
Overall KTT activities	Positive	Positive	Positive	Positive
Forms of KTT activities				
Research	—	Positive	Positive	—
Education	—	Positive	Positive	—
Technical infrastructure	—	No effect	Positive	—
Consulting	—	No effect	No effect	—

a) 'Positive' or 'negative' effects refer to the respective coefficients of the OLS (Ordinary Least Squares)-estimates of the propensity equation under the condition that they are statistically significant at the 10%-test level.[51]

51) S. Arvanitis et al.: *University-Industry Knowledge and Technology Transfer in Switzerland: What University Scientists Think about Co-operation with Private Enterprises*, Research Policy, Volume 37, Issue 10, p. 1865; *Is There Any Impact of University-Industry Knowledge Transfer on Innovation and Productivity?–An Empirical Analysis Based on Swiss Firm Data*, Review of Industrial Organization, Volume 32, Issue 2, p. 77, 2008.

exercise a positive influence on labour productivity, not only through a direct effect but also through an indirect effect by raising the elasticity of R&D intensity with respect to labour productivity.

13.4.4
KTT Strategies Determinants and their Impact on Innovation Performance[52]

There are different channels through which university knowledge can reach enterprises. In order to be able to exploit the advantages of KTT activities, a firm has to formulate and apply a strategy for KTT. This part-study is dedicated to infer such strategies from the KTT activities firms report to be engaged in.

Previously, several firm strategies to transfer knowledge from universities have been identified. The relationship between strategies and the innovation performance of a firm has been investigated. The strategies were constructed based on 19 different forms of transfer (see Table 13.6) and defined as a combination of different forms of KTT. They comprise not only research collaboration, they also refer to softer transfer relationships, such as firm representatives attending conferences, the joint use of technological infrastructure, or consulting activities.

More concretely, based on a cluster analysis of the 19 different forms of transfer, it was possible to identify three dominant ways to make use of public research activities. For a first group of firms, none of the 19 possible single forms of KTT activities is very important on the average. This was labelled 'strategy A'; that is, the firms are in loose contact with public research organizations.

For a further group of KTT active firms only KTT forms are important that are related to informal, educational and consulting activities. In this case, KTT does not immediately focus on the R&D activities of a firm. This was labelled 'strategy B'; that is, the firms are engaged in non-core contacts with public research organizations.

A third and last group of KTT active firms focused on KTT forms (e.g. joint laboratories, sabbaticals, R&D cooperation) that are directly related to the R&D activities of the firm. This was labelled 'strategy C'; that is, the firms are undertaking core transfer activities such as R&D contracts.

13.4.4.1 Firm Characteristics, KTT Strategies, and Innovation Performance

A first interesting question is which type of firm pursues which type of strategy. Given some theoretical notions about the importance of the resource base of a firm, it is likely that the resource endowment – especially the knowledge endowment – determines to a great extent the firm's behaviour or marks its spectrum

---

[52] S. Arvanitis and M. Woerter, *Firms' transfer strategies with universities and the relationship with firms' innovation performance*, Industrial and Corporate Change, Volume 18, Issue 6, p. 1067, 2009.

of behaviour.[53),54),55),56)] Firms develop strategies that enable them to adapt their knowledge resources to perceived needs. Universities are an important source of external knowledge, and which strategy can be used for knowledge transfer from the universities is strongly related to the knowledge already available inside a firm and the ability of the firm to process and utilize external knowledge. In 1989, Cohen and Levinthal called this ability the absorptive capacity of a firm. They emphasized a firm's own R&D efforts for developing a strong absorptive capacity.[57)] In fact, in many studies the absorptive capacity is an important determinant for KTT activities.

Moreover, the choice of a transfer strategy can be influenced by perceived obstacles. Although the latter are related to firms' resource endowments and capabilities, they may also reflect external circumstances beyond the control of a firm, for example deficiencies of the science institution or difficulties to find contact persons. Consequently, variables related to the absorptive capacity of firms or their obstacle profiles are likely to be important determinants of their choice of a transfer strategy.

Further factors that are expected to influence the strategy choice besides firm size are the number of contacts to universities, contacts with foreign universities, and the importance of mediating services, such as TOs. These three factors reflect the extent of incoming knowledge spillovers.

The econometric analysis of the propensity to (any of) the three identified KTT strategies showed that absorptive capacity matters for the choice of a strategy (see Table 13.29). Firms pursuing strategy A appear to have a greater absorptive capacity than firms that pursue strategy B. This indicates that firms choose the loose contacts strategy A not because of a lack of absorptive capacity, but presumably because of the danger that firm-specific knowledge could be disclosed or because of the firms' own deficiencies (e.g. a lack of interest in public research). Moreover, these results may indicate that firms following strategy B are more likely to substitute in-house R&D activities for external knowledge through KTT with universities compared to firms following strategy A. No difference has been found between strategy C and strategy B with respect to absorptive capacity.

Firms following strategy B are less confronted with their own deficiencies than firms conducting strategy A. The obstacle profile differs considerably if we compare strategy C with strategy A. There are two groups of obstacles: (i) risk-related obstacles (organizational/institutional impediments; property rights problems); and (ii) cost-related impediments (follow-up work in order to implement public R&D results, lack of firm financial resources for transfer activities, lack of financial resources of science institutions for cooperation on an equal

---

53) B. Wernerfelt, *A Resource-based View of the Firm*, Strategic Management Journal, Volume 5, N. 2, p. 171, 1984.
54) J. Barney, *Firm Resources and Sustained Competitive Advantage*, Journal of Management, Volume 17, N. 1, p. 99, 1991.
55) E. Penrose, *The Theory of the Growth of the Firm*, Third Edition, Oxford University Press, 1995.
56) J. Barney et al., *The Resource-based View of the Firm: Ten years after 1991*, Journal of Management, Volume 27, Issue 6, p. 625, 2001.
57) W. Cohen and D. Levinthal, *Innovation and Learning: The Two Faces of R&D*, The Economic Journal, Volume 99, N. 397, p. 569, 1989.

Table 13.29 Determinants of KTT strategies 2005.[a]

	Strategy B	Strategy C
Absorptive capacity		
Frequency of R&D activities	Negative	No effect
Share of employees with tertiary-level education	No effect	No effect
Incoming knowledge spill-overs		
Number of contracts to universities	Positive	Positive
Knowledge and technology transfer with foreign universities	No effect	Positive
Mediating institutions (TOs, etc.)	No effect	Positive
Obstacles		
Risks	No effect	Positive
University deficiencies	No effect	Negative
Firm deficiencies	Negative	Positive
Financial obstacles	No effect	Positive
Lack of information	No effect	No effect
Firm size (employees)		
500–999	No effect	Positive
≥1000	Positive	No effect

a) 'Positive' or 'negative' effects refer to the respective coefficients of the multinomial probit-estimates of the propensity to strategy B and strategy C, respectively (compared to strategy C as reference strategy) under the condition that they are statistically significant at the 10%-test level.[58]

basis with firms, financial shortcomings on part of the firm and/or on part of the university partner, costs), that seem to hinder firms following strategy C more strongly than those pursuing strategy A. Taking the variable for cost-related impediments as a proxy for the costs of a strategy, it can be concluded that strategy C is more costly than strategy A. In contrast, there is no discernible difference with respect to this obstacle between strategy B and strategy A.

Two further obstacle categories – university deficiencies (a lack of scientific staff for transfer activities, R&D orientation of science institutions is uninteresting for firms, possible R&D results cannot be commercialized, firms' R&D questions are not interesting for science institutions) and firm deficiencies – have been found to be less of a problem for firms with strategy C than for those following strategy A. This means that deficiencies on part of the universities, as well as deficiencies on part of the firms themselves that hinder KTT activities, are less important for firms following strategy C than for those pursuing strategy A. Thus, the mismatch between firms' research interests and capabilities and the research interests and capabilities of universities and/or public research institutions, as detected by the obstacle variables, university deficiencies and firm deficiencies, respectively, seems to diminish with increasing intensity of transfer contacts. This hints at some learning effects with respect to abilities and

---

58) S. Arvanitis and M. Woerter, *Firms' transfer strategies with universities and the relationship with firms' innovation performance*, Industrial and Corporate Change, Volume 18, Issue 6, p. 1067, 2009.

thus related advantages of transfer activities with universities and/or public research institutions.

Further, it has been found that there is a positive relationship between the number of contacts to universities, a measure of the intensity of interactions with universities, and a firm's choice for strategy B. Contact with foreign universities and the importance of mediating services showed a positive relationship with strategy C. The results clearly demonstrated that firms with a broad network of university contacts (including contacts to foreign universities) that utilize the mediating services of specialized transfer offices follow strategy C, and to a small extent strategy B, rather than strategy A. This indicates that there is a significant relationship between these characteristics of the transfer process and the strategic orientation of a firm in terms of transfer forms.

A second question that naturally comes to the mind is how successful firms pursue a strategy in terms of their innovation performance. In order to investigate this question, an econometric model for the relationship between two innovation measures (the propensity to file a patent and the sales share of new products) and the three strategies A, B and C is built. It emerges that strategies A, B, and C show different effects on the two innovation performance measures. The patent propensity is significantly positively correlated with strategy C compared to firms that pursue strategy B, but also compared to firms that pursue strategy A. Thus, firms that follow strategy C are more likely to patent than firms apply strategy A or strategy B. The sales share of new products is also positively correlated with strategy C compared to strategy A, but not compared to strategy B. For both innovation indicators, significant differences between strategy A and strategy B are not found.

In summary, transfer strategy matters for the transfer success in terms of the innovation performance of a firm. Strategy C clearly shows advantages over strategy A or strategy B and, consequently, from a firm's point of view it is worth considering the type of KTT strategy to be applied in order to transfer publicly available knowledge in an efficient manner. Of course, there are good and better ways to do this. Basically, core KTT contacts focusing on various R&D-related activities (strategy C) seem to be more promising than rather loose contacts, if the involved firm has a high absorptive capacity, envisages a broad KTT network, and utilizes the services offered by transfer-supporting institutions. These conditions are more likely to be found in larger firms, particularly because they are also related to higher costs.

13.4.5
Exploration and Exploitation[59]

The exploration or exploitation of acquired knowledge is an issue of strategic importance for any innovative enterprise that is interested in long-term investment in innovation.

---

59) S. Arvanitis and M. Woerter, *Exploration or Exploitation of Knowledge from Universities: Does It Make a Difference?* KOF Working Papers, N. 322, ETH, 2012.

The seminal studies of March (1991) and of Levinthal and March (1993) on the exploration and exploitation of knowledge provided the background for the investigation of the characteristics of firms that explore new knowledge and/or exploit existing knowledge in cooperation with universities.[60],[61] More concretely, the distinction between exploration and exploitation specifically to knowledge acquired from universities has been applied. In this sense, our interest was to investigate which type of firm acquires additional external knowledge from universities for the purpose of exploring new knowledge, and which for the purpose of exploiting existing knowledge. Consequently, two research questions have mainly been investigated. First, as already mentioned, which type of firm pursues knowledge acquisition from universities for the purpose of exploring and generating new knowledge, and which type of firm focuses on exploiting existing knowledge? Second, if there are discernible differences as to important firm characteristics between the firms that focus on exploration and those that concentrate on exploitation, would this be also relevant for the impact of KTT activities on the innovation performance of a firm and, as a consequence, for the assessment of such activities from the point of view of economic policy? The expectation was that exploitation-orientated activities would show performance effects already in the short run, contrary to exploration-focused activities, for which mostly only long-run effects would be expected.

Given the inherent tension between exploration and exploitation, an analysis of the factors that drive these contradictory activities appears to be an interesting research topic. As Lavie *et al.* in their literature survey write: 'There has been little attempt to uncover why some organizations emphasize exploration, while others pursue exploitation. Empirical research has produced limited or mixed evidence on the causes of exploration and exploitation'.[62] In this sense, it is interesting to investigate which type of firm acquires additional external knowledge from universities for the purpose of exploring new knowledge and which for the purpose of exploiting existing knowledge. Existing literature has identified a series of factors as important antecedents of exploration and exploitation that briefly reviewed in the next paragraphs.

The absorptive capacity of a firm is doubtless a very important factor for having the ability to assess the importance of external knowledge, to transfer it successfully, and to apply it in combination with its internally created know-how.[63],[64]

---

60) J.G. March, *Exploration and Exploitation in Organizational Learning*, Organization Science, Special Issue: Organizational Learning, Papers in Honour of and by J.G. March, Volume 2, Issue 1, p. 71, 1991.
61) D. Levinthal and J.G. March, *The Myopia of Learning*, Strategic Management Journal, Volume 14, Issue S2, p. 95, 1993.
62) D. Lavie et al., *Exploration and Exploitation within and across Organizations*, Academy of Management Annals, Volume 4, Issue 1, p. 109, 2010.
63) W. Cohen and D. Levinthal, *Innovation and Learning: The Two Faces of R&D*, The Economic Journal, Volume 99, N. 397, p. 569, 1989.
64) W. Cohen and D. Levinthal, *Absorptive Capacity: A New Perspective on Learning and Innovation*, Administrative Science Quarterly, Special Issue: Technology, Organizations, and Innovation, Volume 35, N. 1, p. 128, 1990.

Absorptive capacity is related primarily to exploration activities.[65],[66] A high absorptive capacity enables firms to detect and explore for their own purpose new emerging technologies (e.g. nanotechnology) and associated market opportunities. Of course, the further exploitation of existing knowledge (along a given technological trajectory) also requires the use of external knowledge, and therefore the ability to absorb such knowledge, but to a lesser extent as in case of exploration.

In the empirical model, the absorptive capacity was measured through three variables: (i) the share of employees with tertiary-level formal education; (ii) the existence of a R&D department (binary variable yes/no); and (iii) the existence of R&D collaborations (besides the universities; binary variable yes/no).

The relationship between technological diversification on exploration and exploitation of knowledge is ambiguous. On the one hand, it is plausible that firms with a wide portfolio of technological activities are more likely to become engaged in exploration because they have lower exploration costs than highly specialized firms. Their broad knowledge spectrum could enable them to detect earlier than other firms opportunities for new knowledge. On the other hand, specialization could also decrease the marginal costs of exploring new knowledge, especially if it is within a technological paradigm or trajectory. However, it is also likely that decreasing search costs within a paradigm come along with increasing costs for knowledge exploring activities in other fields of investigation. Since specialized firms accumulate knowledge and experiences, it is also likely that they would have advantages in exploiting existing knowledge. Consequently, the effect of technological diversification on the exploration and exploitation of knowledge in collaboration with universities is not, a priori, clear. Technological diversification was measured by the number of technologies, in which a firm is active.

The ability to appropriate the revenues of innovations depends on the appropriability conditions; that is the effectiveness of protection from imitation through available formal (e.g. patents) and informal means (e.g. time lead, product complexity). When appropriability is low, investment in exploration would be discouraged to a larger extent than investment in exploitation, the outcomes (e.g. new products) of which can be more easily protected by available informal means. As Lavie *et al.* write: 'Under such conditions (of a weak appropriability regime), the value of exploration is diminished so that organizations may withhold their investment in exploration and focus on exploitation'.[67] The appropriability conditions are industry-specific. The appropriability has been approximated by the industry affiliation of a firm.

---

65) D. Lavie et al., *Exploration and Exploitation within and across Organizations*, Academy of Management Annals, Volume 4, Issue 1, p. 109, 2010.
66) U. Lichtenthaler and E. Lichtenthaler, *A Capability-based Framework for Open Innovation: Complementing Absorptive Capacity*, Journal of Management Studies, Volume 46, Issue 8, p. 1315, 2009.
67) D. Lavie et al., *Exploration and Exploitation within and across Organizations*, Academy of Management Annals, Volume 4, Issue 1, p. 109, 2010.

The relationship between competitive pressures and the incentives to exploit and/or explore knowledge in collaborations with universities is ambiguous. The Schumpeterian view would suggest that a certain level of market power is necessary for a firm to be motivated to bear the risks of exploration. But there are limits to this effect. 'In the long run, however, the use of power to impose environments is likely to result in atrophy of capabilities to respond to change'.[68] A second line of argumentation suggests that competitive pressures would enhance incentives for exploration that can drive change and reveal new sources of competitive advantage, thus improving a firm's relative position *vis-à-vis* competitors.[69],[70] Three measures of competition, namely, the intensity of price competition, the intensity of non-price competition, and the number of principal competitors in the main sales market of a firm, have been used.

Also, the relationship between firm size and exploration/exploitation of knowledge is ambiguous. As Lavie *et al.* mention: ' . . . conflicting findings exist concerning the impact of organizational size on the tendency to explore versus exploit'.[71] On the one hand, with increasing size firms tend to become less flexible and less adaptable to changes, leading to knowledge exploitation along existing trajectories, while restricting explorative search for new opportunities. On the other hand, larger organizations may have better access to internal and external resources that allow a more effective and less expensive search for new knowledge. We sympathize with the latter hypothesis, which is in accordance also with the empirical evidence for overall innovation activities. Firm size is measured by the number of employees.

Looking at firm age, it is known that younger firms are less strongly dedicated to existing technologies along known technological trajectories than older, established enterprises. Thus, young organizations are more likely to invest in exploration.[72] Older firms rely more on their existing knowledge and experience, so that they are more inclined to engage in exploitation rather than exploration.[73]

13.4.5.1 Empirical Evidence

The firm survey of 2011 also contained questions about the outcomes of the transfer activities. Consequently, information was gathered on whether the knowledge exchange with universities has brought out: (i) the initiation of a new R&D project (yes/no); (ii) the development of new products (yes/no); and (iii) the development of new processes (yes/no). The initiation of new projects would indicate an exploration of knowledge and the development of new products,

---

68) D. Levinthal and J.G. March, *The Myopia of Learning*, Strategic Management Journal, Volume 14, Issue S2, p. 95, 1993.
69) Ibidem.
70) D. Lavie *et al.*, *Exploration and Exploitation within and across Organizations*, Academy of Management Annals, Volume 4, Issue 1, p. 109, 2010.
71) Ibidem.
72) Ibidem.
73) F.T. Rothaermel and D.L. Deeds, *Exploration and exploitation alliances in biotechnology: a system of new product development*, Strategic Management Journal, Volume 25, Issue 3, p. 201, 2004.

Table 13.30 Factors determining explorative/exploitative activities.[a]

	Exploration versus Exploitation
Gross investment per employee	No effect
Absorptive capacity	
Share of employees with tertiary-level education	Positive
R&D department	Positive
R&D collaborations	Positive
Technological diversity	Positive
Technological field	
Nanotechnology	Positive
Other characteristics	
Size	No effect
Age	Positive
Foreign-owned	No effect
Competition	
Intensity of price competition	Negative
Intensity of non-price competition	No effect
Number of principal competitors in a firm's main product market	No effect

a) 'Positive' or 'negative' effects refer to the respective coefficients of the multinomial probit-estimates of the propensity to explorative and exploitative activities (reference: exploitative activities) under the condition that they are statistically significant at the 10%-test level.[74]

while new processes based on existing technology would indicate an exploitation of knowledge.

The analysis was based on econometric models that estimate the relationship between the above-mentioned possible determining factors (firm size, firm age, technological diversity, absorptive capacity, competition, etc.) and the probability to explore or exploit knowledge.

Table 13.30 presents the main 'stylized' results. A clear pattern of the differences between exploration-orientated and exploitation-orientated firms was obtained. Firms with a focus on exploration showed a significantly higher absorptive capacity of scientific knowledge. The positive effects of human capital intensity (share of employees with tertiary-level education), R&D cooperation, and the existence of an R&D department are three variables that measure absorptive capacity. These were significantly stronger for explorative than exploitative activities. This clearly confirmed the importance of specific in-house knowledge for the generation of new knowledge jointly with the universities.

No difference could be found with respect to physical capital intensity measured by gross investment per employee. This indicates that capital endowment is

---

74) S. Arvanitis and M. Woerter, *Exploration or Exploitation of Knowledge from Universities: Does It Make a Difference?*, KOF Working Papers, N. 322, ETH, 2012.

not decisive for distinguishing between exploring or exploiting knowledge in collaboration with universities.

Exploration-orientated firms did not focus on any particular type of technology (with the exception of nanotechnology) as compared with exploitation-orientated firms, but they showed a significantly higher degree of technological diversification (in terms of the number of technological fields, in which they are active) than exploitation-orientated firms.

Larger and/or older firms appeared to be more inclined to exploration than smaller and/or younger firms. Older firms seemed to invest more in exploitation activities, even if they might have strong invested interests in existing knowledge, presumably because their financial means allow them to pursue better a strategy of technological diversification than younger firms. However, the age effect is not very robust.

Competition clearly matters for exploration and exploitation activities. Price competition was more relevant for exploitation-orientated firms. This indicates that exploration is costly, and firms that are exposed to tough price competition do not dispose of sufficient financial means to fund such uncertain research projects. They have to focus more on innovation activities that create cash-flow in a short time frame. No difference could be found with respect to non-price competition and the number of competitors; proxy for market concentration.

13.4.6
Technological Proximity Between Firms and Universities and TT[75]

One of the most important impediments of collaboration between universities and enterprises is the existence of diverging interests as to the technological priorities due to differing technological profiles. As a consequence, a precondition for collaboration and knowledge transfer should be a compatibility of the technological bases of the partners.

It is widely assumed that the compatibility of the technological base of potential research partner is a necessary condition for collaboration. In fact, if we look at well-known economic concepts (e.g. absorptive capacity of a firm and technology trajectories),[76],[77] which describe the ability of a firm to perceive, process and apply external knowledge and/or to change its innovation behaviour in order to further develop their technology base and to develop and commercialize new products, it is likely that collaborations amongst actors with similar technology/knowledge bases are more likely than collaborations amongst partners with a very different knowledge background. However, if firms want to substantially

---

75) M. Woerter, *Technology Proximity between Firms and Universities and Technology Transfer*, Journal of Technology Transfer, Volume 37, Issue 6, p. 828, 2012.
76) W. Cohen and D. Levinthal, *Innovation and Learning: The Two Faces of R&D*, The Economic Journal, Volume 99, N. 397, p. 569, 1989.
77) G. Dosi, *Technological Paradigms and Technological Trajectories. A suggested interpretation of the determinants and directions of technical change*, Research Policy, Volume 11, Issue 3, p. 147, 1982.

modify or extend their knowledge base, the complementarities of knowledge bases might also provide important incentives for KTT activities. In any case, technological proximity seems to be important and it is desirable to know more about its function in the field of KTT.

13.4.6.1 Empirical Evidence

In order to investigate the relationship between technological proximity and the propensity to knowledge and technology transfer the information from three sources was put together: (i) patent data about the technological activities of firms; (ii) information about the technological activities of universities; and (iii) survey data on KTT activities between firms and universities. Based on a comprehensive mapping of technological activities of Swiss firms and the public research sector, the KTT potential between the private and public research sector in Switzerland was identified.

After mapping the technology activities of private enterprises to the technology activities of universities in Switzerland, four areas of technological proximity were detected, as follows.

High Potentials

The following technological fields were frequently found in private enterprises as well as universities:[78]

- Human necessities, that is agriculture (a01), medical or veterinary sciences or hygiene (a61).
- Performing operations/transporting, that is physical or chemical processes (b01), hand tools, workshop equipment, manipulators (b23), vehicles in general (b60).
- Chemistry, that is organic chemistry (c07), organic macromolecular compounds (c08), biochemistry, microbiology (c12).
- Physics, that is measuring (counting), testing (g01), computing, calculating, counting (g06).
- Electricity, that is basic electric elements (h01), and electric communication technique (h04).

The expectation was that in these fields, for which the technological proximity between firms and universities is high, the collaboration potential and therefore the propensity to KTT activities would be at strongest. Detailed econometric analysis showed that a high technology proximity between universities and private enterprises increases the probability of transfer activities, and makes it more likely to have more than one university link (intensity of KTT activities). This was observed particularly for the technological fields of agriculture (a01), organic macromolecular compounds (c08), optics (g02), and electric communication technique (h04) and for smaller firms (fewer than 300–500 employees).

78) The labels in brackets denote the patent classes according to the International Patent Classification.

Low Potentials
The following technological fields were not found frequently either in private enterprises or in universities:

- Human necessities, that is headwear (a42).
- Performing operations/transporting, that is generating or transmitting mechanical vibrations (b06).
- Chemistry, that is manufacturing of fertilizers (c05), explosives, matches (c06), sugar industry – polysaccharides (c13), skins, hides, pelts, leader (c14), combinatorial technology (c40).
- Textiles, paper, that is robes, cables other than electric (d07).
- Mechanical engineering, that is storing or distributing gases or liquids (f17), steam generation – physical or chemical apparatus (f22).
- Physics, that is instrument components (g12).

The econometric analysis yielded – independent of firm size – not significant or negative significant correlations between the respective technology fields and the probability to have comprehensive transfer contacts. This result is quite coherent, if we consider that both sides did not emphasize research in these fields and thus did not accumulate any considerable knowledge, which is a precondition for a mutual interest in knowledge exchange.

Not Used Potentials
The following technological fields were found frequently in universities, but not in private enterprises:

- Human necessities, that is sports, games, and amusements (a63).
- Fixed constructions, that is building – layered materials, layered products in general (e04).
- Physics, that is optics – making optical elements or apparatus (g02), controlling, regulating (g05), educating, cryptography, display, advertising, seals (g09), information storage (g11).
- Electricity, that is basic electronic circuitry (h03), electric techniques not otherwise provided for (h05).

As expected, the econometric analysis yielded for most technological fields of not used potentials significantly positive or not significant results. This finding indicated that firms did not have comprehensive research activities in these fields but tried to build in-house capabilities through transfer activities with universities. This is evidence for the explorative character of these transfer contacts.[79] For the few technological fields for which no significant or significantly negative results

---

79) J.G. March, *Exploration and Exploitation in Organizational Learning*, Organization Science, Special Issue: Organizational Learning, Papers in Honour of and by J.G. March, Volume 2, Issue 1, p. 71, 1991.

were found, the interpretation was that firms did not have the absorptive capacity to make use of public research activities or that they simply did not want, for example for security reasons, to have transfer activities in such technological fields.

This group is of particular interest for policy makers because it helps to identify technological fields in which collaboration could be possible. Hence, universities show considerable research activities but firms seem to be at the moment less inclined to invest in these fields either through in-house R&D or knowledge transfer from the universities.

Lone Stars

The following technological fields were found frequently in private enterprises but not in universities:

- Human necessities, that is furniture, domestic articles and appliances, coffee mills, spice mills, suction cleaners in general (a47).
- Performing operations/transporting, that is working of plastics (b29), conveying, packing, storing, handling thin or filamentary material (b65), hoisting, lifting, hauling (b66).
- Chemistry, that is dyes, paints, polishes, natural resins, adhesives (c09).
- Mechanical engineering, that is engineering elements/units, measures for producing and maintaining effective functioning of machines or installations, thermal insulation (f16).
- Physics, that is watch industry (g04).
- Electricity, that is generation, conversion, or distribution of electric power (h02).

According to the econometric findings, private enterprises patenting in these technology fields showed – independent of firm size – no tendency to KTT activities in general; none of these technological fields was significant. This result is quite intuitive, if it is considered that firms might have problems to find adequate science partners in these fields because universities conduct little or no research in such areas.

13.5 Conclusion[80]

To conclude, it is important to discuss some important implications of the analysis of several aspects of KTT activities for policy makers. In particular, it is

---

80) We refrain here from a discussion on the concepts of technology policy in general. An important part of it is also a policy with respect to technology transfer between universities and the business sector. For a discussion on the Swiss technology policy, see: B. Bozeman, *Technology Transfer and Public Policy: A Review of Research and Theory*, Research Policy, Volume 29, Issue 4–5, p. 627, 2000; C. Chaminade and C. Edquist, *Rationales for Public Policy Intervention in the Innovation Process: Systems of Innovation Approach*, in R.E. Smits *et al.*, The Theory and Practice of Innovation Policy: An International Research Handbook, Cheltenham (Edward Elgar), p. 95, 2010; S. Arvanitis and H. Hollenstein, *Kompaktwissen: Innovationsaktivitäten der Schweizer Wirtschaft: Determinanten, Auswirkungen, Förderpolitik*, Rüegger Verlag, 2012.

important to compare the impediments reported by enterprises with those reported by scientific institutions; knowing the impediments of collaboration from both sides helps to improve transfer across the interface between the two sides (firms, universities).

Since it is known that transfer activities support the innovativeness and labour productivity of firms, it is useful to develop policy measures to ease transfer (especially for smaller firms) by taking into account the different functions of private enterprises and universities in society. From a policy point of view, all four categories of technological fields – high potentials, low potentials, not used potentials and lone stars – may be of interest.

Thus, it seems obvious that a lack of transfer activities in some fields of high potential poses a communication and information challenge to transfer policy makers. Firms may not be well informed about research activities in related fields at universities or their research goals, the time schedules and research questions may be too different, and thus firms will refrain from transfer activities. Secrecy may be a further problem, especially in market-related research.

A lack of transfer activities in fields of low potential is quite understandable, since there seems to be neither an academic research interest nor any commercial interest for such technological fields. As a consequence, such fields are of little or no policy interest, at least in the short term. However, fields of low potential can pose long-term strategic challenges where the government aims at strengthening the capabilities in such technology fields, for instance, due to a perceived long-term benefit for society.

Lone stars may have problems in finding adequate national academic partners for their research activities, which would pose an information challenge to policy makers or a research strategy challenge to universities, if such technology fields are also of academic interest. Of course, universities do not have to take over research goals from the private sector that do not appear interesting from an academic point of view.

Unused potential indicates a lack of absorptive capacity or a lack of commercial potential. This may indicate that research in those fields is simply of less commercial value, at least in the short term; however, the transfer potential of technology fields should be investigated in more detailed empirical analyses. Furthermore, this study suggests that the monitoring of the technological orientation of firms and universities, and a regular comparison of their profiles could be shown as a valuable policy instrument as it would provide policy makers with a sound basis for technology-orientated policy initiatives.

The analysis showed that enterprises and universities do not see their relationship in the same way. There are discernible differences that can be traced back to different goals and different ways of doing things.

A remarkable finding is the quite similar assessment of firm and university deficiencies with respect to KTT activities, as evaluated by the firms both in 2005 and 2011. In these cases, 49% (in 2005) and 53% (in 2011) of KTT-active firms asserted their deficiencies as a serious impediment, while 42% (in 2005) and 41% (in 2011) of firms found university deficiencies a severe obstacle for

KTT activities. This assessment can be interpreted as a hint that there are in fact some problems at the interface between university and industry, given that they are anticipated by firms to the same extent independent of the source of deficiencies (firm or university). Directly policy-relevant are results referring to the relative importance and effective influence of science institutions deficiencies, as seen from the firms. In particular, the two most relevant single obstacles of this category – that R&D orientation of science institutions is uninteresting for firms and that possible R&D results cannot be commercialized – indicate possible weaknesses from the part of the science institutions on the interface between science and business. These findings can serve as a starting point for the improvement of the effectiveness of policy measures aimed at the enhancement of KTT between universities and the business sector. However, these findings should not be interpreted in a way that science institutions have to adapt their research goals in order to be more interesting for private firms; rather, it is more about finding a way to better use the potential of transfer activities.

However, while firm and university deficiencies are the major problematic domains for firms, institutes identify even more serious problems in both periods in the domain of costs/risks uncertainty (primarily, the R&D budgets of potential partners are too low and industry has different ideas on costs and productivity), as well as with respect to (possible) problems in teaching and research (primarily teaching requires too much time and hindrance to academic publication activities) as a consequence of the involvement in KTT activities. Rather remarkable is the fact that relatively few institutes (about 10% in both periods) found administrative and approval procedures to be a problem. Finally, there were significantly fewer institutes in 2011 as compared to 2005 (18% versus 28%, respectively) that found a lack of information to be a problem. The discernibly higher firms' awareness of the mediating services of institutions such as the CTI and the TOs during the period 2008–2010 is presumably the result of additional efforts of these institutions to reach firms and inform them about their services, and the explanation for institutes being confronted with fewer information problems regarding possible KTT activities. On the whole, however, there is still sufficient scope for technology and/or university policy to offer remedies for the types of problems mentioned above.

Indirectly policy-relevant are the results with respect to exploration/exploitation. The clear pattern of the differences between exploration- and exploitation-orientated firms obtained in the econometric estimations shows that firms with a focus on exploration have a significantly higher knowledge absorptive capacity. This means that exploration-orientated firms would be able to utilize university knowledge more effectively in the long-run than firms that focus on (the rather short-term) benefits from the completion of current R&D projects. For science institutions, this means that two groups of KTT-partners can be distinguished with different knowledge requirements, and thus different consequences with respect to the utilization of acquired knowledge.

The presented analysis applies primarily to Switzerland, but a better exploitation of innovative knowledge by firms and an improved transfer of R&D

information from scientific institutions to companies is an issue for most other industrialized countries. The removal of obstacles of any type is necessary in a world dominated by rapid international exchanges and great technological challenges. Each country needs to carry out regular monitoring of respective structures that might have to be adapted in order to achieve country-specific goals of higher innovation performance.

14
Conclusion

Marilena Streit-Bianchi

Since the dawn of humanity the wish to understand *why* and *how* has been the driving force of the human quest. The scientific community of the nineteenth century was based on rationalism, individualism and empiricism, and from this science has developed into a collaborative and multidisciplinary enterprise in order to deal with the cost and challenges of most research explorations (matter, antimatter, cosmology, genome, etc.). Today, contemporary science and discoveries have transitioned from specialist papers to the front pages of the newspapers. Moreover, the economic sphere has begun to play an important role, and fundamental science has become increasingly subject to economic pressures, as have the types of research carried out. The amount of money invested is continuously questioned, and the public demands to be informed on planned developments and their consequences. As a result, the communication of science has become much-valued.

Big Data is no longer limited to High-Energy Physics (HEP). Much has been achieved by dismantling preconceived barriers and working at the limit of what is achievable in physics and in the natural sciences. This has been made possible during the past 50 years thanks to the combination of new technologies which have opened the doors to new areas of human understanding. Big Data is synonymous with big challenges; a big adventure to achieve the exploration of the unknown through Big Data analysis. Consequently, what once was thought unreachable is today within our grasp.

Education is an undisputed outcome of research institutions and universities. Fundamental research is not a luxury, as has been clearly demonstrated in the chapters of this book. It is – in an economically driven world – a necessity from which new and unexpected developments and discoveries may also originate in the future. However, what remains unknown is the *when* and *how* this acquired knowledge may be applied. These are factors which depend not only solely on the cost of the necessary applied research and development, but also on the integration of new technologies into devices and on inter-disciplinary fertilization.

This book has demonstrated how much the availability of cheaper methods and more manageable instrumentation (e.g. miniaturization and integration in electronics, software and hardware developments in informatics, protein sequencing-related analytical methods, and imaging instrumentation) have paved the road for new directions of research, and had a major impact on society at large.

For instance, the World Wide Web (WWW), the touch screen and Positron Emission Tomography (PET) all originated from CERN, and each has greatly modified human behaviour. It has taken more than 40 years for some technologies to become pervasive in everyday life. Today, businesses invest in knowledge management in order to enhance the motivation of their employees. A strong commitment to the organization, a culture of belonging and a unity of purpose are important driving forces, not only for enterprise but also for scientific institutions and large science projects. Thus, we can also apply knowledge management theories to scientific institutions such as CERN. Studies carried out in this field, as reported in Parts I and III of this book, provide new insights into the epistemology of scientific knowledge. Empirical evidence is given that a better interaction between scientific institutions and industrial enterprises improves knowledge and technology transfer. Exploration-orientated and exploitation-orientated firms have different capacities to absorb knowledge and also different focus requirements (long-term versus short-term benefits). This must be taken into account by science institutions when embarking on collaborative activities of transfer and joint R&D with industry. Ontologies used in scientific models enable knowledge sharing and use by a variety of actors, and in diverse contexts, which together can bring innovation.

The HEP Large Hadron Collider (LHC) project has enforced some visible structural and functional changes in the physics community, and its management. This has been due to the size of the project, the time scale (the conception, realization and final upgrading took over 30 years), the disparate location of the production, testing and assembly of the various machine elements, and the rapid technological turnover. Managerial skill has become an important asset, while knowledge transmission and communication tools have played key roles, and technical documentation and safety procedures have had to be more structured and digitized. In addition, the interaction between scientists, engineers and industrialists has become increasingly more close.

One could speculate that in order to facilitate Knowledge and Technology Transfer (KTT) it may be necessary to create specific independent development projects for cutting-edge R&D technologies. In such a framework, scientists would be allowed to use part of their time (e.g. 15%) to collaborate with industry for the conception, realization and testing of new devices. Funding should be found outside the next main HEP projects. As a result, additional KTT would speed up the implementation of new technologies for the benefit of society in general.

As illustrated in this book, the monitoring of the Earth and its climate, the development of accelerators and detectors for medical applications, and of networking databases for diagnostics and the life sciences, will impact on energy production, the environment, and healthcare. Major steps forward in development have been always accompanied by paradigm shifts rather than by the simple evolution of techniques. Yet, major advances are more than just novelty; rather, they rely on the application of new theories and new knowledge. The internationalization of science also means peace, and this aspect has been taken into account at CERN since its early times. Heterogeneous social environments are needed to sustain collective endeavours in research so that we may achieve a better and sustainable world.

It is clear that we owe much that we take for granted in our daily lives to the work carried out in fundamental research, and many of the examples presented in this book are sure to surprise the reader. We often hear the questions 'What is the value of research?', and 'Why should we invest so much in projects that we cannot be sure will yield results?' History, and the examples presented here, provide a part of the answers to these questions, as inevitably we always find uses for new ideas, new thinking and new tools, where research is the driving force. Who knows what life-changing technologies will appear tomorrow based on the scientific research being conducted right now at institutions all over the world?

Author Index

a

Acemoglu, D. 238
Aghion, P. 238
Albert, M. 234
Amaldi, U. 168
Anderson, C. 243, 245, 252
Anderson, C. D. 163
Angell, M. 246
Argyris, C. 24
Argyris, J. H. 146
Arrow 6
Arvanitis, S. 255–306
Aspect, Alain 218
Atkinson, R. D. 240
Attié, D. 202
Autio, E. 44, 46, 49

b

Barak, F. 256
Barney, J. 292
Bartlett, N. 163
Batchelor, G. K. 140
Bates, D. 66
Baur, G. 165
Beath 285
Beck, F. 104
Beck, Verlag C.H. 7
Bell, John 218
Benkler, Y. 246–248
Bennet, A. 6, 8, 9, 25
Bennet, D. 8, 9, 25
Berger, S. 234
Berners-Lee, T. 61, 62, 67
Bertolucci, Sergio 1–2
Berwert, A. 256
Bézier, A. 149
Bézier, P. 149
Bianchi-Streit, M. 42, 307–309
Bird, I. 84

Bishop, Robert 173–194
Blaug, M. 233
Block, F. L. 245
Bohr, Niels 162, 216
Boisot, M. 25, 43
Boltzmann, Ludwig 217
Bomsel, O. 243
Boom, Daan 5–60
Boutang, Y. Moulier 241, 242
Bozeman, B. 257, 258, 260, 279, 302
Brand, Stewart 190
Brayton, George 120
Bressan, B. 5–56, 92, 161, 163, 168
Broût, Robert 160
Bruneel, J. 275
Bruneo, D. 96
Bunge, M. 37
Burgaleta, J. I. 133
Burlamaqui, L. 246

c

Cailliau, R. 61–80
Calloo, J. 168
Campbell, Michael 195–214
Camporesi, T. 45, 46, 52
Carlisle, R. B. 231
Cassiman, B. 280
Chamberlain, O. 163
Chaminade, C. 302
Christensen, C. M. 234
Clausius, Rudolf 217
Coase, Ronald 239, 247
Cohen, W. 280, 292, 295, 299
Coilly, N. 231
Collinson, R. P. G. 153
Collison, C. 19
Crick, Francis Harry Compton 219
Crowley-Milling, M. C. 107, 109

From Physics to Daily Life: Applications in Informatics, Energy, and Environment, First Edition.
Edited by Beatrice Bressan.
© 2014 Wiley-VCH Verlag GmbH & Co. KGaA. Published 2014 by Wiley-VCH Verlag GmbH & Co. KGaA.

Cruz, E. 46, 52
Cukier, K. 252

d
D'Aspremont, C. 279
da Vinci, L. 141
Davenport, T. 252
Davenport, T. H. 19, 22, 32, 33
Davies, H. 92
De Bondt, R. 279
de Casteljau, P. 149
de Geus, A. 5, 24
De Giovanni, D. 234
de Montalk, J. P. Potocki 154
de Saint-Simon, C. H. 231
Decter, M. 260, 270, 274
Deeds, D. L. 297
Demsetz, H. 248
Deutsch, David 218
DeWitt, D. P. 128, 135, 139
Díaz, J. 207
Dirac, Paul 162, 217
Dore, R. 234
Dosi, G. 246, 256, 299
Drucker, P. F. 32, 44

e
Edquist, C. 302
Edvinssons 23
Einstein, Albert 215–219
Engels, F. 160, 222, 231
Ernst 20
Everett, Hugh 218
Ezell, S. J. 240

f
Faraday, Michael 216
Ferris, T. 216
Feynman, Richard 215, 217, 218
Foray, D. 242
Freedman, R. A. 119
Fumagalli, A. 241

g
Galilei, Galileo 216
Ghirardi, Giancarlo 218
Giavazzi, F. 234
Gibbs, Josiah Willard 217
Gillam, M. 251
Gillies, J. 68
Gilsing, V. 278
Ginsparg, Paul 223
Gladwell, M. 25

Glaser, Donald A. 112
Glashow, Sheldon Lee 220
Gleick, J. 14, 217
Gore, Albert Arnold (Al), Jr. 223
Goswami, D. Y. 123, 129
Greenlee, P. 280
Grimaldi, M. 28

h
Hämäläinen, A. 35
Hamel, G. 6
Hamm, S. 251
Hampton-Turner, C. 234
Hardin, G. 242
Haroche, Serge 192
Hawking, Stephen 215
Hayek, F. A. 237
Heijne, Erik 196, 199
Heisenberg, Werner 216
Hellman, T. 285
Helpman, E. 238, 245
Hendee, W. R. 168
Hertz, Heinrich 216
Hess, J. L. 146
Heuer, Rolf-Dieter xxii
Hey, T. 251
Higgs, Peter W. 160, 222
Hinkley, N. 165
Hollenstein, H. 289, 302
Holsapple 6
Holzapfel, R. 74
Horvath-Hrsg, F. 256
Howitt, P. 238
Huber, H. 256
Huuse, H. 46

i
Ichbiah, Jean 154
Incropera, F. P. 128, 135, 139
Irvine, M. 169

j
Jacquemin, A. 279
Jakubek, J. 204
Jameson, Antony 141–158
Jantsch, E. 14
Janz, B. D. 33
Jefferson, Thomas 244
Jensen, R. 285
Johnson, E. A. 106
Johnson, S. 249, 250

Jones, B. 81–102
Jovanovic, B. 245

k
Kautsky, K. 232
Kayyali, B. 252
Keller, M. R. 245
Kelly, J. E., III 251
Kelly, K. 250
Kelsey, S. 146
Kelvin, Lord 159
Keynes, J. M. 235, 236
Kim, J. 6, 252
Klapisch, R. 164
Klevorick, A. K. 289
Knight, F. H. 235
Knorr Cetina, K. 43
Kochenderfer, M. J. 154
Kornai, J. 237
Kowalewsky, M. 231
Kragh, H. S. 162
Krugman, P. 238
Kurki-Suonio, K. 33, 36, 39–41
Kurki-Suonio, R. 33, 34
Kurzweil, R. 250, 251
Kuznets 6

l
Laure, E. 84
Lavie, D. 295–297
Lazaridis, M. 221
Lengert, M. 91
Lenin, V. I. 232
Levinthal, D. 280, 292, 295, 297, 299
Lichtenthaler, E. 296
Lichtenthaler, U. 296
Liebowitz, S. J. 242
Likert, R. 262–264, 268–270, 272, 274, 275, 277
Lipsey, R. G. 245
Lopatka, J. E. 243
Lopez, Cayetano 119–140
Lovegrove, K. 123
Lovelace, Ada 154
Lowe, J. F. 104, 111
Lübke, J. 206
Lucarelli, S. 241
Lucas, R. 238
Lucas, R. E. 238

m
Machlup 6
Madsen, Niels 159–172
Majchrzak, A. 30
Mansfield, E. 260
March, J. G. 295, 297, 301
Marconi, Guglielmo 216
Margolis, S. E. 242
Markram, H. 251
Martin, B. R. 169
Marx, K. 6, 231, 236, 241
Maury, S. 164
Maxwell, James Clerk 216, 221
Mayer-Schönberger, V. 252
Mazzucato, M. 246
McArthur, J. W. 236
McDermott 21
McGinn, C. 7
McKinsey 240
Méndez-Vilas, A. 207
Michelson, A. A. 159, 169
Mitchell, M. 14
Mokyr, J. 6, 7
Moore, Gordon, E. 178, 200
Morishima, M. 234
Morley-Fletcher, E. 229–254
Mowery, D. C. 260

n
Negroponte, N. 241
Newton, I. 159, 215
Ng, H. H. 111
Nonaka, I. 6, 9, 12, 14, 18, 22, 31
Nordahl, O. P. 46
Nordberg, M. 43, 46

o
O'Dell 21
Oelert, W. 166
Osuna, R. 133
Otlet, P. 69

p
Page, W. H. 243
Palmer, L. 35
Parcell, G. 19
Parker, D. J. 168
Pehrson, R. J. 154
Pellow, N. 71
Penrose, E. 292
Petsko, A. G. 183
Phelps, E. 233, 234, 240
Polanyi, M. 6, 26
Pordes, R. 84
Porter, M. E. 6
Prahalad, C. K. 6

Prescott, E. C. 238
Prigogine, I. 14
Probst, G. 19
Procz, S. 211
Prodi, R. 234
Prusak, L. 19, 22
Puchkoff, S. J. 111

q
Quesnay, F. 230
Quevedo, Fernando 215–228

r
Régnier, Ph. 231
Ricardo, David 230, 243
Riege, A. 21
Rifkin, J. 243, 244
Risenfeld, R. 150
Robin, S. 257
Rodriguez-Vidal, J. 189
Romer, Paul 238
Romer, P. M. 6, 238, 239
Romero-Alvarez, M. 123
Rosenberg, N. 234
Rothaermel, F. T. 297
Rousseau, P. L. 245
Rubbia, Carlo 66, 119, 164
Rüegger, Verlag 256
Ruttan, V. W. 245

s
Sachs, J. D. 236
Sadava, D. 193
Sainsbury, David 234
Sakharov, Andrei 224
Salam, A. 220
Schein, E. H. 21
Schmied, H. 42, 46
Schön, D. A. 24
Schrödinger, E. 216, 219
Schubert, T. 257
Schumpeter, W. J. A. 233
Seitz, K. 234
Sendall, Mike 67
Senge, P. M. 6, 24
Shannon, Claude 217
Shapiro, C. 280
Shiga, D. 168
Simons, James 224
Shumpeter 6
Shy, O. 243
Sieloff, C. G. 7
Sikharulidze, I. 211

Simon, H. A. 233
Skidelsky, R. 235, 237
Sliver, N. 252
Smith, A. 230
Smith, A. M. O. 146
Smits, R. E. 302
Solow, R. M. 6, 238
Spence, M. 279
Srinivasan, H. 112
Stengers, I. 14
Stewart, T. A. 6, 23
Stirling, Robert 120
Streit-Bianchi, M. 42
Stumpe, B. 103–118
Sullivan 23
Sutton, C. 104
Sveiby 23

t
Takeuchi, I. H. 9, 12, 18
Thiele, C. 103
Thierstein, A. 256
Thurow, L. 244
Thursby, M. 285
Thursby, M. C. 284
Torp, C. P. 108
Townsend, D. W. 167, 168
Trompenaars, F. 234
Turck, J. A. V. 143
Turner, M. J. 145

v
van der Graaf, H. 208
van der Meer, Simon 119, 164
Vercellone, C. 241
Verne, J. 115
Verspille, K. J. 150
Veugelers, R. 280
Vock, P. 256
von Hayek, F. 237
von Neumann, John 217
von Westfalen, Baron Ludwig 231
Vykydal, Z. 204

w
Warsh, D. 6, 238
Warwick, G. 154
Watkins, P. 164
Watson, James Dewey 219
Weber, Max 233
Weinberg, Steven 220
Weitzman, M. 238
Wernerfelt, B. 292

Wheeler, John Archibald 218
Wilhelm, B. 256
Wilkins, Maurice Hugh
 Frederick 219
Willig, R. 280
Wineland, David J. 192
Woerter, M. 255–306
Wolfram, Stephen 224
Wooters, William 218
Wright, Wilbur 142

y
Young, H. D. 20, 119
Young, Malcolm 94

z
Zarza, E. 123
Zimmerman, B. J. 14
Zinkl, W. 256
Zurek, Wojciech 218

Index

a

absorptive capacity 280
acquired knowledge 49
acquisition, of skills 55
ADA language 154
adaptability 5, 6
Adobe 79
advanced resource connector (ARC) 93
aeroplane design 141
aerosol chemistry 184
African university of science and technology (AUST) 220
airborne collision avoidance systems 154
Airborne Collision Avoidance System X (ACAS X) 154
airborne software 154
Airbus 380 143
– CFD simulation 147
aircraft movement at peak traffic 155
Air Data Monitor (ADM) 154
airguns 98
air traffic control chart 155
Air Traffic Control (ATC) system 155
A large ion collider experiment (ALICE) 199
American Physical Society (APS) 217, 218
analytical skill 8
antimatter 159
– and CERN 162–164
– energy 167
– research 169
Antiproton Cell Experiment (ACE) 169
Antiproton Decelerator (AD) 164
– at CERN 165
Apple 97
Apple's version of operating system ('Mavericks') 79
astronomy, grid 86

Atomic Spectroscopy And Collisions Using Slow Antiprotons (ASACUSA) 165
a toroidal LHC apparatus (ATLAS) 82, 199

b

backup 79
battle-field awareness systems 154
Bayh–Dole Act 260
Beath's analysis 285
Big Data
– analysis 307
– analytics 180, 251–253
– applications in medicine 252
– computational modelling in medicine 252
– healthcare singularity 251
– 'long-tail medicine' drugs 253
biomimicry 190
Boeing 747 143
Boeing Company 146, 147
bounded rationality concept 233
brain drain 1
B-Spline (Basis Spline) 150
bump bonds 200
– interconnect technology 213
bureaucratic regulatory environments 240

c

CAD applications in aircraft design, and manufacturing 152
CAD software 149, 150
cancer therapy 'toolbox' 168
capacitor 104
carbon cycles 184
career development 52
Cargo Smoke Detector System (CSDS), 154
Ceefax 70
central processing unit (CPU) 108
central receiver systems, STE technologies 121
– reflecting elements, helisotats reflect 121

From Physics to Daily Life: Applications in Informatics, Energy, and Environment, First Edition.
Edited by Beatrice Bressan.
© 2014 Wiley-VCH Verlag GmbH & Co. KGaA. Published 2014 by Wiley-VCH Verlag GmbH & Co. KGaA.

– solar flux incident 122
CERN 1
– antimatter facility 164–166
– centre of knowledge production/
  dissemination 1
CERN SPS Accelerator, novelty for control
  room
– central processing unit (CPU) 108
– cheaper touch detector circuits 108
– computer-based message transmission
  systems 106
– computer-controlled brake 107
– computer-controlled knob 107
– consoles for super proton synchrotron
  (SPS) 106
– fixed buttons 106
– general-purpose consoles 106
– general view, CERN accelerator control
  console 110
– intelligent touch terminal 108
– intelligent touch terminal for CERN
  antiproton accumulator (AA) 109
– joint european torus (JET), 109
– microchips, development 107
– microcomputer 108
– minicomputers 106
– mutual capacitance multitouch screen 107
– network-orientated document abstraction
  language (NODAL) 107
– novelty for the control room 106–109
– personal computer (PC) 108
– phase-locked loop circuits 108
– self-capacitance transparent touch
  screen 107
– small table-top cabinet 108
– touch terminal from NESELCO. 109
– TV raster scan displays 107
Citizen Science networks 173
Cloud 78
– computing 100, 182, 183
– formation 184
– services 92
CMS (compact muon solenoid) 82
cognitive capitalism 241
collaborations
– pooling of resources and exploitation of latest
  CMOS processes 214
– and social media platforms 29
– university collaborations, France firms
  with 257
– university collaborations, Switzerland firms
  with 257
combination 10

commercial CFD software 147
communications 55
– modern tools of 55
– scattered software segments 182
communities of innovation (CoI) 28
communities of practice (CoP) 28
community planners 173
compact disc read-only memory (CD-
  ROM) 61
compact linear fresnel concentrators, STE
  technologies 122
– hybrid design 122
– link between parabolic-troughs and central
  receiver systems 122
compact muon solenoid (CMS) 199
competencies, for labour market 55
complementary metal oxide semiconductor
  (CMOS) 201
comprehension skill 8
CompuServe 70
computed tomography (CT) 205
Computer-Aided Design (CAD) system
  149–151
Computer-Aided Manufacturing (CAM)
  system 149–151
computer application 70
Computer-Augmented Design And
  Manufacturing (CADAM) system 150
computer-based message transmission
  systems 106
computer-based simulation 177
computer-controlled valve 112
computers
– automatic check-in 141
– based simulation 177
– computing in structural and aerodynamic
  analysis 145–148
– electronic reservation 141
– microprocessor timeline 145
– multi petaflops performance 178
– supercomputers' timeline 145
– transform aspect
– – aviation and aerospace 141
– visual presentation of early history 143, 144
Computer Science 80
computing and networking (CN) 67
computing hierarchy 182
concentrated solar radiation 134
conceptualization
– direction 38
– hierarchical levels 38
– – spiral of 39
– leads to terminology or language 34

– perception process 38
conducted expeditions 231
control sequence applications
– software developer 110
convection 184
cosmic radiation 196
cost-effective, peer production 229
creative destruction 233, 234
culture
– of managing knowledge 22
– for sharing and collaboration 21

d
daemons 77
data handling division 66
data-intensive healthcare 251–253
data-rich environment 180
deforestation 174
demise of socialism 237
democratizing scientific research 223
detector
– gas detector readout 208
– hybrid pixel (*See* hybrid pixel detector)
– LEEM 211
– LHCb vertex detector 213
– Medipix3 212, 213
– PIXcel 208
– sensitive 206
– type of 195
digital fly-by-wire (FBW) system 153
disasters 174, 189
– Big Data analytics 180
– challenges associated with management 174
– cost of hurricane evacuation 180
– forecasting tools 174
– knowledge of planetary change 174
– natural catastrophes 175
– prediction and uncertainty of extreme events 186–189
– – challenges 186
– – current global models 187
– – Multiple Synchronous Collapse 187, 188
– storm track forecasting, improvements in 180
discovery procedure 237
DNA molecule 219
Doppler radar 191
Douglas Commercial-3 (DC-3) 141, 142

e
Earth 183
– carbon and nitrogen cycles 184
– elementary layers 183, 184
– hydrological cycle 184
– intrinsic complexity 184
– Milankovitch cycles 184
– oceans 184
– special phenomena 184
Earth Science applications, grid 86
Earth Simulation Centre 173
Earth–Sun System 184
Earth System 176, 177, 179, 184
– computational modelling 185
– Coupled Model Inter-comparison Project 5 (CMIP5) 192
– extreme events, prediction and uncertainty of 186–189
– impact on cities and bioregions 189, 190
– Integrated Assessment Models (IAMs) 192
– modelling the whole-earth system 191–193
– multiple nonlinear feedback loops 187
– simulation 182
– towards urban resilience 190, 191
Earth System Grid Federation (ESGF) 99
Earth System Models 187, 189, 192
– features sequentially integrated into 186
economic death 240
economic discourse 232
economic growth 229, 239
– real engine of 237
economic planning 236
economic reasoning 247
economic slowdown 237
economies of scale 232
ecosystem of technologies 250–253
ECP. *See* electronics and computing for physics (ECP)
edge effects 177
EDS. *See* electronic data systems (EDS)
EEC. *See* Electronic Engine Control (EEC)
efficiency 5
EFPs. *See* European Framework Programmes (EFPs)
EGI. *See* European Grid Infrastructure (EGI)
EGI-DS. *See* European Grid Initiative Design Study (EGI-DS)
electromagnetic force 160
electromagnetic waves 216
electronic data systems (EDS) 88
Electronic Engine Control (EEC) 154
electronics and computing for physics (ECP) 67
e-mail 8
emergency responders 173
energy budget, of the Universe 161

energy, STE plants
- thermal storage 137, 138
- - extension of electricity delivery period 139
- - mediums and their characteristics 138
- - photovoltaic plants 137
- - renewable energy systems 137
- - salt mixture of potassium nitrates 137
- - salt mixture of sodium 137
- - Spanish Gemasolar plant 137
- - working fluid 137
enterprise content management (ECM) 28, 29
enterprise portal 29
environment
- for collaborative knowledge sharing/ICTs 28, 29
equity 240
ESA. See European Space Agency (ESA)
ESRF. See European synchrotron radiation facility (ESRF)
ESTELA. See European solar thermal electricity association; European Solar Thermal Electricity Association (ESTELA)
e-Therapeutics 94
EUDET Collaboration 208
European Framework Programmes (EFPs) 268
European Grid Infrastructure (EGI) 91
European Grid Initiative Design Study (EGI-DS) 91
European Solar Thermal Electricity Association (ESTELA) 122
European Space Agency (ESA) 101
European synchrotron radiation facility (ESRF) 205
exosphere 184
exploitation 231
externalization 10

f

financial upheavals 6
fixed-broadband subscriptions 113
flexibility 6
Fluid Mechanics, STE plant
- conduction losses in cylindrical tube 139
- fluid velocity 139
- heat convection coefficient 140
- heat transfer coefficient 139
- plant efficiency 138
- Reynolds number (Re) 140
- thermal conductivity 139
- turbulent flow 139
fluid velocity 140
Fly-By-Wire (FBW) systems 151
- digital 153
- and on-board systems 151–153
Fourier's law 135
four mode of knowledge conversion 9, 12, 13, 40
Future and Emerging Technologies (FET) 251

g

gallium arsenide (GaAs) 200
gas electron multiplier (GEM) foils 202
GBAR. See Gravitational Behaviour of Antihydrogen at Rest (GBAR)
general theory of relativity 217
geocluster 98, 99
geoengineering 178
Geographic Information Systems (GIS) 191
geoscientists, petroleum reservoirs knowledge 97
Germany
- firms with university collaborations 257
GIS. See Geographic Information Systems (GIS)
gLite 93
Global Distribution Systems (GDSs) 156
global economy 5
Global Positioning System (GPS) 103
- system 162, 165
Global Positioning System Sensor Unit (GPSSU) 154
global system for mobile (GSM) phone 70
global warming 174, 186
Google 69
Google Earth 77, 181
Google's applications 77
GPS. See Global Positioning System (GPS)
graphene flakes 210
Gravitational Behaviour of Antihydrogen at Rest (GBAR) 166, 169
Grid
- basic principles 87
- communication channels 89
- computing 182
- enabling grids for e-science (EGEE)
- - to EGI transition 90, 91
- environment 99
- fusion 86
- GridVideo 96
- hurdles 94
- infrastructure 91
- - development 86
- from laboratory to market 97
- lessons learned/anticipated evolution 91, 92
- life sciences research 86
- needs 82–85
- production infrastructure 85, 86
- reservoir simulation: 98, 99

– scientific simulation, modelling and data mining 95
– seismic imaging 98, 99
– sharing resources 94
– societal impact 99–101
– stock analysis application 95, 96
– three-year phases 88–90
– total group geoscience research centre (GRC) in UK 97, 98
– transferring technology 86, 92–94
– WLCG project 86, 87
GridVideo 96
ground-based computer systems 155, 156
groundwater 174

h

HBP. *See* Human Brain Project (HBP)
head-on-head proton collisions 199
healthcare systems, knowledge-based redesign 230
heat convection coefficient 140
heat transfer coefficient 139
heat transfer mechanisms 120
HEP. *See* High-Energy Physics (HEP), for detectors
Hertz's experiments 216
high-energy accelerator research organization 109
High-Energy Physics (HEP), for detectors 1, 31, 196, 213, 307
– bump bonding interconnect technology 213
– developing, chips 213
– LHCb vertex detector 213
– Medipix3 chips 213
– Medipix3 Collaboration 213
– Smallpix chip 213
– Timepix 213
high-performance computing (HPC) 222
– advancement in 178, 179
– resources 95
high-technology developments 47
HPC. *See* high-performance computing (HPC)
HP ProCurve 88
Human Brain Project (HBP) 251
human creativity, remaining scarce resource 246
hybrid pixel detector
– High-Energy Physics 197–199
– – bump bonds on Medipix2 readout chip 199
– – bunch of particles, LHC 198
– – compact muon solenoid (CMS) 199
– – electron microscopy image 198
– – head-on-head proton collisions 199
– – large hadron collider beauty (LHCb) 199
– – A large ion collider experiment (ALICE) 199
– – LHC experiments 198
– – noise-free images, radiation 198
– – protons 197
– – readout chips 198
– – signal-to-noise ratio 198
– – a toroidal LHC apparatus (ATLAS) 199
– imaging, medipix chips 199–205 (*See also* Medipix, applications)
– – ALICE pixel studies 201
– – beta-ray imaging 200
– – chip, achievement 200
– – complementary metal oxide semiconductor (CMOS) 201
– – cosmic ray 202
– – electric field 202
– – EUDET collaboration 202
– – European synchrotron radiation facility (ESRF) 205
– – full-scale prototype 203
– – gallium arsenide (GaAs) 200
– – gas electron multiplier (GEM) foil, image 202
– – mammography 200
– – medipix1 chip 204
– – microelectronics technology 200
– – Moore's law 200
– – Pixelman 205
– – prototypes 201
– – silicon wafers 200
– – smartphones, price 200
– – three-dimensional (3D) image 203
– – time over threshold (ToT) 203
– – Timepix chip 203
– – USB-based readout interfaces 204
– – USB readout system 204
– – user-friendly readout software 205
– – x-ray images 200
– origins of 196, 197
– – cosmic ray detector 196
– – electronics chips 197
– – illustration of 197
– – radiation sensor 196
– – sensor 197
– – tiny square of silicon 196
– – two-dimensional (2D) array 197
hydrological cycle for clouds 184
hyper-competition 6
hypertext markup language (HTML) files 61
hypertext pre-processor (PHP) 73

i

ICT support systems 28, 29
IMAX theatre 181
industrialization 174, 231
Inertial Navigation System 154
Information and Communications Technology (ICT) 245
– environment 8
information gains 248
information robots applications 29
information technology 55
information theory 217
initial public offerings (IPOs) 240
innovative developments 53
Institute of Electrical and Electronics Engineers (IEEE) 90
Intellectual Property 214
– rights 244
interactive and immersive 4D visualizations 180–182
Inter-Governmental Panel on Climate Change (IPCC) 99, 223
internalization 10, 11
international space station (ISS) 210
Internet 179, 182, 183
– search 78
intuitive knowledge 7
invention 36
ionosphere 184
ion sputtering technologies 104
IPCC. *See* Inter-Governmental Panel on Climate Change (IPCC)
iron law of wages 230
irradiation of tumours 168

j

JavaScript 72
joint European torus (JET) 109

k

knowledge 7
– acquisition 47
– based economy 5
– explicit 8, 9
– generating activities 245
– implicit 8
– management 8, 41
– – case study of CERN 41–46
– – core processes 18, 19
– – ICT support systems for 29
– – LHC case study survey 47–56
– – ultimate aim of value creation 31
– matrix 20
– outputs/outcomes 19–31
– – creating a learning organization 24, 25
– – creating an environment for collaborative knowledge sharing/ICTs 28–30
– – creating culture for knowledge sharing 20–22
– – creating knowledge 25–27
– – institutionalizing knowledge management 22, 23
– – maximizing intellectual capital 23, 24
– – networking 27, 28
– – transforming knowledge into organizational wealth 30, 31
– scientific
– – protocols 7
– sharing barriers 21
– specialization (*See* knowledge specialization)
– tacit 8, 9
– theory of 7, 8
– transfer in social process 51
– worker 31–33
– – concept formation 37–39
– – elements categorize 32
– – hierarchical levels of conceptualization 37–39
– – individual learning process 33–35
– – scientific, technological and social processes 36, 37
knowledge and technology transfer (KTT) 214, 255, 308
– absorptive capacity of firm 295
– categories of single forms 264
– degree of specialization 265
– determinants of 293
– econometric analysis 300, 301, 302
– econometric analysis of propensity 292
– empirical model 296
– energy technologies 284
– exploitation-orientated firms 298, 299
– exploration/exploitation 294–299
– – factors determining 298
– – inherent tension 295
– firms' percentage 271
– forms and partners 262–267
– hypothesis of 289
– impact of 270–272
– – economic 259
– – innovation and labour productivity 288–291
– – innovation performance 291–294
– lack of financial resources 274
– mediating institutions and motives 268–270
– obstacles to 272–276

– – categories of 273
– – ranking of importance 278
– R&D-active firms 267
– science institution 264, 265
– share of firms 261, 262
– size-dependency 262
– in Swiss economy 261–276
– – determinants of 280, 281
– – determinants of propensity 286, 287
– – drivers of 279, 280
– – empirical evidence 281–284, 287, 288
– – exploration of 278, 279
– – incidence and forms 276
– – institutions and obstacles 276–278
– – university point of view 284–286
– technological fields of 267
– technological proximity 299–302
– – of firms 259
knowledge creation 11
– acquisition, and transfer model 39–41
– – Kurki-Suonio's approach 39, 40
– – Nonaka's approach 39, 40
– within learning process, basic processes 36
– at ontological dimension 13
– two dimensions 11
– – dimension/phases 13–18
– two spirals 12
– – epistemological dimension 12
– – ontological dimension 13
'knowledge libraries' 2
– key elements 2
– rules of library 2
knowledge management, and science for peace 219–224
– applied mathematics 222
– applied science 225
– applied subjects 219
– basic science discoveries 222
– basic science educational programmes for younger generations 225
– challenges for humanity 220
– culture of science 220
– democratizing scientific research 223
– enthusiastic research workers 224
– essence of scientific and technical education 224
– establishing, scientific culture 221
– high-performance computing (HPC) 222
– high-quality personalities 224
– humanistic culture 224
– human quality

– – curiosity 219
– – sense of wonder 219
– inter- governmental panel on climate change (IPCC) 223
– invention, World Wide Web (WWW) 222
– large hadron collider (LHC) experiment 222
– no-nonsense quantitative approaches 224
– promotes, scientific studies 220
– quantitative biology 222
– scientific career 219
– scientific investigations 223
– socio-political events 220
– study of theoretical physics 220
– support of fundamental science 221
– support science in developing countries 221
– theoretical physicists 221, 224
– the world academy of sciences (TWAS) 220
knowledge specialization
– edge effects 177
– institutions focusing on 177
– numerical models 177
– and splintering 176
– Top500 Supercomputing Sites 177
– Yokohama Earth Simulator 177
KTT. *See* knowledge and technology transfer (KTT)

l

labour force 231
laptops 179
large electron positron (LEP) collider 67
Large Hadron Collider (LHC) 1, 81, 196, 222
– ALICE 44, 199, 201
– ATLAS 25, 43, 44, 82, 199, 204
– CMS 44, 82, 199
– computing model 84
– – grid 88
– LHCb 44, 165, 199, 213
– TOTEM 44
Large Hadron Collider beauty (LHCb) 199
Large Hadron Collider (LHC) project 308
law of reflection 124, 125
learning process 37
– axiomatic-deductive approach 38
– empirical–inductive approach 38
– perceptional approach 38
lidar 191
line mode browser (LMB) 71
Litton LTN-51 154
Lockheed SR-71 142
low-energy antiproton ring (LEAR) 164
low-energy electron microscopy (LEEM) 210

m

MAKE 6, 19, 23, 32
mammography 200
managerial techniques 234
many integrated core (MIC) 90
marginalism 232
market failure 245
market/non-market institutions
– appropriate set of 239–241
market profit 241
markets uncertainty, reduction 248
Mars Exploration Rover experiments 211
Massachusetts Institute of Technology (MIT) 154
Medipix, applications
– art meets science 212, 213
– – micro-focus x-ray source 212
– – x-ray images, wild flowers 213
– biology 206, 207
– – low-contrast organisms 206
– chemistry 210
– – fully unforeseen application 210
– – graphene flakes 210
– – low-energy electron microscopy (LEEM) 210
– – scattered electrons 210
– – silicon sensor 210
– – Timepix, use 210
– dosimetry in space 210, 211
– – conventional LEEM detector 211
– – international space station (ISS), 210
– – Timepix-based detector 211
– education 211, 212
– – astronaut Christofer Cassidy, image 212
– – Langton ultimate cosmic ray intensity detector (LUCID) 212
– – TechDemoSat satellite 212
– – USB-based readout system 212
– – USB-based Timepix detector 211
– gas detector readout 208, 209
– – amorphous silicon layer 208
– – beta-particles, strontium-90 209
– – CMOS readout 208
– – 3D reconstruction, particle tracks 208
– – GridPix 208, 209
– medical x-ray imaging 205, 206
– – computed tomography (CT) 205
– – CT images of mouse head, comparison 206
– – Medipix3 silicon sensors 205
– – number of photons 205
– – paradigm shift 205
– – photon-counting technique 205
– radiation monitoring 209, 210
– – four spin-off companies 209
– – imaging system 209
– – Medipix2-based system 210
– x-ray materials analysis 207, 208
– – incarnation of the PIXcel detector, image 208
– – PIXcel detector 208
Medipix Collaborations 214
mesosphere 184
Messerschmitt ME-264 142
micro-focus x-ray 212
microprocessors
– timeline 145
– transform aspect 141
Milankovitch cycles 184
mobile phones 179
– subscriptions 113
mobile traffic forecast 114
modelling and simulation
– advanced simulation methods 178
– compute cycles 178
– numerical models 178
– as a platform for collaboration 177, 178
– prediction centre with global capabilities 178
Moore's Law 178, 200, 218
motivation
– due to increase in expertise 54
MPEG3 (moving picture experts group audio layer 3) 76
multicultural, and multifield interaction 52

n

NASA STRucture ANalysis (NASTRAN) software 146
National Aeronautics and Space Administration (NASA) 146
National Institute of Standards and Technology (NIST) 100
natural catastrophes 175
Navier–Stokes equations 147
NEC SX-6, supercomputer 177
networked information economy 246
network-orientated document abstraction language (NODAL) 107
networks 27, 28, 242, 243
– externalities 243
– research organizations 28
Newton's law 135, 159
NeXT software 71
– browser-editor 63
NeXT system 62, 72
Nicomachean Ethics 7
nitrogen cycles 184

Non-Governmental Organizations 173
numerical computation 181
numerical simulations 178
NURBS (Non-Uniform Rational B-Spline) 150

o
oceans 184
office computing systems 66
Open Access 2
Open Innovation 2
Open Science 2
Open Source 2
operating systems (OS) 71
opportunity 53
optical issues 120
organizational environment 6
organizational knowledge 9
organizational learning 49
OS. *See* operating systems (OS)
overcoming uncertainty
– different organizational modes for 247, 248

p
PALS. *See* Positron Annihilation Lifetime Spectroscopy (PALS)
PANalytical 208
parabolic-trough collectors, STE technologies 121
– concentrate, solar radiation 121
– focal line 121
– receiver pipe 121
– specific enthalpy, working fluid 121
peer production
– information and allocation gains of 248–250
Peer Review 2
PEPT. *See* Positron Emission Particle Tracking (PEPT)
PET. *See* Positron Emission Tomography (PET)
Pixelman, user-friendly readout software 205
Planck's law 128
planetary changes 176
plasmasphere 184
policy makers 173
Positron Annihilation Lifetime Spectroscopy (PALS) 168
Positron Emission Particle Tracking (PEPT) 168
Positron Emission Tomography (PET) 167, 217, 308
positron-emitting radioactive source 167
Positronium (Ps) 168
post-scarcity 242–244

practical extraction and reporting language (PERL) 73
Primary Flight Computer (PFC) 154
product obsolescence 6
profits
– accumulation without innovation 236, 237
– creative destruction 233, 234
– disappearance of 232
– economic growth, real engine of 237
– endogenous technological change 238, 239
– entrepreneurs 232
– risk and uncertainty 235, 236
project development
– flow of information in team 50
property system implementation costs 247
Protection of the Intellectual Property 2
Protect Respond Inform Secure Monitor (PRISM) programme 8

q
quantum mechanics 169, 216, 218
quantum superposition 218

r
radiation detection
– circuit technologies 196
religious knowledge 7
research infrastructures 1
research laboratories 13, 57, 80, 98, 245
– powerful tool to re-unite Europe through science 1
reservoir simulation 99
Reynolds number (Re) 140
– find, correct fluid velocity 140
– piping designer 140
– ratio of inertial forces 140

s
SABRE (Semi-Automated Business Research Environment) 155
Satellite Communications (SATCOM) system 154
scalable vector graphics (SVG) 72
scarcity to abundance 243
Schumpeter's assumptions 235
science institutions
– innovation collaborations 257
science for peace 219–225
scientific organizations 55
scientific outcome, determined in project management 54
scientific process 36

scientific research, governments' support 245, 246
Second Life 181
seismic data processing 98
selective and anti-reflective coatings, in optical issues STE Plant 124–128
- anti-reflective coatings 126
- coating, opaque substrate 126
- concentration ratio 126
- high solar flux 124
- ideal and commercial reflectance behaviour, solar receiver 127
- interference 126
- parabolic-trough concentrators 126
- ratio, energy flux 126
- reflection and refraction, light wave 127
- refraction index, coating material 126
- selective coatings 126
- solar absorbers 128
- three-dimensional (3D)-concentrators 126
- wavelength range 126
- waves superimpose 126
sensor networks 180
service level agreements (SLAs) 101
signal-to-noise ratio 198
SI International System of Units 133
SIMATIC WinCC system 90
SimCity 181
simulation
- based research, as potent tool
- - within scientific research community 176
- computer-based 177
- model
- - generic STE plant 131
- numerical 178
SLC. See social learning cycle (SLC)
Smallpix chip 213
Smart devices 179
smart eco-city architectures 191
smart phone 115
SMS messages 8
social behavioural trend analyses 7
socialization 10
socialization, externalization, combination and internalization (SECI) model 9, 10
- and knowledge creation spiral 9–11
social learning cycle (SLC) 43
social media 7
social network 78
social process 36
software development system 70
solar concentrators, optical issues STE Plant 124
- cost-effective manner 124
- flux density per surface unit 124
- law of reflection 124
- parabolic concentrator 125
- solar collectors 124
- solar radiation by reflection 124
- thermal energy conversion 124
solar systems 177
- Coronal Mass Ejections (CMEs) 185
- influence of 184–186
- many body gravitational effects 185
- Space Weather phenomenon 185
solar thermal electricity (STE) plants
- aerial views 123
- dependence of STE plant efficiency 132
- four STE Technologies 120–123
- - antireflective 120
- - central receiver systems 121, 122
- - compact linear fresnel concentrators 122
- - fluid mechanics 120
- - heat transfer mechanisms 120
- - links with physics 120
- - main components and subsystems 121
- - optical issues 120
- - parabolic-trough collectors 121
- - power block 120
- - selective coatings 120
- - solar concentrators 120
- - solar radiation 120
- - stirling dishes 122
- - thermal efficiency, solar receivers 120
- - thermal energy 120
- - thermodynamic issues 120
- issues related to heat transfer 134–137
- - concentrated solar radiation 134
- - conversion, radiation into heat 134
- - Fourier's law 135
- - heat engine 134
- - Newton's law of cooling 135
- - pipes and components in STE plant employing parabolic-trough collectors 135
- - specific enthalpy 134
- - steam turbine operating 134
- - Stefan–Boltzmann constant 136
- - stirling dishes 134
- - thermal insulation 134
- - thermal losses 134
- - thermal storage system 134
- - typical receiver tube for parabolic-trough collector 136
- optical issues 124
- - selective and anti-reflective coatings 124–128

– – solar concentrators 124
– – thermography 128–130
– thermodynamic issues 131–133
– – Carnot efficiency 131
– – dependence of STE plant efficiency 132
– – intermetallic compounds 133
– – plant efficiency *versus* receiver temperature 131
– – problem 133
– – ratio, electricity produced and solar radiation 131
– – selective coatings 133
– – simulation model 131
– – working fluid 133
society 3, 8, 17, 27, 30, 41–47, 55–57, 59, 64, 73, 101, 161, 169, 173, 185, 191, 216, 218, 222, 223, 225, 229, 231, 232, 235, 244, 303, 308
Spanish Gemasolar plant 137
spark chamber
– detects, cosmic particles 195
– example of 196
Spitfire 141
Stefan-Boltzmann law 129
stirling dishes, STE technologies 122, 123
– convert, solar radiation into mechanical energy 122
– European solar thermal electricity association (ESTELA) 123
– solarized Stirling engine 122
stratosphere 184
string theory 223
stylized model 257, 259
– for knowledge and technology transfer 258
subscription model 79
supercomputer 182
Supermarine Spitfire 141, 142
super proton synchrotron (SPS) 103
surgical simulation 181
Swiss National Science Foundation (SNSF) 268
synoptic rationality assumption 233

t
tablet 115
technical skills 55
technological developments 6
technological process 37
Technology, Entertainment, Design (TED) 190
technology products 37
tech-savvy 76
Teletel 70
theoretical physicist 217

theoretical research history, importance of 216–219
– black hole, physics theory 218
– cryptography 217
– data compression, storage and communication 217
– electromagnetic waves 216
– general theory of relativity 217
– genetics revolution 219
– global positioning system 217
– Hertz's experiments 216
– information technology, development 217
– information theory 217
– mechanical view, and impact on establishment 216
– medical applications 216, 217
– nanotechnology ideas 217, 218
– quantum computer 218
– quantum mechanics 216, 218, 219
– stimulated emission 219
theory for development
– CERN, crucial role in 215
– theoretical scientific subjects and impact on funds 215
thermal insulation materials 135
thermal storage system 134
thermodynamic
– cycle 120
– issues 120
thermography in optical issues STE Plant 128
– black-body approximation 128
– contact-methods 129
– glass transmissivity for two thicknesses 132
– Planck's law 128
– ratio of radiant energy 129
– reduce, thermal losses 129
– Stefan–Boltzmann constant 129
– Stefan–Boltzmann law 129
– temperature measurements 129
– thermal emission 129
– thermal-image camera 130
– thermal image, infrared camera 130
– thermal radiation 128
– Wien's displacement law 128
thermosphere 184
time over threshold (ToT) 203
Timepix chip 203
– clock signal 203
– Mars Exploration Rover experiments 211
– 150 peer-reviewed scientific papers 203
– silicon sensor 210
Top500 Supercomputing Sites 176

touch screens
- birth of
- - cathode-ray tube 104
- - CERN proton synchrotron (PS) control room 105
- - consoles 105
- - contact of object 103
- - cost-efficient 104
- - display device 103
- - electronic circuit 104
- - electronic device 103
- - idea 104
- - ion sputtering technologies 104
- - novel device 104
- - prototypes 104
- - screen technology 103
- - self-capacitance 104
- - self-capacitive transparent touch-button capacitor 105
- - separate overlay 103
- - technology, evolution 105
- - traditional methods 104
- - transparent capacitors 104
- - transparent multitouch-button screen 105
- early knowledge transfer, attempts 111, 112
- - Intelligent Oyster 111
- - medical equipment 112
- - touch watches 111
- - treatments 112
- evolution turned into revolution 113–115
- - Global mobile subscription forecasts 114
- - machine-to-machine percontinent 114
- - mobile phone subscriptions 113
- - mobile traffic forecasts 114
- - use of portable telephones 113
- global positioning system (GPS) 103
- human behaviour 115
- - modern smart phones, use 115
- - real-time commercial transactions 115
- - smart phones, use 115
- - technical problems 115
- interaction process 103
- replacement for mechanical buttons 110, 111
- - features, modern touch screen 110
- technology 114
touch-sensitive devices 110
touch watches 111
Traffic Alert and Collision Avoidance System (TCAS) 154
transfer agent 257

transfer offices (TOs) 268
transfer recipient 259
transfer strategy 292
transforming knowledge, into organizational wealth 30
troposphere 184

u
Uniform Interface to Computing Resources (UNICORE) 93
UNISURF system 149
United Nations' Intergovernmental Panel for Climate Change (IPCC) 192
Universal Serial Bus (USB) 204
universities 28, 40, 42, 46, 65, 73, 95, 255–305
urban development 190
US Federal Aviation Administration (FAA) 154

v
victory of capitalism 237
video on-demand (VOD) service 96
virtualisation of economy 241
visual computing 181
visual simulation 181

w
wealth of nations 230
weather reporting 180, 181
Web programs 71
Web's developers 75
Web server 62
- first 71
Wien's displacement law 128
Windows Control Centre (WinCC) 90
wireless communication 216
work-related wikis 30
the world academy of sciences (TWAS) 220
World Bank 6, 20, 21
Worldwatch Institute 190
World-wide LHC Computing Grid (WLCG) project 82
- grid sites, worldwide distribution of 85
World Wide Web (WWW) 61
- CERN's role 65
- - commercial services 70
- - computer-based document 66
- - definition 65
- - Director General 66, 67
- - Hyper-G 69
- - large hadron collider (LHC) 67, 68
- - for money 68, 69
- - networked information systems 70

– conferences 74
– Consortium 75, 76
– first page 62, 63
– history of 64
– – approach 64
– – matter of age 64
– nature of computing 76
– – changed of state 77, 78
– – digital world 78, 79
– – interface 78
– – spy software 77
– programming language 72, 73
– public domain 73, 74
– Web development 72
WWW. *See* World Wide Web (WWW)

X
X-rays
– computed tomography (CT) images 205
– detector 206
– photons 200
– radiation 168
– tube with tungsten anode 213
x-teams 28